William T. Blanford, Eugene W. Oates

The Fauna of British India

Including Ceylon and Burma. Birds - Vol. 2

William T. Blanford, Eugene W. Oates

The Fauna of British India
Including Ceylon and Burma. Birds - Vol. 2

ISBN/EAN: 9783337237905

Printed in Europe, USA, Canada, Australia, Japan

Cover: Foto ©berggeist007 / pixelio.de

More available books at **www.hansebooks.com**

THE FAUNA OF BRITISH INDIA,

INCLUDING

CEYLON AND BURMA.

Published under the authority of the Secretary of State for India in Council.

EDITED BY W. T. BLANFORD.

BIRDS.—Vol. II.

BY

EUGENE W. OATES.

LONDON:
TAYLOR AND FRANCIS, RED LION COURT, FLEET STREET.

CALCUTTA: | BOMBAY:
THACKER, SPINK, & CO. | THACKER & CO., LIMITED.

BERLIN:
R. FRIEDLÄNDER & SOHN, 11 CARLSTRASSE

1890.

ALERE FLAMMAM

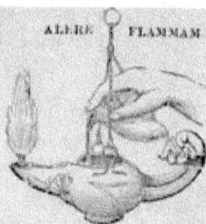

PRINTED BY TAYLOR AND FRANCIS,
RED LION COURT, FLEET STREET.

PREFACE.

THE appearance of the second volume of 'Birds' with fewer pages than are contained in other volumes belonging to the 'Fauna of British India' requires a brief explanation.

When the 'Birds' were undertaken by Mr. Oates in 1888, he knew that it would not be possible to complete them within the two years of furlough to which he was entitled, but it was hoped both by him and by myself that he would obtain additional leave of absence. This has not proved to be the case, and within the time available Mr. Oates has only found it practicable to finish the Passerine Order, comprising about five ninths of all the species of birds found in India. As will easily be understood by those who have been engaged in similar scientific work, constant application has been necessary in order to accomplish this within the period mentioned.

The first volume of the 'Birds' having appeared in December of last year, there was, when Mr. Oates left England in August last, considerably less than another volume ready in manuscript. To have waited for a full volume to be prepared would have entailed considerable delay, and, under the circumstances, it has been thought best to publish at once a second volume of less bulk at a reduced price, and to leave the remaining birds to be

described in a thicker third volume, the cost of which will
be proportionately greater, so that the price of the two
volumes together will remain unaltered. By this means
descriptions of all Indian Passerine birds, which are more
numerous than those of all other orders together, and
which afford the greatest difficulties in identification, are
placed at once in the hands of Indian ornithologists, whilst
Mr. Oates's work is kept distinct from that of any other
writer. I can only express my regret that Mr. Oates has
been unable to finish the work he has so well begun.

The present is the second volume of the 'Fauna of British
India' published in the current year, Mr. Boulenger's
'Reptilia and Batrachia' having been issued in August.
The only part now wanting to complete the Vertebrata of
the Indian Fauna, besides the third volume of Birds, is the
second half of the volume containing Mammalia; and this
half-volume, the greater portion of which is written, will, I
hope, be completed early in 1891. It is not probable that
the Birds can now be finished next year, but I propose to
undertake the third volume as soon as the Mammalia are
completed.

                                    W. T. BLANFORD.

October, 1890.

SYSTEMATIC INDEX.

Fig. 1.— *Terpsiphone paradisi.*

Family MUSCICAPIDÆ.

The intrinsic muscles of the syrinx fixed to the ends of the bronchial semi-rings; the edges of both mandibles smooth, the upper one simply notched; hinder aspect of tarsus smooth, composed of two entire longitudinal laminæ; wing with ten primaries; tongue non-tubular; nostrils clear of the line of forehead, the lower edge of the nostril nearer to the commissure than the upper edge is to the culmen; plumage of the nestling mottled or squamated; nostrils covered more or less by long curly hairs; rectrices twelve; tarsi short; an autumn moult only.

The *Muscicapidæ*, or Flycatchers, constitute a large family of birds, which is well represented in India. Some are resident: but the majority are migratory to a greater or less extent.

The Flycatchers may be known by the mottled plumage of the

nestling, and by the presence of numerous hairs stretching from the forehead over the nostrils. These hairs lie horizontally, and in all cases reach beyond the nostrils, and not unfrequently nearly to the end of the bill. They are not to be confounded with the rictal bristles, which are stiff and strong and lie laterally, nor are they to be confounded with the lengthened shafts of the frontal feathers, which in some of the Thrushes resemble hairs. These latter, moreover, are never horizontal, nor do they extend over the nostrils except in cases where this is brought about by accident, such as careless preparation of the preserved specimen.

The amount of mottling or squamation in the plumage of the nestling varies considerably, but is present in every species in a more or less marked degree. This character is perhaps least developed in the genus *Terpsiphone*, the most typical of Fly-catchers so far as structural characters are concerned; but even in this genus the mottled breast is unmistakable.

The Flycatchers may further be recognized by their very feeble tarsi and feet, which quite incapacitate them from walking on the ground; and this character will by itself be sufficient to separate them from the Thrushes, in which the tarsi are long and the feet strong.

Young Flycatchers moult into adult plumage in most cases the first autumn: but *Terpsiphone* differs in this respect, the males retaining an intermediate plumage for two or more years.

Those Flycatchers which have abandoned their migratory habits and have become resident are well differentiated by generic characters; but others which are still migratory resemble each other structurally very closely, and generic characters by which to separate them into convenient groups are not easy to be found. In the following key, therefore, I have had recourse to types of colour, which appear to work well and to bring allied birds together into natural groups.

The Flycatchers feed on insects, which they either catch on the wing, starting from a perch to which they usually return several times, or by running with the aid of their wings along the limbs of trees. They seldom or never descend to the ground. The majority construct their nests in holes of trees or banks, and some of the species build very beautiful cup-shaped nests in the branches of trees. Few of these birds have any song, and on the whole the Flycatchers are remarkably silent. They are found solitary or in pairs, and they are frequently familiar birds.

Key to the Genera.

a. Tail considerably shorter than wing.
   a'. Second primary equal to the fifth.
      a''. Closed wings not reaching beyond
         middle of tail.................... MUSCICAPA, p. 4.
      b''. Closed wings reaching nearly to tip
         of tail .... HEMICHELIDON, p. 5.

b'. Second primary very much shorter than fifth.

  c''. Frontal feathers of ordinary structure, not concealing the nostrils.

    a'''. Rictal bristles short and few in number, generally less than six.

      a⁴. Sexes different.

        a⁵. In both sexes base of tail white, upper tail-coverts black, upper plumage brown or rufescent, never blue nor black . . . . SIPHIA, p. 7.

        b⁵. Males with whole upper plumage blue or black : lower plumage never entirely blue or green. Females brown or rufescent above, never combined with black upper tail-coverts and white on tail.

          a⁶. Bill wide at base and strong. CYORNIS, p. 11.

          b⁶. Bill narrow throughout, and feeble . . . . NITIDULA, p. 27.

        c⁵. Both sexes with the entire plumage suffused with blue or green . . . . STOPAROLA, p. 27.

      b⁴. Sexes alike ; plumage plain brown or rufous throughout.

        d⁵. First primary never less than half second.

          c⁶. Bill laterally compressed ; lower mandible pale . . . . . . MUSCITREA, p. 30.

          d⁶. Bill flattened ; lower mandible dark . . . . ANTHIPES, p. 31.

        e⁵. First primary much less than half second . . . . ALSEONAX, p. 34.

    b'''. Rictal bristles very long and numerous, about ten on each side.

      c⁴. Tail much rounded ; first primary much longer than half second . . OCHROMELA, p. 37.

      d⁴. Tail quite even ; first primary much less than half second . . . . CULICICAPA, p. 38.

  d''. Frontal feathers lengthened and very dense, concealing the nostrils.

    c'''. Bill carinated and narrow ; both sexes with a brilliant neck-spot . . NILTAVA, p. 39.

    d'''. Bill broad and flattened ; no neck-spot in either sex . . . . PHILENTOMA, p. 42.

b Tail as long as, or longer than, wing.

  c. Head crested . . . . TERPSIPHONE, p. 44.

  d. Head not crested.

    e''. Tail about equal to wing.

      e'''. Length of culmen about twice the breadth of bill at forehead . . . . . . HYPOTHYMIS, p. 48.

      f'''. Length of culmen about equal to breadth of bill at forehead . . . . . . CHELIDORHYNX, p. 51.

    f''. Tail considerably longer than wing . . RHIPIDURA, p. 52.

Genus **MUSCICAPA**, Brisson, 1760.

The genus *Muscicapa* contains the Spotted Flycatcher, a common summer visitor to England and Europe.

In this genus the sexes are alike: the culmen of the bill is about as long as twice the breadth of the bill at the forehead; the rictal bristles are few and moderate in length; the wing is long and pointed, the first primary being very small and the second very long and equal to the fifth; the tail is square; and the plumage is streaked. *M. grisola* is migratory.

557. Muscicapa grisola.   *The Spotted Flycatcher.*

Muscicapa grisola, *Linn. Syst. Nat.* i, p. 328 (1766); *Sharpe, Cat. B. M.* iv, p. 151; *Scully, Ibis,* 1881, p. 437; *Oates in Hume's N. & E.* 2nd. ed. ii, p. 1.

Butalis grisola (*Linn.*), *Blyth, Cat.* p. 175; *Hume & Henders. Lah. to Yark.* p. 185; *Hume, S. F.* iii, p. 467, v, p. 495; *id. Cat.* no. 289 bis; *Biddulph, Ibis,* 1881, p. 52; *Barnes, Birds Bom.* p. 163.

Coloration. Upper plumage brown, the forehead, crown, and nape with black centres; wing-coverts, secondaries, and tertiaries dark brown, rather broadly edged with pale fulvous; primaries and the

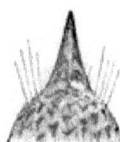

Fig. 2.—Bill of *M. grisola.*

primary-coverts more narrowly edged with the same; tail dark brown, obsoletely edged paler; lores greyish white; a buff ring round the eye; sides of the head brown; cheeks whitish, with an irregular dark moustachial streak below; lower plumage white, the breast and the sides of the throat streaked with brown; the sides of the body less distinctly streaked.

The young have the upper plumage pale fulvous, with brown or blackish margins; the wings broadly edged, the lesser coverts broadly tipped, with buff; the lower plumage whitish, variegated with dark brown.

Legs and feet black; bill blackish, yellow at the base of the lower mandible; iris dark brown (*Butler*).

Length about 6; tail 2·5; wing 3·3; tarsus ·6; bill from gape ·8.

Distribution. Common from May to September in Gilgit, where this species breeds at elevations over 8000 feet. This bird visits the plains in the autumn, and is found at that season in Sind, Raj-putana, Guzerat, Cutch, and Kattywar. Its eastern limit in the

Himalayas appears to be Simla, where it has been obtained in September.

This Flycatcher has an extensive range, being found, according to season, over the greater part of Europe, Africa, and South-western Asia.

Habits, &c. This bird breeds in Gilgit, but nothing beyond this is on record about its nidification in India. In Europe it makes its nest on a branch of a tree near the trunk, in a shallow hole in a tree, or on a branch of a fruit-tree or creeper trained against a wall. The eggs are pale bluish or greenish, marked with reddish brown, and measure about ·75 by ·57.

Genus **HEMICHELIDON**, Hodgs., 1844.

The genus *Hemichelidon* contains two species of Flycatchers which are permanent residents in the Himalayas, a considerable number descending to the lower ranges and plains in the winter.

In *Hemichelidon* the bill viewed from above is almost an equilateral triangle, sharp-pointed, pinched in towards the tip, and very depressed: the rictal bristles are moderate: the wing is long, reaching nearly to the end of the tail, the first primary very minute and the second equal to the fifth; the tail is square. In this genus the sexes are alike, and the plumage brown or ferruginous.

Key to the Species.

a. General colour of plumage brown ...... *H. sibirica*, p. 5.
b. General colour of plumage ferruginous .. *H. ferruginea*, p. 6.

558. **Hemichelidon sibirica**. *The Sooty Flycatcher.*

Muscicapa sibirica, *Gm. Syst. Nat.* i, p. 936 (1788).
Hemichelidon fuliginosa, *Hodgs. P. Z. S.* 1845, p. 32; *Blyth, Cat.*
p. 175; *Horsf. & M. Cat.* i, p. 137; *Jerd. B. I.* i, p. 458; *Stoliczka,*
J. A. S. B. xxxvii, pt. ii, p. 28; *Brooks, J. A. S. B.* xli, pt. ii,
p. 75; *Hume & Henders. Lah. to Yark.* p. 184, pl. iv.
Hemichelidon sibirica (*Gm.*), *Hume, N. & E.* p. 205; *id. Cat.*
no. 295; *Sharpe, Cat. B. M.* iv. p. 120; *Oates, B. B.* i, p. 275;
id. in Hume's N. & E. 2nd ed. ii, p. 1.

Dang-chim-pa-pho, Lepch.

Fig. 3.—Bill of *H. sibirica.*

Coloration. Upper plumage brown, the feathers of the head with darker centres and those of the wings more or less edged paler;

tail plain brown; a ring of white feathers round the eye; lores
mixed white and brown; sides of the head brown; chin, throat,
breast, and sides of the body smoky brown, dashed with grey
in places; an indistinct white patch on the lower throat; abdomen,
vent, and under tail-coverts white, the last mixed with brown.
After the autumn moult, the margins to the wing-feathers are
broader and more rufous.

The young have the crown and nape streaked with fulvous
white and the upper plumage spotted and streaked with fulvous;
the lesser wing-coverts are tipped, and the greater coverts and
quills margined broadly, with fulvous; the lower plumage is much
whiter than in the adult.

Upper mandible dark brown, lower yellowish; iris brown; legs
brownish black.

Length about 4·5; tail 2; wing 2·8; tarsus ·5; bill from
gape ·55.

Distribution. A permanent resident in the Himalayas, from
Afghanistan and Kashmir to Sikhim, occurring as high as
13,000 feet in the summer. In the winter this species is found
along the lower ranges of these mountains, and it has been
observed at Shillong, Manipur, and generally throughout Pegu and
Tenasserim, extending into the Malay peninsula. It is widely
spread over China and Eastern Siberia in summer. This Flycatcher
appears to be entirely absent from the plains of India.

Habits, &c. Breeds in Kashmir in June. A nest found by
Major Cock was placed against the side of a tree-trunk. The
eggs are pale green mottled with pale reddish, and measure about
·65 by ·46.

559. Hemichelidon ferruginea. *The Ferruginous Flycatcher.*

Hemichelidon ferruginea, *Hodgs. P. Z. S.* 1845, p. 32; *Blyth, Cat.*
p. 175; *Horsf. & M. Cat.* i, p. 137; *Hume, N. & E.* p. 207;
Sharpe, Cat. B. M. iv, p. 122; *Oates, B. B.* i, p. 276; *id. in
Hume's N. & E.* 2nd ed. ii, p. 2.
Alseonax ferrugineus (*Hodgs.*), *Jerd. B. I.* i, p. 460; *Hume, Cat.*
no. 299; *id. S. F.* xi, p. 106.

Dung-chim-pa-pho, Lepch.

Coloration. Forehead and crown of head dark brown; back,
scapulars, lesser wing-coverts, rump, and upper tail-coverts reddish
brown, changing to chestnut on the latter two parts: median and
greater coverts brown, edged and tipped with chestnut; quills
dark brown, the later secondaries and tertiaries edged with reddish
brown; tail reddish brown; a distinct ring of feathers round the
eye whitish or pale buff; lores and ear-coverts mixed rufous and
brown: lower plumage pale rufous, deepening to chestnut on the
abdomen, under tail-coverts, and flanks, the breast infuscated, the
lower part of the throat whitish, and the sides of the throat mottled
with brown.

The young have the forehead, crown, and nape black, boldly

streaked with fulvous; the upper plumage chestnut, mottled with black; wings and tail more rufous than in the adult; chin and throat white; lower plumage uniform pale chestnut.

Bill dusky, fleshy-yellow at the base beneath; legs pale whitish-fleshy; iris dark brown (*Jerdon*).

Length about 5; tail 2; wing 2·7; tarsus ·5; bill from gape ·65.

Distribution. A permanent resident in the Himalayas from Nepal to the extreme east of Assam, from about 4000 to 8000 feet. This species has been procured in winter at Shillong, in the Khási hills; in Pegu and in Tenasserim. Hume obtained it in April in Manipur, where he is of opinion that it breeds. This Flycatcher extends to China.

Habits, &c. Hodgson figures the nest, made of moss and lichens and placed upon the surface of an old stump of a tree. The eggs of this bird appear to be buff freckled with reddish, and to measure ·69 by ·5.

Genus **SIPHIA**, Hodgs., 1837.

The genus *Siphia* contains four Indian birds, one of which is the type of the genus, and the other three are closely allied species, which have been placed by various ornithologists in *Muscicapa*, *Siphia*, or *Erythrosterna*. I consider the four species now noticed to be absolutely congeneric both in structure and in style of coloration. They have no close relationship with *Muscicapa*, in which the sexes are alike and the wing very lengthened, and I prefer to associate them together in the genus *Siphia*, which is equal to *Erythrosterna* but of older date.

In *Siphia* the sexes are differently coloured, the base of the tail in both sexes is white, the upper tail-coverts black, and the back brown or rufous. The bill is small, and the rictal bristles moderate; the wing is of moderate length, but sharply pointed, and the first primary is shorter than half the second; the tail is square.

The male nestlings soon lose their spotted plumage, and assume the plumage of the adult female in September. It is not, however, till towards the end of the winter that they commence to put on the characteristic red colouring of the adult male, and consequently the mass of birds which visit India are in the garb of the female till near the time for their departure to summer-quarters.

Key to the Species.

a. Throat chestnut, not extending to the chin or breast .... *S. strophiata*, p. 8.
b. Chin, throat, and breast chestnut: crown of different shade to back .... *S. parva* ♂, p. 9.
c. Chin and throat chestnut, breast ashy: crown and back of the same shade .... *S. albicilla* ♂, p. 10.
d. Chin, throat, breast, and upper abdomen chestnut, surrounded by a black band .. *S. hyperythra* ♂, p. 10.
e. No chestnut on lower plumage .... { *S. parva* { ♀ et ♂ { *S. albicilla* { juv. { *S. hyperytha*

560. **Siphia strophiata.** *The Orange-gorgeted Flycatcher.*

Siphia strophiata, *Hodgs. Ind. Rev.* i, p. 651 (1837); *Blyth, Cat.* p. 171; *Horsf. & M. Cat.* i, p. 283; *Jerd. B. I.* i, p. 479; *Stoliczka, J. A. S. B.* xxxvii, pt. ii, p. 32; *Blanf. J. A. S. B.* xli, pt. ii, p. 47; *Godw.-Aust. J. A. S. B.* xlv, pt. ii, p. 72; *Anders. Yunnan Exped., Aves,* p. 620; *Hume & Dav. S. F.* vi, p. 232; *Sharpe, Cat. B. M.* iv, p. 455; *Hume, Cat.* no. 319; *Scully, S. F.* viii, p. 278; *Oates, B. B.* i. p. 280.

Siphia rufigularis, *Scully, S. F.* viii, p. 279 (1879).

Sij hya, Nep.; *Phatl-tograk-pho,* Lepch.

Fig. 4.—Bill of *S. strophiata.*

Coloration. *Male.* Upper plumage olive-brown, tinged with fulvous on the back and rump; upper tail-coverts black; lores, cheeks, chin, and throat black; forehead and a short eyebrow white; ear-coverts and feathers above the eye deep slaty; a large oval patch below the throat bright chestnut; breast and sides of the neck slaty; abdomen, vent, and under tail-coverts white; flanks olive-brown; lesser wing-coverts slaty; the other coverts and all the quills brown, edged with fulvous; tail blackish; the pair next the middle pair with a patch of white on the outer web; the others with a larger white patch on both webs; under wing-coverts and axillaries light buff.

Female. Similar in style of coloration to the male, but the orange gular patch paler and smaller, the white on the forehead of less extent, and the black of the face and throat replaced by slaty.

The young bird is brown all over, closely streaked and mottled with fulvous; the tail is marked with white as in the adult, but there is no indication of the gular patch.

Bill black; gape fleshy-whitish; iris dark brown; feet dark horny-brown; claws black (*Scully*).

Length nearly 5·5; tail 2·3; wing 3; tarsus ·8; bill from gape ·65.

It is not unusual for the female to have the throat to some extent orange-rufous, and it was to a specimen exhibiting this peculiarity that Scully assigned the name of *Siphia rufigularis.*

Distribution. The Himalayas from Eastern Kashmir to the Daphla hills in Assam up to 12,000 feet in summer, and descending to the lower valleys in winter; the Khási and Nága hills; Manipur; the neighbourhood of Bhámo; Arrakan; Muleyit mountain in Tenasserim. This species extends into China.

Habits, &c. Nothing is known of the nidification of this Flycatcher. According to Jerdon it frequently alights on the ground to pick up an insect, and occasionally makes a dart at one in the air, returning after each sally to its perch.

561. Siphia parva. *The European Red-breasted Flycatcher.*

Muscicapa parva, *Bechst. Naturg. Deutschl.* iv, p. 505 (1795); *Sharpe, Cat. B. M.* iv, p. 161.
Erythrosterna parva (*Bechst.*), *Blanf. J. A. S. B.* xxxviii, pt. ii, p. 174; *id. S. F.* v, p. 484; *Hume, J. A. S. B.* xxxix, pt. ii, p. 116; *id. Cat.* no. 323 bis; *Barnes, Birds Bom.* p. 167.

Coloration. Male. When in fresh plumage, after the autumn moult, the forehead, lores, and cheeks are grey, speckled at times with blackish; a ring of white feathers round the eye; sides of the head bluish ashy; crown and nape ashy brown; remainder of the upper plumage fulvous-brown; upper tail-coverts black; wing-coverts, secondaries, and tertiaries brown, edged with fulvous-brown; primaries and primary-coverts edged more narrowly with the same; chin, throat, and breast bright chestnut; remainder of the lower plumage white, tinged with buff on the sides of the body; the two middle pairs of tail-feathers wholly black; the others with the basal two thirds more or less white.

Female. The whole upper plumage brown, tinged with fulvous, the crown being of the same colour as the back; wings and tail as in the male; upper tail-coverts black; feathers on the eyelids white; sides of the head rufous-brown; lores whitish; lower plumage dull white, suffused with pale fulvous-ashy on the breast and sides of the body.

The young are spotted on the upper plumage and breast with fulvous. After the autumn moult young males commence to assume some red on the breast, and they become fully adult by the spring.

Iris blackish brown; legs and feet black; bill brown above, brownish-flesh below (*Butler*).

Length about 5; tail 2·1; wing 2·6; tarsus ·65; bill from gape ·6.

Distribution. A winter visitor to a great portion of the Indian peninsula, being found to the east as far as the Bhutan Doars at the base of the Himalayas and Singbhoom in the plains, and to the south as far as Mysore and the Nilgiris.

This species is found in Central and South-eastern Europe during the summer. Its distribution out of India is very difficult to trace, as this Flycatcher has been confounded with the next by many ornithologists, Seebohm going so far as to unite *S. parva, S. albicilla*, and *S. hyperythra* into one species. I have seen no example of *S. parva* from any portion of the Himalayas *, and I doubt if it ever crosses those mountains, the specimens said to

---

* Stoliczka (J. A. S. B. xxxvii, pt. ii, p. 32), however, records *Erythrosterna leucura* from Kotgarh. The species he obtained may have been *S. parva* or more likely *S. hyperythra.*

have been procured from Central Asia having probably found their
way thither from the west. This species is found in India from
October to April.

Habits, &c. This bird breeds in Europe, making a nest of moss
lined with grass and hairs, either against the trunk of a tree or in
a hollow of the trunk. The eggs are pale green, marked with
pinkish brown, and measure about ·65 by ·53.

562. Siphia albicilla. *The Eastern Red-breasted Flycatcher.*

Muscicapa albicilla, *Pall. Zoogr. Rosso-Asiat.* i, p. 462, *Aves,* tab. i
(1811); *Sharpe, Cat. B. M.* iv, p. 162; *Oates, B. B.* i, p. 278.
Erythrosterna leucura (*Gm.*), *Blyth, Cat.* p. 171; *Horsf. & M. Cat.*
i, p. 297; *Jerd. B. I.* i, p. 481.
Erythrosterna albicilla (*Pall.*), *Anders. Yunnan Exped., Aves,* p. 621;
Hume & Dav. S. F. vi, p. 233; *Scully, S. F.* viii, p. 280; *Hume,*
Cat. no. 323.

The White-tailed Robin Flycatcher, Jerd.; *Turra,* Hind.; *Chat-ki,*
Beng.

Coloration. Male. Similar to the male of *S. parva,* but having
only the chin and throat chestnut, and not the breast, which is
ashy; it differs also in the crown being, in freshly moulted birds
in good plumage, of the same colour as the back, and in the ear-
coverts being brown instead of bluish ashy.

Female. So similar to the female of *S. parva,* as to be undistinguishable from it.

I have not been able to examine nestlings of this species, but
there is no reason to think that they differ from those of *S. parva.*
The youngest birds I have seen are like the females, but with some
fulvous tips to the wing-coverts.

Bill dark brown, yellowish at the gape; mouth yellow; iris
hazel-brown; legs and claws black; eyelids grey.

Length about 5; tail 2·1; wing 2·7; tarsus ·65; bill from
gape ·6.

Distribution. Visits the Eastern portion of the Empire from
October to April, extending on the west as far as Nepal in the
Himalayas and Dinapore in the plains, and southwards to Tenas-
serim. This species summers in Eastern Siberia and Northern
China.

Habits, &c. The nest and eggs of this bird do not appear to be
known. This Flycatcher frequents groves of trees, running among
the larger branches and constantly flitting its tail up and down
and partially expanding it.

563. Siphia hyperythra. *The Indian Red-breasted Flycatcher.*

Siphia hyperythra, *Cabanis, J. f. Orn.* 1866, p. 391; *Oates in Hume's*
N. & E. 2nd ed. ii, p. 2.
rythrosterna parva (*Becht.*), *Brooks, J. A. S. B.* xli, pt. ii, p. 76,
xliii, pt. ii, p. 245.

Erythrosterna hyperythra (*Cab.*), *Holdsworth, P. Z. S.* 1872, p. 442, pl. 17; *Hume, N. & E.* p. 217; *Brooks, S. F.* iii, p. 236; *Hume, S. F.* vii, p. 376; *id. Cat.* no. 323 ter.

Muscicapa hyperythra (*Cab.*), *Sharpe, Cat. B. M.* iv, p. 163; *Legge, Birds Ceyl.* p. 428.

Coloration. Male. The whole upper plumage dark ashy brown; the tail-coverts black; wings and coverts dark brown, edged with the colour of the back; tail black, with the same distribution of white as in *S. parva* and *S. albicilla*; sides of the head dark ashy brown like the crown; chin, throat, breast, and upper part of the abdomen rich chestnut, separated from the head and neck by a broad black band produced down the sides of the breast; remainder of the lower plumage white, tinged with rufous on the flanks and under tail-coverts.

Female. Very similar to the females of *S. parva* and *S. albicilla*, but darker above.

Some young males in May are acquiring the black pectoral band and show indications of rufous on the throat and breast.

Iris hazel-brown; bill above brown, pale next the forehead; gape and lower mandible fleshy yellow, with the tip dusky; inside of mouth yellow; legs and feet deep brown; soles yellowish (*Legge*).

Length about 5; tail 2; wing 2·7; tarsus ·75; bill from gape ·6.

Distribution. Summers in Kashmir and winters in Ceylon. This species has not yet been procured in the intervening countries during the periods of migration.

Habits, &c. Brooks remarks that this Flycatcher breeds in Kashmir between 6000 and 7000 feet elevation, but he failed to find the nest.

Genus CYORNIS, Blyth, 1843.

I place in the genus *Cyornis* fourteen species of Flycatchers in which the sexes are different, and which appear to be congeneric in structure, habits, and style of coloration.

The females of some of the species of this genus are amongst the most difficult of birds to discriminate, and they remained in great confusion till Sharpe brought them into order with the aid of Hodgson's types and drawings.

In *Cyornis* the bill is about half the length of the head, depressed, and rather broad at the base; the rictal bristles are moderate; the wing in most of the species is sharply pointed, and the first primary generally small; the tail is square or nearly so.

In this genus all the males are blue or black on the upper plumage, and the females brown or rufescent.

All the species are true Flycatchers, catching their prey on the wing or by running along branches.

Key to the Species.

a. Base of tail white.
  a'. White on tail extending nearly to tips of
    feathers....................... *C. cyaneus*, p. 13.
  b'. White on tail confined to base.
    a''. Upper plumage blue.
      a'''. Breast orange-chestnut.
        a⁴. No white frontal band .... *C. hodgsoni* ♂, p. 14.
        b⁴. A white frontal band .... *C. hyperythrus* ♂, p. 15.
      b'''. Breast white, pale buff, or fulvous-
        grey.   [p. 16.
        c⁴. No white supercilium .... *C. leucomelanurus* ♂,
        d⁴. A white supercilium.......... *C. superciliaris* ♂, p. 17.
    a''. Upper plumage black .... *C. melanoleucus* ♂, p. 18.
b. No white on tail.
  c'. Upper plumage blue.
    c''. Crown and rump cobalt-blue, back
    dull blue.
      c'''. Axillaries and under wing-coverts
      white .... *C. sapphira* ♂, p. 20.
      d'''. Axillaries and under wing-coverts
      chestnut .... *C. oatesi* ♂, p. 20.
    d''. Crown, rump, and back of the same
    blue.
      e'''. Whole lower plumage white.... *C. astigma* ♂, p. 19.
      f'''. Chin, throat, and breast dark blue;
      abdomen white .... *C. pallidipes* ♂, p. 22.
      g'''. Whole lower plumage pale blue.. *C. unicolor* ♂, p. 22.
      h'''. Breast ferruginous or chestnut.
        e⁴. Chin and throat blue.......... *C. rubeculoides* ♂, p. 23.
        f⁴. Throat ferruginous like breast.
          a⁵. First primary about half length
          of second; wing under 3 .. *C. tickelli*, p. 25.
          b⁵. First primary much shorter
          than half second; wing over 3. *C. magnirostris* ♂, p. 26.
  d'. Upper plumage brown or rufescent.
    e''. Breast chestnut or ferruginous.
      i'''. First primary about half length of
      second; sides of head grey .... *C. pallidipes* ♀, p. 22.
      k'''. First primary much shorter than
      half second; sides of head fulvous
      or rufous.
        g⁴. Wing considerably under 3 .... *C. rubeculoides* ♀, p. 23.
        h⁴. Wing over 3 .... *C. magnirostris* ♀, p. 26.
    f''. No chestnut on breast.
      l'''. Wing considerably over 3.
        i⁴. First primary much shorter than
        half second ; yellowish - buff
        patch on throat .... *C. oatesi* ♀, p. 20.
        k⁴. First primary about equal to half
        second; no patch on throat .. *C. unicolor* ♀, p. 22.
      m'''. Wing considerably under 3.
        l⁴. Second primary equal to ninth;
        wing 2·5; tail 1·4 .... *C. hyperythrus* ♀, p. 15.
                                    [p. 16.
        m⁴. Second primary shorter than the
        secondaries; wing 2·4; tail 2.. *C. leucomelanurus* ♀,

n^4. Second primary between the
sixth and seventh or equal to
seventh.

c^5. Wing 2·8; tail 2·2 . . . .   *C. hodgsoni* ♀, p. 14.
d^5. Wing 2·3 or 2·4; tail 1·7 or 1·8.
  a^6. Chin, throat, and breast pale
   buff or sordid white.
    a^7. Upper tail-coverts and
     outer webs of tail-fea-
     thers suffused with blue.   *C. superciliaris* ♀, p. 17.
    b^7. Upper tail-coverts and
     outer webs of tail-fea-
     thers fulvous . . . .   *C. astigma* ♀, p. 19.
    c^7. Upper tail-coverts and
     outer webs of tail-fea-
     thers bright ferruginous.   *C. melanoleucus* ♀, p. 18.
  b^6. Chin, throat, and breast
   orange-chestnut . . . . . .   *C. sapphira* ♀, p. 20.

564. **Cyornis cyaneus.**   *The White-tailed Blue Flycatcher.*

Muscitrea cyanea, *Hume, S. F.* v, p. 101 (1877); *Hume & Dav. S. F.*
  vi, p. 207; *Hume, Cat.* no. 266 bis.
Trichastoma leucoproctum, *Tweedd. P. Z. S.* 1877, p. 336; *Hume,*
  S. F. vii, p. 318.
Niltava leucoprocta (*Tweedd.*), *Oates, B. B.* i, p. 293.
Pachycephala cyanea (*Hume*), *Gadow, B. M. Cat.* viii, p. 224.

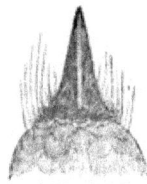

Fig. 5.—Bill of *C. cyaneus.*

Coloration. Male. Lores and front line of the forehead black;
forehead, crown, nape, and some of the lesser coverts cobalt-blue;
the remaining wing-coverts and the whole upper plumage deep
blue; quills dark brown, edged with blue; the four middle tail-
feathers dull blue, the next pair white on both webs, with a broad
black tip, the next two pairs nearly entirely white on the inner
web, the outermost pair narrowly white on the inner web; sides
of the head and neck, chin, throat, and breast dull blue; abdomen
and sides of the body bluish grey; vent and under tail-coverts
pure white; under wing-coverts ashy. The disposition of white
in the tail varies somewhat.

Female. Lores and front line of the forehead reddish brown;
upper plumage olive-brown, tinged with rufous; wing-coverts and
quills dark brown, broadly edged with bright rufous; tail brown,
tinged with rufous and with the same distribution of white on it
as in the male; ear-coverts olive-brown, with pale shafts; chin,
throat, breast, and sides of the body rufous-olive; a large patch of

white on the fore neck; middle of the abdomen whitish; vent and under tail-coverts pure white; under wing-coverts and axillaries rufous-ashy.

Bill black; legs light brown; iris deep brown (*Limborg*).

Length about 7; tail 3; wing 3·6; tarsus ·9; bill from gape 1.

Distribution. Muleyit mountain in Tenasserim up to 5000 feet. This species has also been found on the mountains of Perak in the Malay peninsula.

Habits, &c. A forest bird, found constantly on trees, and never descending to the ground.

565. Cyornis hodgsoni. *The Rusty-breasted Blue Flycatcher.*

Siphia erythacus, *Jerd. & Blyth, P. Z. S.* 1861, p. 201 (*nec Blyth*); *Jerd. B. I.* i, p. 480; *Hume, S. F.* ii, p. 458; *Godw.-Aust. J. A. S. B.* xliii, pt. ii, p. 158; *Hume, S. F.* v, p. 137; *Hume & Dav. S. F.* xi, pp. 233, 510; *Hume, Cat. no.* 322; *id. S. F.* xi, p. 115.

Siphia hodgsonii, *Verr. N. Arch. Mus.* vi, *Bull.* p. 34 (1870), vii, p. 29; *David. op. cit.* ix, pl. 4, fig. 4.

Erythrosterna sordida, *Godw.-Aust. J. A. S. B.* xliii, pt. ii, p. 158 (1874); *Hume, S. F.* iii, p. 392.

Poliomyias hodgsoni (*Verr.*), *Sharpe, Cat. B. M.* iv, p. 203; *Oates, B. B.* i, p. 286.

The Rusty-breasted Flycatcher, Jerd.

Coloration. Male. The whole upper plumage slaty blue; lores, cheeks, under the eye, and the upper tail-coverts black; sides of the head and neck slaty blue; wing-coverts brown edged with cyaneous; quills black edged with brown; tail black, the base of all the feathers except the middle pair white; chin, throat, breast, and abdomen orange-chestnut; lower abdomen, vent, flanks, and under tail-coverts pale ferruginous.

Female. Upper plumage olive-brown, tinged with fulvous on the upper tail-coverts; tail brown, edged on the basal half with fulvous-brown; wing-coverts and quills brown edged with fulvous-brown, and the greater coverts tipped with the same; lores whitish; a pale ring round the eye; sides of the head olive-brown tinged with rufous; lower plumage ashy brown, the abdomen whitish.

The young bird is not known.

The legs and feet vary from dusky liver-brown to plain dark brown; the bill in one bird entirely black, in other two blackish, horny grey on base and lower ridge of rami of lower mandible; iris deep brown (*Hume*).

Length about 5·5; tail 2·2; wing 2·8; tarsus ·65; bill from gape ·55.

C. lutcola is an allied species found outside our limits and differs chiefly in having a considerable amount of white on the wing-coverts of both sexes. It was to this species that Blyth first applied the name *Siphia erythaca* (J. A. S. B. xvi, p. 126, 1847). Subsequently he and Jerdon reapplied this name to the Indian species, for which, under these circumstances, the term *erythaca* cannot be used.

I have examined the types of *E. sordida* obligingly sent me by
Godwin-Austen and I find them to be the females of the present
species.

Distribution. Specimens of this species have been obtained in
Sikhim (March and April); Shillong (no date); Japvo peak, Nága
hills, at 6000 feet (January); Manipur (February and April); Karen
hills near Toungngoo at 4000 feet (January); pine-forests, Salween
(February); Muleyit mountain (January and February). This
Flycatcher extends into China.

566. **Cyornis hyperythrus.** *The Rufous-breasted Blue Flycatcher.*

Dimorpha superciliaris, *Blyth, J. A. S. B.* xi, p. 190 (1842, *nec Jerd.*).
Muscicapa hyperythra, *Blyth, J. A. S. B.* xi, p. 885 (1842).
Muscicapula rubecula, *Blyth, J. A. S. B.* xii, p. 940 (1843).
Siphia superciliaris (*Bl.*), *Blyth, Cat.* p. 172; *Jerd. B. I.* i, p. 480;
  Blanf. J. A. S. B. xli, pt. ii, p. 159; *Hume, Cat.* no. 321; *id. S. F.*
  xi, p. 115.
Digenea superciliaris (*Bl.*), *Horsf. & M. Cat.* i, p. 293; *Hume, N. & E.*
  p. 216.
Muscicapula hyperythra (*Bl.*), *Sharpe, Cat. B. M.* iv, p. 206.
Cyornis hyperythrus (*Bl.*), *Oates in Hume's N. & E.* 2nd ed. ii, p. 2.

The Rufous-breasted Flycatcher, Jerd.

Coloration. Male. Forehead, lores, chin, and cheeks black; a
frontal band extending to the posterior part of the eye white; the
whole upper plumage, sides of the head, and wing-coverts slaty
blue; quills brown, the primaries and secondaries edged with rufous;
tail brown suffused with slaty blue, and the bases of all the feathers
except the middle two pairs white; throat and breast orange-
chestnut, paling below the breast and becoming white on the
abdomen and under tail-coverts; sides of the body tinged with
brown; axillaries white.

Female. Forehead, lores, and a conspicuous ring round the eye
fulvous; sides of the head fulvous-brown with paler shafts; upper
plumage olive-brown, the wings edged with ferruginous and the
tail suffused with rufescent olive-brown; lower plumage ochraceous,
pale on the throat and abdomen, bright on the breast, the flanks
infuscated.

The young are streaked with fulvous above and on the sides of
the head, and the lower parts are fulvous with black margins on
the feathers of the breast.

Legs and feet very pale silvery to fleshy pink, the terminal joints
of the toes and the claws slightly brownish; bill black; iris deep
brown (*Hume*).

Length about 4·5; tail 1·6; wing 2·3; tarsus ·7; bill from gape
·5.

Distribution. The Himalayas from Garhwál to Sikhim; the Khási
and Nága hills; Manipur. This species breeds in the Himalayas
and also in the Khási hills, whence I have seen specimens obtained
in July and August.

Habits, &c. According to Hodgson this species makes a nest of

moss under the roots, or near the base, of a tree. The eggs are said
to be pale grey or brownish white marked with brownish red, and
to measure about ·68 by ·44.

567. Cyornis leucomelanurus. *The Slaty-blue Flycatcher.*

Digenea leuconelanura, *Hodgs. P. Z. S.* 1845, p. 26; *Horsf. & M.
Cat.* i, p. 294; *Hume, N. & E.* p. 216; *Sharpe, Cat. B. M.* iv,
p. 459, pl. xiii.
Digenea tricolor, *Hodgs. P. Z. S.* 1845, p. 26; *Horsf. & M. Cat.* i,
p. 294.
Siphia tricolor (*Hodgs.*), *Blyth, Cat.* p. 172; *Jerd. B. I.* i, p. 478;
Brooks, S. F. v, p. 471; *Hume, Cat.* no. 318.
Siphia leucomelanura (*Hodgs.*), *Blyth, Cat.* p. 172; *Jerd. B. I.* i,
p. 479; *Stoliczka, J. A. S. B.* xxxvii, pt. ii, p. 32; *Brooks, J. A.
S. B.* xli, pt. ii, p. 76; *Blanf. ibid.* p. 159; *Godw.-Aust. J. A.
S. B.* xlvii, pt. ii, p. 15; *Hume, Cat.* no. 320.
Siphia minuta, *Hume, Ibis,* 1872, p. 109; *id. S. F.* vii, p. 376; *id.
Cat.* no. 318 bis.
Digenea cerviniventris, *Sharpe, Cat. B. M.* iv, p. 460 (1879); *Salv.
Ann. Mus. Civ. Gen.* (2) vii, p. 388.
Cyornis leucomelanurus (*Hodgs.*), *Oates in Hume's N. & E.* 2nd ed.
ii, p. 3.

The Brown-winged Flycatcher, The Slaty Flycatcher, Jerd.

Coloration. Male. Upper plumage and the margins of the wing-
coverts and tertiaries dull blue; forehead and eyebrow greyish blue;
lores and sides of head black; upper tail-coverts and tail black,
the basal half of all the tail-feathers, except the middle pair, white;
quills brown, edged with pale rufous; chin and throat white, and
the remainder of the lower plumage pale fulvous-grey; sometimes
the whole lower plumage including the chin and throat is a pale
buff.

Female. Whole upper plumage olive-brown, tinged with rufous
on the rump; upper tail-coverts and tail ferruginous; a fulvous
ring round the eye; lores and sides of the head mixed fulvous and
brown; wings brown, margined with pale rufous; chin, throat,
and middle of the abdomen whitish; remaining lower plumage
ochraceous.

The young nestling has the upper plumage brown streaked with
fulvous and the lower plumage fulvous.

Bill black; legs dark brown; iris brown (*Cockburn*).

Length about 4·5; tail 2; wing 2·4; tarsus ·75; bill from gape ·5.

In Sikhim males the chin and throat are generally white and the
lower plumage pale. Birds from Shillong and Manipur become
much darker; and to such a dark bird from the latter locality Sharpe
gave the name *Digenea cerviniventris.* I do not think the two
forms are more than races.

Distribution. The Himalayas from Murree and Kashmir to Sib-
sagur and Sadiya in Assam; the Khási hills; Manipur; Karennee.
This species is found up to 7000 or 8000 feet.

Habits, &c. The nest of this Flycatcher is a massive little cup

of moss, fur, and wool placed in a hollow at the side of the trunk of
a tree. Brooks found the nest in Kashmir at the commencement
of June. The eggs, four in number, are pale buff clouded with
rufous, and measure about ·62 by ·48. Many males of this species
breed while still in immature plumage, that is in the plumage of
the female.

568. Cyornis superciliaris. *The White-browed Blue Flycatcher.*

Muscicapa superciliaris, *Jerd. Madr. Journ. L. S.* xi, p. 16 (1840).
Dimorpha albogularis, *Blyth, J. A. S. B.* xi, p. 190 (1842).
Muscicapa ciliaris, *Hodgs. in Gray's Zool. Misc.* p. 84 (1844).
Muscicapa hemileucura, *Hodgs. in Gray's Zool. Misc.* p. 84 (1844).
Muscicapula acornans, *Hodgs., Blyth, J. A. S. B.* xvi, p. 127 (1847).
Erythrosterna acornans (*Hodgs.*), *Blyth, Cat.* p. 171; *Jerd. B. I.* i,
 p. 483; *Hume, Cat.* no. 325.
Muscicapula hemileucura (*Hodgs.*), *Horsf. & M. Cat.* i. p. 296.
Muscicapula albogularis (*Bl.*), *Horsf. & M. Cat.* i, p. 297.
Muscicapula superciliaris (*Jerd.*), *Blyth, Cat.* p. 172; *Horsf. & M.
 Cat.* i, p. 296; *Jerd. B. I.* i, p. 470; *Stoliczka, J. A. S. B.* xxxvii,
 pt. ii, p. 30; *Hume, N. & E.* p. 213; *Ball, S. F.* v, p. 415; *Hume,
 Cat.* no. 310; *Sharpe, Cat. B. M.* iv, p. 204; *Barnes, Birds Bom.*
 p. 106.
Muscicapula ciliaris (*Hodgs.*), *Hume, Cat.* no. 311 bis.
Cyornis superciliaris (*Jerd.*), *Oates in Hume's N. & E.* 2nd ed. ii,
 p. 4.

The White-browed Blue Flycatcher, The Brown Flycatcher, Jerd.

Coloration. Male. The whole upper plumage, lesser and median
wing-coverts, ear-coverts, cheeks, and sides of the neck dull blue:
greater coverts and quills dark brown, edged with pale blue; tail
black, edged with blue, the basal half of all the feathers except the
median pair white; lores black; a broad supercilium from the eye
to the nape white; a broad collar across the breast, interrupted in
the middle, dull blue like the back; the whole lower plumage
white.

Female. After the autumnal moult the upper plumage is olive
brown, the forehead tinged with fulvous, the crown with minute
dark spots; the upper tail-coverts tinged with blue; wing-coverts
and tertiaries edged and tipped with fulvous; the other quills more
narrowly edged with the same; tail brown, with a tinge of blue on
the outer webs; lores and sides of the head fulvescent; lower
plumage pale buff, turning to white on the lower part of the
abdomen and under tail-coverts.

When the plumage becomes worn, the female is frequently found
with the back and rump suffused with blue.

The nestling is ashy brown above, with numerous buff spots, and
the wing-coverts tipped with the same; lower plumage pale buff,
closely mottled with brown; the outer webs of the tail-feathers
suffused with blue. In the young male the white on the tail is
present from the earliest age. The spotted plumage is soon lost,
and the adult plumage quickly acquired.

Bill black ; legs and feet dull purplish black ; iris deep brown (*Hume*).

Length about 4·5 ; tail 1·9 ; wing 2·6 : tarsus ·6; bill from gape ·6.

Distribution. The Himalayas, from Kashmir and the Hazára country to Sikhim. In summer this species is found up to 12,000 feet ; but in winter it descends to the lower ranges, and many birds find their way to the plains, whence I have examined specimens procured at Allahabad, Etáwah, Jhansi, Saugor, Raipur, Seoni, and Khandesh. It probably does not occur east of the longitude of Calcutta. Two birds procured in Karennee by Wardlaw Ramsay were entered in my 'Birds of Burmah' as belonging to this species. On reexamining these specimens, I find that they are without doubt females of *C. astigma*. Other localities recorded for this species are Ajanta by Jerdon and Ahmednagar in the Deccan by Fairbank.

Habits, &c. Breeds throughout the Himalayas from April to June, laying five eggs in a cup-shaped nest of moss in a hole of a tree or between two stones in a wall. The eggs are pale green, profusely marked with reddish, and measure about ·62 by ·48.

569. **Cyornis melanoleucus.** *The Little Pied Flycatcher.*

Muscicapa maculata, *Tick. J. A. S. B.* ii, p. 574 (1833, *descr. null.*), *nec P. L. S. Müll., nec Gmel.*

Muscicapula melanoleuca, *Hodgs., Blyth, J. A. S. B.* xii, p. 940 (1843); *id. Cat.* p. 172.

Erythrosterna pusilla, *Blyth, J. A. S. B.* xviii, p. 813 (1849); *id. Cat.* p. 171 ; *Jerd. B. I.* i, p. 482 ; *Brooks, S. F.* iii, p. 236, v, p. 471 ; *Hume, Cat.* no. 324.

Muscicapula maculata (*Tick.*), *Horsf. & M. Cat.* i, p. 296; *Sharpe, Cat. B. M.* iv, p. 297 ; *Oates, B. B.* i, p. 291.

Erythrosterna maculata (*Tick.*), *Jerd. B. I.* i, p. 483; *Brooks, S. F.* iii, p. 277 ; *Hume, Cat.* no. 326; *Barnes, Birds Bom.* p. 167 ; *Hume, S. F.* xi, p. 117.

The Rufous-backed Flycatcher, The Little Pied Flycatcher, Jerd. ; *Tuni-ti-ti,* Lepch.

Coloration. Male. The whole upper plumage, including the lores, cheeks, ear-coverts, and sides of the neck, black ; a very broad superciliary streak, reaching to the nape and widening posteriorly, white : the whole lower plumage white : wings black, the later secondaries edged with white on the outer webs : greater wing-coverts white ; tail black, the basal two-thirds of all the feathers except the middle pair white ; the bases of some of the feathers of the rump white.

Female. Resembles closely the female of *C. superciliaris*, but may be recognized by the bright ferruginous colouring of the upper tail-coverts and the pale colour of the lower plumage.

Specimens of females from Manipur and south of that place are generally much darker than those from the Indian peninsula.

The young closely resemble those of *C. superciliaris*, and are, in fact, not distinguishable from them.

Bill, legs, and claws black ; iris deep brown.

Length about 4·5 ; tail 1·7 ; wing 2·4 ; tarsus ·6 ; bill from gape ·55.

Distribution. The Himalayas, from Nepal to the extreme east of Assam, up to 7000 feet ; the eastern portion of the Empire, from the Rajmehal hills, Manabhoom, Singbhoom, and Midnapore, through Bengal and Assam, down to Tenasserim and Karennee, extending to Sumatra and Java. This species breeds in the Himalayas and also in the Khási hills : and from the latter locality I have seen young nestlings procured in June. To the other parts above mentioned it appears to be a winter visitor. The nest of this Flycatcher does not yet appear to have been found.

570. **Cyornis astigma.** *The Little Blue-and-white Flycatcher.*

Muscicapa astigma, *Hodgs. in Gray's Zool. Misc.* p. 84 (1844) ; *Gray, Cat. Mamm. etc. Coll. Hodgs.* p. 90, *App.* p. 155 (1846).

Muscicapula astigma (*Hodgs.*), *Horsf. & M. Cat.* i, p. 297 ; *Jerd. B. I.* i, p. 474 ; *Godw.-Aust. J. A. S. B.* xliii, pt. ii, p. 157, xlv. pt. ii, p. 201, xlvii, pt. ii, p. 22.

Muscicapula astigma (*Hodgs.*), *Sharpe, Cat. B. M.* iv, p. 205 ; *Hume, Cat.* no. 311 ; *id. S. F.* xi, p. 112.

Muscicapula superciliaris (*Jerd.*), *apud Oates, B. B.* i, p. 292.

Coloration. Male. Resembles the male of *C. superciliaris.* Differs in having no white on the tail-feathers, the supercilium

Fig. 6. - Bill of *C. astigma.*

absent or very faintly indicated, and the white on the throat and breast somewhat narrower.

Female. Resembles the female of *C. superciliaris.* Differs, generally speaking, in having the upper tail-coverts more fulvous than the other parts of the upper plumage, and in never having any tinge of blue on these parts, as is almost always the case in *C. superciliaris.* The plumage may be termed olive-brown on the sides of the neck and breast, and not buff. The females of the two species are, however, difficult to separate unless series of both are examined.

The young are precisely similar to those of *C. superciliaris.*

Bill and legs black ; iris brown (*Cockburn*).

Length about 4·5 ; tail 1·9 ; wing 2·6 ; tarsus ·6 ; bill from gape ·6.

The two female specimens of a Flycatcher procured by Wardlaw Ramsay in Karennee, and referred by me to *C. superciliaris* in the ' Birds of Burmah,' are now, I find, females of *C. astigma.*

c 2

Distribution. Nepal and Sikhim : the plains of Bengal, whence I have seen specimens collected by Brooks at Mudhupur and Assensole on the E.I. Railway in winter : the Khási hills, where this species is a constant resident and breeds ; probably Manipur, where Hume procured a specimen which he doubtfully refers to this species, but which specimen I have not been able to discover in his collection ; Karennee, where Wardlaw Ramsay obtained two specimens.

571. Cyornis sapphira. *The Sapphire-headed Flycatcher.*

Muscicapula sapphira, *Tickell, Blyth, J. A. S. B.* xii, p. 939 (1843) ; *Jerd. Ill. Ind. Orn.* pl. 32 ; *Blyth, Cat.* p. 173 ; *Horsf. & M. Cat.* i, p. 295 ; *Jerd. B. I.* i, p. 471 ; *Sharpe, Cat. B. M.* iv. p. 205 ; *Anders. Yunnan Exped., Aves,* p. 619 ; *Hume, Cat. no.* 312.
Siphia superciliaris ♀, *apud Godw.-Aust. J. A. S. B.* xlv, pt. ii, p. 201.

Coloration. Male. Forehead, crown, nape, and upper tail-coverts bright ultramarine-blue ; lores and a band from the eyes to the nostrils black ; sides of the head and neck, upper breast, back, rump, and visible portions of wing-coverts deep purplish blue ; quills black, narrowly edged with blue ; tail black, the outer webs suffused with bright blue ; chin, throat, and upper breast chestnut ; remainder of lower plumage pale bluish white ; under wing-coverts and axillaries white.

Female. Upper plumage olive-brown tinged with rufous ; the upper tail-coverts ferruginous ; wing-coverts and quills dark brown, edged with rufous ; forehead and sides of the head bright fulvous-brown, with a ring of fulvous round the eye ; chin, throat, and breast pale orange-chestnut ; remainder of lower plumage whitish, the flanks and sides of the body tinged with fulvous ; under wing-coverts and axillaries white.

The female of this species, with its red throat and breast, cannot be confounded with the females of *C. superciliaris, C. melanoleucus,* or *C. astigma.*

The nestling is densely spotted above and on the coverts with fulvous, and mottled with brown below.

At the autumn moult young males assume the plumage of the female, but have the wings, tail, rump, and upper tail-coverts like the adult male.

Bill black ; legs ashy brown : iris brown (*Cockburn*).

Length 4·5 ; tail 1·8 ; wing 2·3 ; tarsus ·6 ; bill from gape ·55.

Distribution. A permanent resident in Sikhim, probably extending into Nepal ; found also in the Gáro and Nága hills, and in the hills to the east of Bhámo.

572. Cyornis oatesi. *The Rufous-bellied Blue Flycatcher.*

Cyornis vivida, *Swinh. apud Hume & Dav. S. F.* xi, p. 229 ; *Hume, Cat. no.* 300 bis ; *id. S. F.* xi, p. 111.

Niltava vivida (*Swinh.*), *apud Sharpe, Cat. B. M.* iv, p. 463 (part.) ;
Oates, B. B. i. p. 286.
Niltava oatesi, *Salvadori, Ann. Mus. Civ. Gen.* (2) v, pp. 514 (1887),
578 (1888).

Coloration. Male. Forehead and lores deep black : crown, nape,
rump, upper tail-coverts, lesser and median wing-coverts shining
cobalt-blue ; back and scapulars dark bluish black ; winglet and

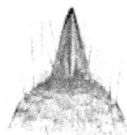

Fig. 7.—Bill of *C. oatesi.*

primary-coverts black ; greater coverts and quills black, edged with
blue ; tail black, the outer webs suffused with cobalt-blue ; region
of eye and ear-coverts black, the latter bordered posteriorly by a
band of cobalt-blue running up to the nape; chin and throat
black suffused with blue ; remainder of lower plumage, axillaries,
and under wing-coverts chestnut.

Female. Forehead, lores, round the eye, cheeks, chin, and upper
throat rufous, speckled and irregularly barred with brown ; under
tail-coverts and a large patch on the throat, the axillaries, and under
wing-coverts clear yellowish buff; remainder of lower plumage
ashy olive suffused with buff ; crown, nape, and sides of neck
ashy brown ; remainder of upper plumage olive-brown with a
fulvous tinge ; tail brown, suffused with rufous on the outer
webs.

Legs, feet, and claws dark to blackish brown ; soles yellowish ;
bill black ; iris deep brown to reddish chocolate (*Hume*).

Length about 7·5 ; tarsus ·7 to ·8 ; bill from gape about ·8.
Males from Tenasserim have wings varying from 3·7 to 4, tails 3
to 3·7 ; males from Manipur have wings varying from 3·9 to 4,
tails 3·1 to 3·3. In the females the wing is 3·7, and the tail 2·8
to 3·2.

C. vividus from China differs from the present species in being
much smaller, the wing in males varying from 3·3 to 3·5 and tail
2·6 to 2·8 ; in females the wing is 3·3 and the tail 2·5. The male
has the upper parts of a much more brilliant blue. The female
differs merely in size.

I have examined the type of *C. oatesi*, and find it the same bird
as the one which Hume identified with doubt with *C. vividus*, and
of which there are numerous specimens in the Hume collection
from Manipur and Tenasserim.

There is a female specimen of a *Cyornis* in the Hume collection
from Tenasserim, which I cannot identify with any known species.
It differs from the females of *C. oatesi* and *C. vividus*, among other
things, in wanting the conspicuous yellowish-buff patch on the
throat. It is probably the female of an undescribed species, and

it appears to be the same as a specimen noticed by Count Salvadori
(Ann. Mus. Civ. Gen. (2) vii. p. 385, 1889) under the generic name
of *Niltava*. Mr. Fea procured this specimen in Karennee, and
Count Salvadori, with his usual courtesy, forwarded it to me for
examination. Being a female specimen, and, as such, difficult to
deal with till the male is known, the Count has prudently
refrained from naming it *.

Distribution. Muleyit mountain in Tenasserim ; Manipur.

573. Cyornis pallidipes.   *The White-bellied Blue Flycatcher.*

Muscicapa pallipes, *Jerd. Madr. Journ. L. S.* xi, p. 15 (1840).
Cyornis pallipes (*Jerd.*), *Blyth, Cat.* p. 358; *Jerd. B. I.* i, p. 469;
    Hume, S. F. iv. p. 397; *id. Cat.* no. 309; *Butler, S. F.* ix, p. 397;
    Davison, S. F. x, p. 371; *Barnes, Birds Bom.* p. 165.
Siphia pallidipes (*Jerd.*), *Sharpe, Cat. B. M.* iv, p. 441.

Coloration. Male. Forehead and a supercilium ultramarine-
blue ; lores and a space in front of the eye black ; feathers round
the eye bluish black ; whole upper plumage, sides of the head and
neck, chin, throat, breast, and the margins of the wings and tail
indigo-blue; the lesser wing-coverts brighter; abdomen, vent, and
under wing-coverts white.

Female. Lores, in front of the eye, and the point of the chin
white ; an indistinct supercilium grey ; ear-coverts greyish brown ;
forehead, crown, and nape ashy brown ; upper plumage and the
sides of the neck rufescent olive-brown, the upper tail-coverts and
the outer webs of the tail-feathers chestnut; wings brown, margined with pale rufous ; throat and breast orange-chestnut
remainder of the lower plumage white.

Legs and feet fleshy tinged purple ; bill black ; iris dark wood-
brown (*Davison*).

Length about 6·5; tail 2·6 ; wing 3; tarsus ·75 ; bill from
gape ·75.

Distribution. A permanent resident in the South of India on the
Western Ghâts, from about the latitude of Belgaum to Mynall
in Travancore. This species appears to be found up to about 6000
feet.

Habits, &c. Davison calls this Flycatcher a magnificent songster.

574. Cyornis unicolor.   *The Pale Blue Flycatcher.*

Cyornis unicolor, *Blyth, J. A. S. B.* xii, p. 1007 (1843); *id. Cat.*
    p. 173; *Jerd. B. I.* i, p. 465; *Wald. Ibis,* 1876, p. 353; *Godw.-
    Aust. J. A. S. B.* xlv, pt. ii, pp. 71, 195, xlvii, pt. ii, p. 15; *Hume,*

* Count Salvadori thus describes the specimen :—

" ♀ . *Supra olivacea, pileo et cervice griseis ; uropygio et supracaudalibus
brunneo-rufescentibus ; fronte, lateribus capitis et gula rufo tinctis, gula ima
albicaste ; gastræo reliquo griseo-olivaceo, pectore magis olivascente sub-
caudalibus et subalaribus fulvescente albidis ; alis fuscis, externis brunneo-
olivaceo marginatis ; cauda brunneo-rufescente ; rostro nigro ; pedibus
fuscis.* Long. tota 0ᵐ.200 ; al. 0ᵐ.096 ; caud. 0ᵐ.077 ; rostri culm.
0ᵐ.010 ; tarsi 0ᵐ.020."

Str. F. vii, p. 516; id. Cat. no. 303; Salvadori, Ann. Mus. Civ.
Gen. (2) vii, p. 386; Hume, S. F. xi, p. 107; Oates in Hume's N.
& E. 2nd ed. ii, p. 5.
Siphia unicolor (Bl.), Sharpe, Cat. B. M. iv, p. 444.

Coloration. Male. Forehead, a broad eyebrow, and the lesser
wing-coverts ultramarine-blue: lores black, tipped with blue;
sides of the head and neck and the whole upper plumage, with the
exposed parts of the closed wing, light blue; upper tail-coverts
and the margins of the tail-feathers deep blue; lower plumage
pale dull blue, becoming albescent on the abdomen; the under
tail-coverts broadly fringed with white; axillaries pale fulvous.
Female. Lores and a ring round the eye pale rufescent; the
whole upper plumage olive-brown tinged with rufous; the upper
tail-coverts and the margins of the wings, together with the whole
tail, ferruginous; the whole lower plumage earthy brown, tinged
with ochraceous on the sides of the body; abdomen albescent;
axillaries pale fulvous.
The young nestling is brown, densely spotted and mottled with
bright fulvous.
In the male the legs and feet are dull pale purple; bill black;
iris brown. In the female the legs and feet are greyish brown;
upper mandible brown, lower mandible pale bluish horny; iris
brown (*Hume*).
Length about 6·5; tail 2·8; wing 3·3; tarsus ·65; bill from
gape ·75.
Distribution. The Himalayas from Sikhim to the Daphla hills in
Assam; the Khási hills; the Nága hills; Manipur; Karennee;
probably Arrakan.
C. cyanopolius, Blyth, from the Malay peninsula and islands, is
a much brighter bird than *C. unicolor*, and may be kept distinct.
Habits, &c. Mandelli found the nest of this bird in Sikhim in
August, a cup of moss and fern-roots placed in a depression in the
trunk of a tree about 10 feet from the ground.

575. Cyornis rubeculoides. *The Blue-throated Flycatcher.*

Phœnicura rubeculoides, *Vigors, P. Z. S.* 1831, p. 35; *Gould, Cent.*
pl. 25, fig. 1.
Cyornis rubeculoides (*Vig.*), *Blyth, Cat.* p. 173; *Horsf. & M. Cat.* i,
p. 289; *Jerd. B. I.* i. p. 466; *Hume, N. & E.* p. 211; *Hume &
Dav. S. F.* vi, p. 227; *Anders. Yunnan Exped., Aves,* p. 619;
Hume, Cat. no. 304; *Barnes, Birds Bom.* p. 164; *Hume, S. F.* xi,
p. 107; *Oates in Hume's N. & E.* 2nd ed. ii, p. 5.
Siphia rubeculoides (*Vig.*), *Sharpe, Cat. B. M.* iv, p. 445; *Legge,
Birds Ceyl.* p. 424; *Oates, B. B.* i, p. 287.
Cyornis dialilæma, *Salvadori, Ann. Mus. Civ. Gen.* (2) vii, p. 387
(1889).

The Blue-throated Redbreast, Jerd.; *Chatki,* Beng. *Manzhil-pho,*
Lepch.

Coloration. Male. Forehead and a streak over the eye glis-

tening blue; lores and the feathers at the base of the bill black;
ear-coverts dusky blue; upper plumage dark blue, the tail with

Fig. 8.—Bill of *C. rubeculoides.*

black shafts and the inner webs mostly brown; wings dark brown,
each feather narrowly edged with dark blue; lesser wing-coverts
bright blue; chin, throat, cheeks, and sides of the neck dusky
blue; breast and upper abdomen bright ferruginous; lower abdomen and under tail-coverts white; under wing-coverts pale
ferruginous.

Female. Lores albescent; upper plumage olive-brown, tinged
with ferruginous, especially on the forehead, round the eye, and
on the upper tail-coverts; wings and tail brown, edged with ferruginous; chin, throat, and breast ruddy ferruginous; abdomen
and under tail-coverts white.

The young are brown, streaked above with fulvous, the coverts
broadly tipped with fulvous: throat and breast bright fulvous
mottled with brown: abdomen white.

Iris brown; bill black, flesh-coloured at the gape; legs and toes
pale flesh-colour; claws pale horn-colour.

Length 5·7; tail 2·4; wing 2·7; tarsus ·7; bill from gape ·7.

All the males of this species throughout the continent of India,
and from Assam to Manipur have the blue on the throat of considerable extent, and sharply defined from the red breast. In only
a very few instances does the red of the breast run up into the
blue for a short distance.

From Manipur to Tenasserim the males almost invariably have
the red running up into the blue throat, but a considerable amount
of blue is always left between the tip of the red and the angle of
the chin. The amount and character of the blue on the sides of
the throat also varies a good deal in this species. I am therefore
quite unable to recognize the Tenasserim and Karennee race as
distinct from the Indian, as the transition from one type to the
other is very gradual.

Distribution. The whole extent of the Himalayas up to 6000 or
7000 feet; a considerable portion of the plains of India from the
Himalayas to Ceylon; I have failed to find any record of the
occurrence of this species in Sind, the Punjab, Rajputana, Guzerat,
and Cutch; the line of migration from Kashmir is apparently
along the Himalayas to Nepal, and thence to the plains. In India
proper this bird is by no means common, but to the east, throughout Burma down to the extreme south of Tenasserim, it is more
abundant. It is found in the Himalayas during the summer, and

in the plains during the winter; but in some portions of the latter this Flycatcher appears to be a constant resident, Hume stating that it breeds in Manipur, and Davison having found the nest in Tenasserim near Ye at the end of March. In Pegu it appeared to me to be only a winter visitor. In the Malay peninsula this species is replaced by *C. elegans*, in which the throat is a bright blue.

Habits, &c. Breeds in the Himalayas from April to June, constructing a nest of moss and lichens in a hole of a tree, bank, or rock. The eggs, three or four in number, are greenish or brownish stone-colour, marked with purplish brown, and measure about ·73 by ·6.

576. Cyornis tickelli. *Tickell's Blue Flycatcher.*

Cyornis banyumas (*Horsf.*), *apud Blyth, Cat.* p. 173; *Horsf. & M. Cat.* i, p. 290; *Jerd. B. I.* i, p. 456.

Cyornis tickelliæ, *Blyth, J. A. S. B.* xii, p. 941 (1843); *Jerd. B. I.* i, p. 467; *Blanf. Ibis,* 1870, p. 533; *Lloyd, Ibis,* 1872, p. 197; *Hume. N. & E.* p. 212; *id. S. F.* iii, p. 468.

Cyornis elegans (*Temm.*), *apud Blyth, Cat.* i, p. 173.

Cyornis jerdoni, *G. R. Gray, Blyth, Ibis,* 1866, p. 371; *Hume, Cat.* no. 305.

Cyornis tickelli, *Blyth, Anders. Yunnan Exped., Aves,* p. 620; *Hume, Cat.* no 306; *Barnes, Birds Bom.* p. 164; *Oates in Hume's N. & E.* 2nd ed. ii, p. 7.

Siphia tickelliæ (*Bl.*), *Sharpe. Cat. B. M.* iv, p. 447; *Legge, Birds Ceyl.* p. 421; *Oates, B. B.* i, p. 289.

Horsfield's Blue Redbreast, Tickell's Blue Redbreast, Jerd.

Coloration. Male. Resembles the male of *C. rubeculoides*, but is larger and of a duller colour; the rufous of the breast runs up into the throat and chin, leaving only the extreme point of the latter black.

Female. Resembles the male closely. The whole upper plumage, sides of the head, coverts and visible portions of the closed wings and tail dull blue; forehead, eyebrow, and bend of wing shining cobalt-blue; lores and feathers over the nostrils whitish; chin, throat, and breast pale orange; remainder of lower plumage, whitish.

The young are streaked with fulvous, and resemble the young of the other members of the genus.

Iris brown; bill blackish; legs and feet bluish brown, dusky bluish, or bluish grey (*Legge*).

Length nearly 6; tail 2·5; wing 2·8; tarsus ·7; bill from gape ·75.

The few Tenasserim specimens I have seen agree well in colour and size with Indian ones, but at Kossoom, in the Malay Peninsula, a very much smaller race occurs, which may hereafter be thought worthy of separation under a distinct name. I have only been able to examine four specimens, and cannot do more than draw attention to the subject.

Distribution. The whole peninsula of India, except Sind and the extreme north-west portion. This species is absent from the Himalayas. It appears to be a resident wherever found, but may be locally migratory according to season and place. To the east it has occurred in Manipur, the hills east of Bhámo, the Karen hills east of Toungngoo, Karennee, Thoungyah in Tenasserim, extending to Kossoom in the Malay peninsula in a modified form as above noticed. This Flycatcher occurs in Ceylon up to 4000 feet.

Habits, &c. Breeds in May, June, and July, constructing a nest of dead leaves and grass in a hole of a bank or tree, and laying three eggs, which are so thickly speckled as to appear to be of an olive-colour or brownish rufous throughout, and measure about ·76 by ·56.

577. **Cyornis magnirostris.** *The Large-billed Blue Flycatcher.*

Cyornis magnirostris, *Blyth, J. A. S. B.* xviii, p. 814 (1849); *Jerd. B. I.* i, p. 469; *Godw.-Aust. J. A. S. B.* xxxix, pt. ii, p. 100; *Blanf. J. A. S. B.* xli, pt. ii, p. 158; *Hume, Cat.* no. 308.
Siphia magnirostris (*Bl.*), *Sharpe, Cat. B. M.* iv, p. 453; *Oates, B. B.* i, p. 290.

Colouration. Male. Upper plumage, cheeks, ear-coverts, sides of the neck, and wing-coverts deep blue, brilliant on the forehead and over the lores and eyes; feathers at the base of the upper mandible and lores black; chin, throat, and breast chestnut; sides of the breast blue; sides of the abdomen fulvous; middle of the abdomen and the under tail-coverts white; under wing-coverts and axillaries clear buff; tail-feathers dark brown, suffused with blue on the outer webs; greater wing-coverts and quills dark brown, with dull blue margins.

Female. Upper plumage olive-brown, tinged with rufous on the upper tail-coverts; tail brown, suffused with ferruginous; coverts and quills brown, edged with rufous-olive; lores and a ring round the eye fulvous; chin, throat, and breast orange-rufous, paler on the chin and throat; flanks suffused with ochraceous; abdomen and under tail-coverts white.

The young are similar to those of *C. rubeculoides.*

Iris dark brown; legs pale flesh-colour (*Godwin-Austen*).

Length about 6; tail 2·5; wing 3·2; tarsus ·7; bill from gape ·9.

Distribution. Sikhim; Cachar; the Khási hills; the extreme south of Tenasserim. The distribution of this species as known is very incomplete, and it will probably be found spread over the greater part of Assam and Burma. It is a resident in Sikhim; occurs in Cachar in May, and in Tenasserim from December to March.

There are some specimens of this bird in the Hodgson collection, but nothing to show that they came from Nepal. They were probably obtained in Sikhim.

Genus NITIDULA, Jerdon & Blyth, 1861.

The single species of this genus differs from the members of *Cyornis* in having a very narrow slender bill. It has been observed very little, and I cannot find a single note about its habits. In structure this bird is a true Flycatcher, having well-developed hairs over the nostrils.

578. Nitidula hodgsoni. *The Pigmy Blue Flycatcher.*

Nemura hodgsoni, *Moore, P. Z. S.* 1854, p. 76, pl. 62; *Horsf. & M. Cat.* i, p. 300.
Nitidula campbelli, *Jerdon & Blyth. P. Z. S* 1861, p. 201.
Nitidula hodgsoni (*Moore*), *Jerd. B. I.* i, p. 472; *Blanf. J. A. S. B.* xli, pt. ii, p. 159; *Godw.-Aust. J. A. S. B.* xlvii, pt. ii, p. 15; *Hume, Cat.* no. 313; *id. S. F.* xi, p. 112.
Tarsiger hodgsoni (*Moore*), *Sharpe, Cat. B. M.* iv, p. 258.

Coloration. Male. Lores and a frontal band black; sides of the head black, with a bluish tinge; the whole upper plumage bright

Fig. 9.—Bill of *N. hodgsoni.*

blue, the anterior half of the crown ultramarine; wings and tail black, the outer webs edged with blue; the whole lower plumage pale orange-yellow; under wing-coverts and axillaries white.

Female. The whole upper plumage, the exterior margins of the wing- and tail-feathers, and the ear-coverts olive-brown, slightly rufescent on the rump and upper tail-coverts; lores and cheeks fulvous yellow, slightly mottled with brown; the whole lower plumage saffron-yellow, paling on the abdomen and under tail-coverts.

I have not been able to examine a nestling bird; but after the autumn moult the young of both sexes resemble the adult female, and the male begins to assume the adult plumage about March.

Bill black; legs pale reddish; iris dark brown (*Jerdon*).

Length nearly 4; tail 1·4; wing 1·9; tarsus ·6; bill from gape ·5.

Distribution. A resident species in Sikhim up to 7000 feet or higher; Sadiya and Dibrugarh in Assam; the Nága hills.

Genus STOPAROLA, Blyth, 1847.

The genus *Stoparola* is hardly worthy of separation from *Cyornis*. All the members of the genus, however, are green or blue throughout in both sexes, and the type of the genus has a small and very depressed bill which, when viewed from above, forms an equilateral triangle.

Two members of the genus are sedentary and confined to small areas. The third is spread over the Empire and is migratory to a greater or less extent. They are typical Flycatchers in habits.

Key to the Species.

a. No white on tail.
  a'. First primary much shorter than half
         second; tail blue or green .... *S. melanops*, p. 28.
  b'. First primary quite half the length of
         second; tail dark brown .... *S. sordida*, p. 29.
b. Base of tail white........................ *S. albicaudata*, p. 30.

579. **Stoparola melanops.** *The Verditer Flycatcher.*

Muscicapa melanops, *Vigors, P. Z. S.* 1831, p. 171; *Gould, Cent.* pl. 6.
Stoparola melanops (*Vig.*), *Blyth, Cat.* p. 174; *Hume, N. & E.* p. 208; *Sharpe, Cat. B. M.* iv, p. 438; *Hume, Cat.* no. 301; *Oates, B. B.* i, p. 285; *Barnes, Birds Bom.* p. 164; *Oates in Hume's N. & E.* 2nd ed. ii, p. 9.
Hypothymis melanops (*Vig.*), *Horsf. & M. Cat.* i, p. 292.
Eumyias melanops (*Vig.*), *Jerd. B. I.* i, p. 463; *Hume & Henders. Lah. to Yark.* p. 186; *Anders. Yunnan Exped., Aves,* p. 622.

Nil kat-katia. Beng.: *Sibgell-pho,* Lepch.

Fig. 10.—Bill of *S. melanops.*

Coloration. Male. Lores, feathers in front of the eye and the feathers at the base of the upper mandible black; the whole plumage verditer-blue, brightest on the forehead, chin, throat, breast, and upper tail-coverts; under tail-coverts broadly fringed with white; tail blue, the shafts black and the inner webs edged with brown; primaries and secondaries blue on the outer and black on the inner webs; tertiaries wholly blue; upper wing-coverts blue.

Female. The general colour is much duller, but otherwise resembles the male; the chin and throat are mottled with white, the lores are brown, and the under tail-coverts are more broadly fringed with white.

The young are greenish grey, the sides of the head and the whole lower plumage being spotted with fulvous. Occasionally white spots are present on the head and back, and one adult has a white nape-patch.

The *Stoparola spilonota* of Gray (Hand-list, no. 4898), the type of which is still in the British Museum, resembles the present species, but each feather of the rump, upper tail-coverts, and abdomen has a triangular streak of brown. No other specimen resembling it has yet been found, and it is probably an accidental variety. It came from Nepal.

Bill and legs black; iris brown; mouth flesh-colour; claws black.

Length about 6; tail 2·8; wing 3·2; tarsus ·65; bill from gape ·7.

Distribution. The whole Empire with the exception of Sind, Ceylon, the Andamans and Nicobars, and that portion of the peninsula of India south of the Nilgiris. This species breeds throughout the Himalayas up to about 9000 feet, and visits the plains during the winter. It probably breeds in some portions of the plains, and in some of the hill-ranges of the peninsula and Burma, for I have examined specimens killed at Ahmednagar in July, Shillong in the same month, and Momein, to the east of Bhámo, in June. Hume found it breeding in Manipur, and Godwin-Austen on the Khási hills. This Flycatcher extends into China, Cochin China, and the Malay peninsula.

Habits, &c. Breeds from April to July, constructing a nest of moss inside a hole in a tree, wall, or bank, and laying four eggs which are pinky white, sometimes unmarked, at others speckled with reddish. The eggs measure about ·78 by ·57.

580. **Stoparola sordida.** *The Dusky-blue Flycatcher.*

Glaucomyias sordida, *Wald. A. M. N. H.* (4) v, p. 218 (1870); *Hume, S. F.* iii, p. 401.
Stoparola sordida (*Wald.*), *Sharpe, Cat. B. M.* iv, p. 440; *Hume, Cat.* no. 302 bis; *Legge, Birds Ceyl.* p. 419, pl. xviii; *Oates in Hume's N. & E.* 2nd ed. ii, p. 11.

Coloration. Male. Forehead and a short eyebrow bright cobalt-blue; lores and region of the nostrils black; the whole body-plumage ashy grey tinged with blue, brightest on the crown; abdomen, vent, and under tail-coverts albescent; wing-coverts, wings, and tail dark brown, very narrowly margined with ashy blue.

The young are dark brown above, each feather with a streak or oval drop of fulvous in the centre; lower plumage fulvous white, each feather margined with blackish; abdomen albescent.

Iris reddish brown to brown; bill black; legs and feet dark plumbeous, the feet sometimes blackish, much darker than tarsus; claws black (*Legge*).

Length about 6; tail 2·5; wing 3; tarsus ·75; bill from gape ·7.

It is very probable that the female of this species will be found to be slightly duller in colour than the male, but I have seen no sexed female. All the birds in the small series of this species in the British Museum appear to be males, but only one is so sexed.

Distribution. A resident in Ceylon up to 2000 feet.

581. **Stoparola albicaudata.** *The Nilghiri Blue Flycatcher.*

Muscicapa albicaudata, *Jerd. Madr. Journ. L. & S.* xi, p. 16 (1840); id. *Ill. Ind. Orn.* pl. xiv.

Stoparola albicaudata (*Jerd.*), *Blyth, Cat.* p. 175; *Hume, N. & E.* p. 210; id. *Cat. no.* 302; *Sharpe, Cat. B. M.* iv, p. 457; *Davison, S. F.* x, p. 370; *Oates in Hume's N. & E.* 2nd ed. ii, p. 11.

Hypothymis albicaudata (*Jerd.*), *Horsf. & M. Cat.* i, p. 292.

Eumyias albicaudata (*Jerd.*), *Jerd. B. I.* i, p. 464.

Coloration. Male. The whole plumage indigo-blue; the forehead, a short and broad eyebrow, and the edge of the wing ultramarine-blue; the throat suffused with the same; lores and the region of the nostrils black; abdomen bluish brown mottled with white; under tail-coverts bluish brown, broadly fringed with white; wings and tail dark brown edged with blue, all the tail-feathers except the median pair white at the base.

Female. The whole upper plumage dull greyish tinged with olivaceous; upper tail-coverts dark bluish brown; tail blackish, the bases of all the feathers except the median pair white; entire lower plumage dull greyish blue, tinged with olivaceous on the throat and fore neck; coverts and quills dark brown edged with rufescent, the greater coverts tipped with fulvous.

The young have the upper plumage and wing-coverts brown, each feather margined with black and centred with fulvous; wings and tail as in the adult; lower plumage pale greyish brown, barred with black and fulvous.

Bill, legs, and feet black; iris dark brown (*Davison*).

Length about 6; tail 2·6; wing 3·2; tarsus ·75; bill from gape ·7.

Distribution. The Nilgiri and Palni hills up to 7000 feet.

Habits, &c. Breeds from February to May, constructing a nest of moss in a hole in a tree, wall, or bank, and laying three eggs, which are white or pale buff marked with reddish, and measure about ·81 by ·59.

Genus **MUSCITREA**, Blyth, 1847.

The genus *Muscitrea* contains one Indian species the position of which is somewhat doubtful. The British Museum does not contain a nestling bird of this species, but judging from the circumstance that a few birds have the wing-coverts margined with rufous, as is the case with so many young Thrushes and Flycatchers, I incline to the belief that the nestling will prove to be spotted. The presence of numerous long hairs over the nostrils further induces me to place this species in its present position.

In *Muscitrea* the sexes are alike; the bill is strong, deep, and much compressed laterally; the wing is rather long and straight, and the first primary is large, being more than half the length of the second; the tail is square, and the plumage brown.

There is but little on record about the habits of this bird. The one I observed in Pegu was solitary and silent, and was perched on a stalk of elephant-grass.

582. Muscitrea grisola. *The Grey Flycatcher.*

Tephrodornis grisola, *Blyth, J. A. S. B.* xii, p. 180* (1843); *id. Cat.* p. 153; *Jerd. B. I.* i, p. 411.
Muscitrea cinerea, *Blyth, J. A. S. B.* xvi, p. 122; *Hume, S. F.* v, p. 101.
Hylocharis philomela (*Boie*), *Hume, S. F.* ii, p. 201.
Hylocharis occipitalis, *Hume, S. F.* ii, p. 202 (1874).
Muscitrea grisola (*Bl.*), *Hume & Dav. S. F.* vi, p. 206; *Hume, Cat.* no. 266; *Oates, B. B.* i, p. 257.
Pachycephala grisola (*Bl.*), *Gadow, Cat. B. M.* viii, p. 220.

The Arakan Wood-Shrike, Jerd.

Figs. 11 & 12.-- Head and bill of *M. grisola*.

Coloration. Forehead, crown, nape, and lores ashy brown; upper plumage, wings, and tail rufous- or olive-brown, the secondaries broadly edged with rufous; sides of the head pale brown; chin and throat white mottled with ashy; breast pale ashy; remainder of lower plumage and under wing-coverts white.

The young are slightly rufous.

Bill dark brownish black; mouth flesh-colour; iris reddish brown; eyelids plumbeous; legs plumbeous; claws pale horn-colour.

Length about 6·5; tail 2·6; wing 3·3; tarsus ·8; bill from gape ·8.

Some specimens are paler and greyer, others darker and browner, and the plumage varies apparently according to the length of time which has elapsed since the moult in the autumn.

Distribution. Jerdon states that this species has been procured near Calcutta; it occurs in the Andaman Islands, Arrakan, Pegu, and Tenasserim, extending to the Malay peninsula and islands.

Genus ANTHIPES, Blyth, 1847.

I place in the genus *Anthipes* five species of Flycatchers in which the sexes are alike, the plumage brown or rufous, relieved, in the case of three, by a patch of white on the throat, the bill flattened, the first primary large, and the lower mandible dark coloured. They are all very local, and they are not known to migrate.

In addition to the above characters the rictal bristles are long but few in number, and the tail is square.

Key to the Species.

a. Chin and throat white, in strong contrast to
surrounding parts.
 a'. White of chin and throat surrounded by a
 firm black band.
  a''. Forehead and eyebrow fulvous........ *A. moniliger*, p. 32.
  b''. Forehead and eyebrow white .... *A. leucops*, p. 33.
 b. White of chin and throat not surrounded
 by a black band..................... *A. submoniliger*, p. 33
b. Chin and throat buff or whitish, blending with
 surrounding parts.
 c'. General colour of lower plumage orange-
 buff................................. *A. poliogenys*, p. 33.
 d'. General colour of lower plumage white,
 merely tinged with ochraceous on breast
 and flanks .... *A. olivaceus*, p. 34.

583. Anthipes moniliger. *Hodgson's White-gorgeted Flycatcher.*

Dimorpha moniliger, *Hodgs. P. Z. S.* 1845, p. 26.
Anthipes gularis, *Blyth, J. A. S. B.* xvi, p. 122 (1847).
Anthipes moniliger (*Hodgs.*), *Blyth, Cat.* p. 172; *Jerd. B. I.* i,
 p. 477; *Hume, Cat.* no. 317; *Oates in Hume's N. & E.* 2nd ed.
 ii, p. 13.
Digenea moniliger (*Hodgs.*), *Sharpe, Cat. B. M.* iv, p. 460, pl. xiv,
 fig. 1; *Oates, B. B.* i, p. 300.

The White-gorgeted Flycatcher, Jerd.; *Phutt-tugruk-pho,* Lepch.

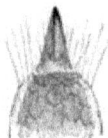

Fig. 13.—Bill of *A. moniliger.*

Coloration. Forehead and a short eyebrow bright fulvous; lores,
ear-coverts, and under the eye greyish brown with white shafts;
the whole upper plumage and sides of the head olive-brown tinged
with rufous on the rump; upper tail-coverts and tail dull ferruginous; coverts and wings brown edged with ferruginous; chin
and throat white, surrounded on all sides by a black band; lower
plumage fulvous-olive, becoming white on the abdomen.

Bill black; legs and claws pale fleshy; iris dark brown (*Jerdon*).
Length about 5; tail 2; wing 2·4; tarsus ·8; bill from gape ·6.

The young nestling is no doubt spotted, but the youngest bird I
have seen merely differs from the adult in not having the black
band round the white of the throat.

Distribution. Sikhim up to about 7000 feet or so. Hodgson's
specimens, now in the British Museum, do not appear to have been
obtained in Nepal but in Sikhim.

Habits, &c. According to Mandelli, this species breeds in Sikhim

from April to June, constructing a nest of moss on the ground in
grass and low jungle. The eggs are described as being white
marked with brownish red, and measuring about ·73 by ·54.

584. Anthipes leucops. *Sharpe's White-gorgeted Flycatcher.*

Anthipes moniliger (*Hodgs.*), *Blyth & Walden, Birds Burm.* p. 103;
  Godw.-Aust. J. A. S. B. xlv, pt. ii, p. 195.
Digenea leucops, *Sharpe, P. Z. S.* 1888, p. 246.

Coloration. Resembles *A. moniliger*, but has the forehead and
eyebrow white, and the sides of the head ashy grey.
Iris bright dark brown; bill slate-brown; legs white tinged
fleshy (*Wardlaw Ramsay*).
Of the same size as *A. moniliger.*
Distribution. The Khási hills; Gonglong, Manipur hills; Karennee at 5000 feet.

585. Anthipes submoniliger. *Hume's White-gorgeted Flycatcher.*

Anthipes submoniliger, *Hume, S. F.* v, p. 105 (1877); *Hume & Dav.
  S. F.* vi, pp. 232, 510; *Hume, Cat. no.* 317 bis.
Digenea submoniliger (*Hume*), *Sharpe, Cat. B. M.* iv, p. 461; *Oates,
  B. B.* i, p. 301; *Sharpe, P. Z. S.* 1888, p. 246.

Coloration. Upper plumage fulvous-brown; forehead, lores, a
broad but short supercilium, and a circle of feathers round the eye
rich golden-fulvous; sides of the head fulvous-brown; chin and
throat white, with a few black feathers on the side of the chin;
breast and flanks olive-brown; abdomen, vent, and under tail-
coverts white; wings and wing-coverts brown, edged with rufous;
tail ferruginous.
Bill black, yellowish on lower mandible; iris deep brown;
legs fleshy white (*Davison*).
Length about 5; tail 2; wing 2·4; tarsus ·85; bill from gape ·6.
Distribution. Tenasserim, where this species has been found on
Muleyit mountain and at the foot of Nwalabo mountain.

586. Anthipes poliogenys. *Brooks's Flycatcher.*

Cyornis poliogenys, *Brooks, S. F.* viii, p. 469 (1879); *Hume, S. F.*
  ix, pp. 195, 295, xi, p. 108.
Siphia cacharensis, *Madarász, Zeitschr. ges. Orn.* 1884, p. 52, pl. i,
  fig. 2.

Coloration. Forehead, crown, and nape greyish brown; back,
rump, scapulars, and the margins of the quills olive-brown; upper
tail-coverts and tail ferruginous; lores and a ring round the eye
light grey; sides of the head brownish grey; chin and throat pale
buff; breast and sides of the body orange-buff; middle of the
abdomen whitish; under tail-coverts pale buff or buffy white.

Iris brown; bill black; edges of the eyelids yellowish; legs pale greyish pink, pale silvery fleshy, pale silvery purplish (*Hume*).

Length about 6; tail 2·4; wing 2·8; tarsus ·75; bill from gape ·8.

This bird resembles the female of *Cyornis rubeculoides*, but may be recognized at a glance by the size of its first primary, which is equal to half the length of the second, whereas in *C. rubeculoides* the first primary is much less than half the second.

Distribution. The Bhutan Doars; Dibrugarh in Assam; Shillong; Cachar; Tipperah; Manipur. Hume records this species from the Sikhim Terai.

587. **Anthipes olivaceus.** *Hume's Flycatcher.*

Cyornis olivacea, *Hume, S. F.* v, p. 338 (1877); *Hume & Dav. S. F.* vi, p. 229; *Hume, Cat.* no. 307 ter.
Siphia olivacea (*Hume*), *Sharpe, Cat. B. M.* iv, p. 457; *Oates, B. B.* i, p. 292.

Coloration. Upper plumage greyish brown, tinged with fulvous on the back and rump; lores and the sides of the head ashy, the shafts of the ear-coverts whitish; lower plumage whitish, the breast and the sides of the body suffused with ochraceous; tail reddish brown edged with ferruginous; wing-coverts and quills brown edged with rufous-brown.

Bill black in the male, brown in the female; iris brown; legs, feet, and claws pinkish white (*Hume & Davison*).

Length 5·8; tail 2·5; wing 3; tarsus ·75; bill from gape ·8.

The young bird, which Hume identified with some doubt with *Hemichelidon ferruginea* (S. F. vi, p. 227), is in my opinion the young of the present species. It resembles very closely the young of *Cyornis rubeculoides* at the same age.

Distribution. The extreme southern part of Tenasserim at Bankasan and Maiawun. This species also occurs in Java and Borneo. It appears to be a resident in Tenasserim, Hume's specimens having been obtained in March, June, and December.

Genus **ALSEONAX**, Cabanis, 1850.

The genus *Alseonax* contains three Indian Flycatchers which are allied to *Cyornis*. In *Alseonax*, however, the sexes are alike and the plumage is brown or rufous as in *Anthipes*. The first primary is very small and all the three species are wide migrants.

Key to the Species.

a. Upper plumage and tail ashy brown with no tinge of rufous . . . . *A. latirostris*, p. 35.
b. Upper plumage olive-brown; upper tail-coverts and the whole tail chestnut . . . . . . *A. ruficaudus*, p. 36.
c. Upper plumage ruddy brown; upper tail-coverts ferruginous; tail brown, suffused with rufous on the outer webs of the feathers only. *A. muttui*, p. 36.

588. **Alseonax latirostris.** *The Brown Flycatcher.*

Muscicapa latirostris, *Raffl. Tr. Linn. Soc.* xiii, p. 312 (1821).
Butalis terricolor, *Hodgs., Blyth, J. A. S. B.* xvi, p. 120 (1847); *id. Cat.* p. 175.
Hemichelidon latirostris (*Raffl.*), *Blyth, Cat.* p. 175; *Horsf. & M. Cat.* i, p. 137.
Muscicapa cinereo-alba, *Temm. & Schleg. Faun. Jap., Aves,* p. 42, pl. 15 (1850).
Alseonax latirostris (*Raffl.*), *Jerd. B. I.* i, p. 450; *Hume & Henders. Lah. to Yark.* p. 185, pl. v; *Hume, S. F.* ii, p. 219; *Sharpe, Cat. B. M.* iv, p. 127; *Legge, Birds Ceyl.* p. 415; *Hume, Cat.* no. 297; *Brooks, S. F.* ix, p. 225; *Oates, B. B.* i, p. 277; *Barnes, Birds Bom.* p. 163.
Alseonax terricolor (*Hodgs.*), *Jerd. B. I.* i, p. 460; *Hume, Cat.* no. 298.

The Southern Brown Flycatcher, The Rufescent-brown Flycatcher, Jerd.; *Zakki,* Hind.

Fig. 11.—Bill of *A. latirostris.*

Coloration. Upper plumage ashy-brown, the feathers of the crown with darker centres; tail dark brown, the outer feathers very narrowly tipped with whitish; wings and coverts dark brown, all but the primaries broadly edged with ashy white; lores and a ring of feathers round the eye white; sides of the head brown; lower plumage white, tinged with ashy on the breast and sides of the body.

The young have the crown blackish streaked with fulvous; the upper plumage and wings with large terminal fulvous spots; the lower plumage like that of the adult but mottled with brown. After the autumn moult and till the following spring the young are very rufous.

Bill black, the base of the lower mandible yellow; mouth orange; iris brown; legs and claws black. The young bird has the whole bill yellow except the tip, which is dusky.

Length rather more than 5; tail 2; wing 2·8; tarsus ·5; bill from gape ·7.

Distribution. The whole Empire except the north-west portion. I have seen no example of this species from Sind, the Punjab, Rajputana, or Guzerat. It occurs in Ceylon and the Andamans. On the Himalayas it is a summer visitor as far west as Chamba, and it is found in the other portions of the Empire chiefly in winter, but some birds appear to be resident in certain parts all the year round, for I have seen a specimen obtained in Ceylon in June.

This Flycatcher has a wide range, being found from Japan and Eastern Siberia to the Philippines and Java. Specimens killed in the Malay peninsula in July and August are contained in the British Museum Collection.

D 2

589. Alseonax ruficaudus. *The Rufous-tailed Flycatcher.*

Muscicapa ruficauda, *Swains. Nat. Lib.* x, *Flycatchers*, p. 251 (1838).
Cyornis ruficauda (*Swains.*), *Jerd. B. I.* i, p. 464; *Stoliczka, J. A. S. B.* xxxvii, pt. ii, p. 50; *Hume, S. F.* iv, p. 396; *id. Cat.* no. 307; *Barnes, Birds Bom.* p. 165.
Siphia ruficauda (*Swains.*), *Sharpe, Cat. B. M.* iv, p. 457.

Coloration. Upper plumage and sides of the neck dull olive-brown; upper tail-coverts and tail chestnut; coverts and quills of wing brown edged with olive-brown; lores and round the eye greyish white; ear-coverts fulvous-brown with pale shafts; the whole lower plumage pale earthy brown.

The young are of the usual spotted character, but the upper tail-coverts and tail are chestnut from the first.

Iris dark brown; upper mandible pale brown, lower fleshy; legs, feet, and claws purplish brown (*Davison*).

Length 5·5 to 6; tail 2·3; wing 3·3; tarsus ·6; bill from gape ·7.

Distribution. Found in summer in Kashmir, Gilgit, and the Himalayas from Murree to the valley of the Bhagirati river up to altitudes of 10,000 feet. This species extends in summer to Afghanistan, where Wardlaw Ramsay observed it breeding but failed to find its nest. In the winter months it has been recorded from Ahmednagar and Mount Abu, and from numerous other localities in the west of India down to Travancore, extending on the east as far as Raipur in the Central Provinces.

Godwin-Austen records this species from Cachar, but probably by an oversight.

590. Alseonax muttui. *Layard's Flycatcher.*

Butalis muttui, *Layard, A. M. N. H.* (2) xiii, p. 127 (1854); *Legge, Ibis*, 1878, p. 203; *Hume, S. F.* vii, p. 513; *id. Cat.* no. 299 ter.
Cyornis mandellii, *Hume, S. F.* ii, p. 510 (1874), iv, p. 396, vii, p. 456; *id. Cat.* no. 307 bis.
Alseonax flavipes, *Legge, S. F.* iii, p. 367 (1875).
Alseonax muttui (*Layard*), *Sharpe, Cat. B. M.* iv, p. 132; *Legge, Birds Ceyl.* p. 417, pl. xviii.
Siphia mandellii (*Hume*), *Sharpe, Cat. B. M.* iv, p. 453, footnote.
Alseonax mandellii (*Hume*), *Sharpe, Cat. B. M.* iv, App. p. 472; *Hume, S. F.* xi, p. 109.

Coloration. Forehead, crown, and nape olive-brown, with indistinct brown shaft-streaks; remainder of the upper plumage ruddy brown, changing to bright ferruginous on the upper tail-coverts; wings brown, the outer webs of the coverts and quills broadly edged with ferruginous; tail brown, suffused with ferruginous on the outer webs; lores and a conspicuous ring round the eyes white; chin and throat white; ear-coverts olive-brown; cheeks, sides of the neck, the whole breast, and the sides of the body yellowish brown; middle of the abdomen, vent, and under tail-coverts pale yellowish white.

Legs and feet pale wax-yellow; claws brown; upper mandible blackish brown, yellowish at tip; lower mandible dull horny-yellow; iris brown (*Hume*).

Length about 5·5; tail 2·1; wing 2·8; tarsus ·55; bill from gape ·65.

I have not been able to examine a young bird.

Distribution. Summers in Sikhim and probably in other parts of the Himalayas and winters in Travancore and Ceylon. This species occurs at Shillong on the Khási hills, where it has been obtained in August, September, and October, and in Manipur, where Hume observed it in April. It may probably be a permanent resident, therefore, in Shillong and in Manipur. Brooks observed it at Mudhupur on the E. I. Railway on passage. Legge notes it from Ceylon in January and June. Altogether our knowledge of the movements of this Flycatcher is very imperfect.

Genus **OCHROMELA**, Blyth, 1847.

The genus *Ochromela* was instituted by Blyth for the reception of a Flycatcher which is remarkable alike for its coloration and its habits.

In *Ochromela* the sexes are slightly dissimilar, but both preserve the characteristic black and orange plumage. The bill is blunt and thick, and the rictal bristles are not only very numerous but very long. The wing is blunt, with the first primary longer than half the length of the second, and the tail is considerably rounded.

In habits this species approaches *Pratincola*, descending to the ground for an instant to capture its prey, and it differs from all other Flycatchers in constructing a large globular nest. It is described by Jerdon as frequenting dense woods.

591. **Ochromela nigrirufa.** *The Black-and-Orange Flycatcher.*

Saxicola nigrirufa, *Jerd. Madr. Journ. L. S.* x, p. 266 (1839).
Ochromela nigrirufa (*Jerd.*), *Blyth, Cat.* p. 173; *Horsf. & M. Cat.* i, p. 289; *Jerd. B. I.* i, p. 462; *Hume, N. & E.* p. 207; *id. Cat.* no. 309; *Davison, S. F.* x, p. 369; *Oates in Hume's N. & E.* 2nd ed. ii, p. 14.
Siphia nigrirufa (*Jerd.*), *Sharpe, Cat. B. M.* iv, p. 455; *Legge, Birds Ceyl.* p. 425.

Fig. 15.—Bill of *O. nigrirufa.*

Coloration. Male. Forehead, crown, nape, hind neck, lores, sides

of the head, and the whole wing black, some of the coverts and
inner quills very narrowly tipped with orange; remainder of the
plumage rich orange, somewhat paler on the abdomen.

Female. Resembles the male, but the black is replaced by
greenish brown, and the orange-coloured parts of the plumage are
paler; the lores and a large space round the eye are rufous
speckled with dusky, and the wings are dark brown.

I have not been able to examine a young bird of this species.

Bill black; iris dark brown; legs and feet dark plumbeous fleshy
(*Davison*).

Length about 5; tail 2; wing 2·5; tarsus ·8; bill from gape ·6.

Distribution. The Hill-ranges of Southern India from the
Wynaad to Cape Comorin at 2500 feet and upwards. This species
appears to be common on the Nilgiri, the Palni, and the Assambu
hills. Its occurrence in Ceylon is doubtful. Colonel McMaster is
of opinion that he observed this bird at Chikalda in Berar.

Habits, &c. Breeds from April to June, constructing a large
globular nest of coarse grass and fern-leaves in a clump of reeds
or on the summit of a stump of a tree. The eggs, two or three in
number, are greyish white marked with brownish red, and measure
about ·7 by ·53.

Genus CULICICAPA, Swinhoe, 1871.

The genus *Culicicapa* contains one species of very wide distribution over the Empire, migratory in the plains, but resident in
the Himalayas and many of the hill-ranges.

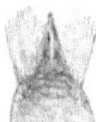

Fig. 16.—Bill of *C. ceylonensis.*

In this genus the sexes are alike and the plumage is grey and
yellow. The bill is very much depressed, and viewed from above
forms an equilateral triangle: the rictals are extremely numerous
and long; the first primary is short and the tail is square.

592. **Culicicapa ceylonensis.** *The Grey-headed Flycatcher.*

Platyrhynchus ceylonensis, *Swains. Zool. Ill.* ser. 1, i, pl. 13 (1820-21).
Cryptolopha cinereocapilla (*Vieill.*), *Blyth, Cat.* p. 205; *Horsf. & M.
Cat.* i, p. 147; *Jerd. B. I.* i, p. 455.
Myialestes cinereocapilla (*Vieill.*), *Hume, N. & E.* p. 205.
Culicicapa ceylonensis (*Swains.*), *Hume, Cat.* no. 295; *Scully, S. F.*
viii. p. 275; *Sharpe, Cat. B. M.* iv, p. 369; *Legge, Birds Ceyl.*
p. 410; *Oates, B. B.* i, p. 274; *Barnes, Birds Bom.* p. 162; *Oates
in Hume's N. & E.* 2nd ed. ii, p. 16.

Zird phutki, Beng.

Coloration. The whole head, neck, and breast ashy, darker on the crown, the feathers of which are obsoletely centred with brown; lower plumage bright yellow; under wing-coverts pale yellow; lores and the edges of the eyelids whitish; back, rump, scapulars, and upper tail-coverts greenish yellow; wings and coverts dark brown, the outer webs of all the feathers edged with greenish yellow; the lesser coverts edged on both webs; tail dark brown, the outer webs of all the feathers except the median pair broadly edged with greenish yellow.

Iris dark hazel; bill brown, paler at the base and gape; mouth yellow; legs yellowish brown; claws horn-colour.

Length about 5; tail 2·2; wing 2·4; tarsus ·55; bill from gape ·55.

Distribution. The whole Empire except Sind, the Punjab, and Rajputana, from which provinces I have not seen any specimens. This species is a permanent resident on the Himalayas up to 8000 feet, and on all the hill-ranges such as the Nilgiris, Khási hills, &c., but to many portions of the plains it is probably only a winter visitor. The most westerly locality on the Himalayas from which I have examined a specimen is Baramula on the Jhelum river in Kashmir. It is found in Ceylon at about 1000 feet elevation, but it apparently does not extend to the Andamans nor to the Nicobars. This Flycatcher ranges as far as Java and Borneo.

Habits, &c. Breeds from March to June, constructing a small nest of moss against a rock or tree-trunk. The eggs, three or four in number, are whitish marked with brown and grey, and measure about ·61 by ·48.

Genus **NILTAVA**, Hodgs., 1837.

The genus *Niltava* contains three species of Flycatchers remarkable for the brilliant plumage of the males. The sexes differ in colour, but both may be recognized by the presence of a bright spot or mark on the side of the neck. *Cyornis oatesi* approaches these birds in having some bright blue on the side of the neck, but the patch is of a different character, being connected with the nape and forming a band rather than a spot. This bright mark is moreover absent in the female.

The Niltavas frequent thick jungle and are less typical in their habits than the species of *Cyornis*, and they are said to eat berries. They appear to be resident on the Himalayas. Mr. Cripps, however, states that *N. sundara* and *N. macgrigoriæ* are seasonal visitors to Dibrugarh in Assam, but he omits to state at what season they visit that district.

In *Niltava* the bill is somewhat compressed laterally and narrow, and the base is covered by a multitude of dense plumelets, which conceal the nostrils; the rictal bristles are moderate in number and in length; the first primary is large, being quite half the length of the second, and the tail is rounded.

Key to the Species.

a. Wing 4 inches or longer .... *N. grandis*, p. 40.
b. Wing not much exceeding 3 inches or less.
    a′. Under wing-coverts and axillaries chest-
      nut or buff .... *N. sundara*, p. 41.
    b. Under wing-coverts and axillaries white
      or ashy white .... *N. macgrigoriæ*, p. 42.

593. **Niltava grandis.** *The Large Niltava.*

Chaitaris grandis, *Blyth, J. A. S. B.* xi, p. 189 (1842).
Niltava grandis (*Blyth*), *Blyth, Cat.* p. 174; *Horsf. & M. Cat.* i,
  p. 288; *Jerd. B. I.* i, p. 476; *Hume, N. & E.* p. 215; *Hume &*
  Dav. S. F. vi, p. 232; *Sharpe, Cat. B. M.* iv, p. 464; *Hume, Cat.*
  no. 316; *id. S. F.* xi, p. 113; *Oates, B. B.* i, p. 297; *id. in Hume's*
  N. & E. 2nd ed. ii, p. 18.

The Large Fairy Blue-Chat, Jerd.; *Margony,* Lepch.

Fig. 17.—Bill of *N. grandis.*

Coloration. Male. Forehead, crown, rump, upper tail-coverts,
lesser and median wing-coverts, and a large patch on each side of
the neck brilliant cobalt-blue; back and scapulars purplish blue;
middle tail-feathers purplish blue, the others brown on the inner
web and blue on the outer; greater coverts and quills black,
narrowly edged with blue; feathers at base of upper mandible,
lores, sides of the head, chin, throat, and upper breast black;
lower breast dull blue; abdomen bluish ashy, the under tail-coverts
fringed with whitish.

Female. Forehead, lores, round the eye, ear-coverts, and cheeks
fulvous, with pale shafts; crown and nape ashy brown; a patch on
each side of the neck bright blue; back and rump fulvous-brown;
tail and wings dark brown, with the outer webs suffused with deep
rufous; the whole lower plumage rich olive-brown, the feathers of
the throat and breast with whitish shafts; the middle of the chin
and throat clear buff, and of the abdomen ashy; under wing-coverts
and axillaries buff.

The young nestling is dark brown, streaked with fulvous; wings
and tail as in the female.

Iris deep brown; in the male the bill is black, the legs and feet
black or very dark plumbeous; in the female the bill is brownish
black, the legs, feet, and claws fleshy-pink (*Hume & Davison*).

Length about 8·5; tail 3·6; wing 4·2; tarsus ·9; bill from gape ·1.

Distribution. The Himalayas from Nepal to the Daphla hills in Assam from 4000 to 7000 feet elevation; the Khási and Nága hills; Manipur; Muleyit mountain in Tenasserim.

Habits, &c. Breeds from April to June, constructing a nest of moss in a cleft of a rock or of a tree-trunk, and laying four eggs, which are pale buff freckled with pinkish brown, and measure about ·9 by ·7.

594. Niltava sundara. *The Rufous-bellied Niltava.*

Niltava sundara, *Hodgs. Ind. Rev.* i, p. 650 (1837); *Blyth, Cat.* p. 174; *Horsf. & M. Cat.* i, p. 288; *Jerd. B. I.* i, p. 473; *Hume, N. & E.* p. 213; *Blyth, Birds Burm.* p. 102; *Hume, Cat. no.* 314; *Sharpe, Cat. B. M.* iv, p. 463; *Oates, B. B.* i, p. 295; *id. in Hume's N. & E.* 2nd ed. ii, p. 20.

The *Rufous-bellied Fairy Blue-Chat,* Jerd.; *Niltau,* Nep.; *Margong,* Lepch.

Coloration. Male. Forehead, lores, sides of the head, chin, and throat deep black; crown of the head, nape, rump, upper tail-coverts, a spot on either side of the neck, and the lesser wing-coverts glistening blue; the remaining coverts and quills dark brown, edged with purplish blue; back and scapulars purplish black; tail black, the outer webs edged with bright blue; the whole lower plumage and the under wing-coverts chestnut.

Female. Upper plumage olive-brown, tinged with fulvous on the rump; upper tail-coverts and tail chestnut; wings brown, edged with rufous; sides of the head and lores mixed fulvous and brown, with paler shafts; lower plumage rich olive-brown, tinged with ochraceous; the chin and throat rather rufous, the abdomen whitish; a large oval patch of white on the fore neck, and a small patch of brilliant blue on each side of it; under tail-coverts pale buff; under wing-coverts and axillaries buff.

Young birds are dark brown streaked with fulvous both above and below. In males some trace of blue, and in females an indication of the white neck-patch, make their appearance at an early age.

Bill black; legs brown; iris dark brown (*Jerdon*).

Length about 6·5; tail 2·7; wing 3·2; tarsus ·8; bill from gape ·7.

Distribution. The Himalayas from Simla to Assam from 5000 to 8000 feet elevation; the Khási hills; Karennee at 4000 feet. Blyth records this species from Arrakan and Tenasserim. I have not been able to examine specimens from these localities. It extends into Western China.

Habits, &c. Breeds in April and May, constructing a nest of moss among the roots, or in a crevice of a trunk of a tree, or sometimes against a rock. The eggs are pale buff freckled with dingy pink, and measure about ·93 by ·71.

595. **Niltava macgrigoriæ.** *The Small Niltava.*

Phœnicura macgrigoriæ, *Burton, P. Z. S.* 1835, p. 152.
Niltava macgrigoriæ (*Burton*), *Blyth, Cat.* p. 174; *Horsf. & M. Cat.*
i, p. 288; *Jerd. B. I.* i, p. 475; *Hume, N. & E.* p. 214; *Wald. in
Blyth, Birds Burm.* p. 102; *Hume & Dav. S. F.* vi, p. 231; *Sharpe,
Cat. B. M.* iv, p. 465; *Hume, Cat.* no. 315; *id. S. F.* xi, p. 113;
Oates, B. B. i, p. 280; *id. in Hume's N. & E.* 2nd ed. ii, p. 21.

The Small Fairy Blue-Chat, Jerd.; Phatt-tagrak-pho, Lepch.

Coloration. Male. Upper plumage bright purplish blue; forehead, supercilium, rump, upper tail-coverts, and a patch on each side of the neck cobalt-blue; lesser wing-coverts brown, tipped with blue; median coverts entirely blue; greater coverts and quills dark brown, edged with blue; median tail-feathers entirely blue, the outer webs of the others blue, the inner dark brown; lores, feathers at base of the upper mandible and those in front of and below the eye black; cheeks, ear-coverts, chin, throat, and breast purple; the breast occasionally ashy like the abdomen; remainder of the lower plumage ashy, becoming albescent on the abdomen and under tail-coverts; under wing-coverts ashy white.

Female. Upper plumage olive-brown, tinged with rufous; tail rufous-brown; wing-coverts and quills brown, edged with rufous-brown; forehead and sides of the head mixed brown and fulvous; a patch of brilliant blue on each side of the neck; lower plumage ochraceous buff, paling on the abdomen and under tail-coverts; under wing-coverts and axillaries pure white.

The young nestling is streaked with fulvous.

Bill black; legs reddish black; iris dark brown (*Jerdon*).

Length nearly 5; tail 2·1; wing 2·6; tarsus ·7; bill from gape ·6.

Distribution. The Himalayas from Garhwál to Assam from about 3000 to 5000 feet elevation; the Khási hills; North Cachar hills; Manipur; Karennee; Northern Tenasserim.

Habits, &c. Breeds from April to June, constructing a nest of moss, sometimes, it is said, on the ground, at other times in the hole of a trunk of a tree. The eggs are described as being white or stone-coloured, freckled with brownish purple or brownish pink, and measure about ·76 by ·53.

Genus **PHILENTOMA**, Eyton, 1845.

The two Flycatchers which constitute the genus *Philentoma* are birds of peculiar coloration, maroon and chestnut entering into its composition.

In this genus the sexes are dissimilar. The bill is very large and coarse, and its base is covered by the frontal plumelets; the wing is rounded, and the first primary is much longer than half the length of the second; the tail is square. Both species are resident, and they have all the habits of the typical Flycatchers, catching insects on the wing.

Key to the Species.

a. Wings and tail blue .... *P. velatum*, p. 43.
b. Wings and tail chestnut .... *P. pyrrhopterum*, p. 43.

596. Philentoma velatum. *The Maroon-breasted Flycatcher.*

Drymophila velata, *Temm. Pl. Col.* no. 334.
Philentoma velatum (*Temm.*), *Blyth, Cat.* p. 201 ; *Horsf. & M. Cat.*
i, p. 392 ; *Hume & Dav. S. F.* vi, pp. 224, 503 ; *Hume, Cat.*
no. 289 ter ; *Sharpe, Cat. B. M.* iv, p. 365 ; *Oates, B. B.* i, p. 263.

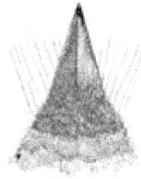

Fig. 18.—Bill of *P. velatum.*

Coloration. Male. General colour indigo-blue ; forehead, lores,
chin, cheeks, ear-coverts, and round the eye black ; throat and
breast rich maroon ; quills black, the outer webs broadly margined
with indigo-blue, and the tertiaries wholly of this colour, but with
the shafts black ; median tail-feathers indigo-blue ; the others
black on the inner web and blue on the outer.
Female. Wholly dull indigo-blue except the wings and tail,
which are coloured as in the male.
The nestling has the whole plumage chestnut except the wings
and tail, which are the same as in the adult, but all the median
and greater wing-coverts are broadly tipped with chestnut. Almost
before it is able to leave the nest blue feathers appear on all parts,
and probably before the first autumn the full adult plumage is
assumed ; but both sexes resemble the adult female at first, the
male donning the maroon breast later on.
Legs and feet bluish- or purplish-black ; bill black, iris lake to
crimson (*Hume & Davison*).
Distribution. Tenasserim from about Muleyit mountain to the
extreme south. This species extends down the Malay peninsula to
Sumatra, Java, and Borneo.

597. Philentoma pyrrhopterum. *The Chestnut-winged Flycatcher.*

Muscicapa pyrrhoptera, *Temm. Pl. Col.* no. 596 (1823).
Philentoma pyrrhopterum (*Temm.*), *Blyth, Cat.* p. 205 ; *Hume &
Dav. S. F.* vi, p. 223 ; *Hume, Cat.* no. 289 bis ; *Sharpe, Cat. B. M.*
iv, p. 366 ; *Oates, B. B.* i, p. 264.

Coloration. Male. The whole head and neck, back, breast, and lesser wing-coverts indigo-blue; lower back and rump rufous-grey; upper tail-coverts, the tail, a portion of the outer webs of the scapulars, the whole of the tertiaries, and the greater portion of the outer webs of the secondaries bright chestnut; remainder of the quills dark brown, the outer webs of the primaries margined with reddish grey; primary-coverts blue, centred with blackish; greater wing-coverts chestnut; lower plumage pale buff, becoming paler on the vent and under tail-coverts.

Female. Forehead, crown, nape, and ear-coverts dark olive-brown; lores and round the eye pale ashy; back, scapulars, and rump rufous-ashy; upper tail-coverts and tail chestnut; lesser wing-coverts olive-brown, tinged with rufous on the margins; greater coverts chestnut; wings dark brown, the outer webs of the primaries rufescent, those of the other quills chestnut; the tertiaries wholly chestnut; lower plumage pale buff, somewhat brighter on the throat and breast.

In the males legs, feet, and claws pale purplish blue; bill black; iris crimson; in the adult female legs, feet, and claws plumbeous olive; upper mandible pale horny-brown; lower mandible fleshy-white; iris dull red; in a young female legs and feet pale horny-red; iris pale red, speckled with white (*Hume & Davison*).

Length about 7; tail 2·8; wing 3·2; tarsus ·65; bill from gape ·9.

I have not been able to examine a young bird of this species, but it probably follows the young of *P. velatum* in its style of coloration.

Distribution. Tenasserim from Nwalabo mountain to its southern extremity, extending to the Malay peninsula, Cochin China, Sumatra, and Borneo.

Genus **TERPSIPHONE**, Gloger, 1827.

In *Terpsiphone* the typical characteristics of the Flycatchers are developed to a greater extent than in any other genus.

The bill is extremely large, depressed, and swollen, and the rictal bristles are very numerous, coarse, and long. The head is crested, and the tail is greatly developed in the mature males.

In *Terpsiphone* the sexes are alike, or closely so, during the first two years, and the prevailing colour of the plumage is chestnut. The female never drops her chestnut garb, but the male after the second autumn, or even later, assumes a white plumage. It thus happens that a pair may be found breeding both being in the chestnut plumage, or a female in chestnut plumage may be found mated with a white male.

The Paradise Flycatchers are found over the whole Empire, both on the Himalayas and in the plains and lesser hill-ranges, and are resident in many portions of the country, but in others they appear to be seasonal visitors, or in great measure so, but their movements are probably of no great extent.

The members of this genus are typical Flycatchers, catching

insects on the wing, never descending to the ground, and of solitary habits. Their notes are very harsh.

Key to the Species.

a. Crest long and pointed, reaching to upper part of back.	
a'. Throat and sides of head ashy........	*T. paradisi*: ♀ at all ages and ♂ before second autumn, p. 45.
b'. Throat and sides of head glossy black..	*T. paradisi* : ♂ after second autumn moult, p. 45.
b. Crest short and rounded, not reaching beyond nape.	
c'. Back deep chestnut; throat and sides of head bluish ashy..............	*T. affinis*: ♀ at all ages, ♂ while in chestnut plumage, p. 47.
d'. Back ashy rufous.	*T. nicobarica*: ♀ at all
a''. Throat and sides of head dark ashy..	ages, p. 48.
b''. Throat and sides of head black	*T. nicobarica*: ♂ while in chestnut plumage, p.48.

598. **Terpsiphone paradisi.** *The Indian Paradise Flycatcher.*

Muscicapa paradisi, *Linn. Syst. Nat.* i, p. 324 (1766).
Muscipeta paradisea, *Jerd. Ill. Orn.* pl. 7.
Tchitrea paradisi (*Linn.*), *Blyth, Cat.* p. 203; *Horsf. & M. Cat.* i, p. 133; *Jerd. B. I.* i, p. 445; *Hume, N. & E.* p. 196; *Hume & Henders. Lah. to York.* p. 184.
Muscipeta paradisi (*Linn.*), *Cripps, S. F.* vii, p. 274; *Hume, Cat.* no. 288; *Scully, S. F.* viii, p. 273; *Barnes, Birds Bom.* p. 158.
Terpsiphone paradisi (*Linn.*), *Sharpe, Cat. B. M.* iv, p. 346; *Legge, Birds Ceyl.* p. 404; *Oates in Hume's N. & E.* 2nd ed. ii, p. 22.

The Paradise Flycatcher, Jerd.; *Shah Bulbul, Hosseini Bulbul, Sultana Bulbul, Taklah, Doodhraj*, Hind.; *Touka pigli pitta*, Tel.; *Wal Kondalati*, Tam.

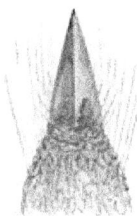

Fig. 19.—Bill of *T. paradisi*.

Coloration. The young of both sexes have the forehead, crown, nape, and crest metallic bluish black : sides of the head, chin, throat, and the neck all round ashy brown, the brown of the throat blending gradually with the pale ashy of the breast, and this again with the white of the remainder of the lower parts ; upper plumage, tail, wing-coverts, tertiaries, and the outer webs of the other quills chestnut.

This plumage is retained till the second autumn by the male, and permanently by the female, which undergoes no further change of any kind. The young male for some time previous to the second autumn becomes gradually blacker on the chin and throat, and sometimes becomes quite black on those parts, as well as on the sides of the head, as in the adult, but the breast remains ashy and is never pure white contrasting with the black throat.

After the autumn moult of the second year the male has the whole head and crest glossy black, the lower parts as before the moult, and the whole upper plumage rich chestnut ; the median tail-feathers grow to a great length, and are retained till May or June, when they are cast.

After the autumn moult of the third year the chestnut plumage is again assumed, and also the long median tail-feathers, but the whole lower plumage from the throat downwards is pure white, the breast being sharply demarcated from the black throat. After this moult a gradual transition to the white upper plumage takes place, the wings and tail being the first parts to be affected, but the change to a complete white plumage is not affected till the moult of the fourth autumn.

After this moult the male bird is fully adult, and permanently retains the white plumage. The whole head, neck, and crest are glossy bluish black ; the whole body-plumage white, the feathers of the back, rump, scapulars, and wing-coverts with black shafts ; tail white, with black shafts and narrow outer margins, except on the middle feathers, where the shaft is black only on the basal third of its length and at the tip ; wings black, with broad white margins on both webs, the later secondaries and tertiaries being almost entirely white.

Bill, gape, and margin of eyelids cobalt-blue, the tip of the bill darker ; iris dark brown ; feet plumbeous blue ; claws dusky (*Scully*).

Length from about 9 to 21 ; tail 4·5 to 16·5 ; wing 3·7 ; tarsus ·65 : bill from gape 1·1.

Distribution. The whole of India proper as far east as Nepal in the Himalayas and the Brahmaputra river in the plains. To the west this species extends into Afghanistan, and to the north into Turkestan. In Kashmir and other parts of the Himalayas it is found in summer up to 9000 feet or so. This Flycatcher occurs in Ceylon. It appears to be everywhere a permanent resident, except in the Himalayas, where it moves to lower levels in winter.

Habits, &c. Breeds from May to July, constructing a small cup-shaped nest of grass, fibres, or moss in the branch of a tree. The eggs, four or five in number, are pink marked with brownish red, and measure about ·81 by ·6.

599. **Terpsiphone affinis.** *The Burmese Paradise Flycatcher.*

Tchitrea affinis, *Hay, Blyth, J. A. S. B.* xv, p. 292 (1846); *id. Cat.* p. 203; *Horsf. & M. Cat.* i, p. 134; *Jerd. B. I.* i, p. 448.
Tchitrea paradisi (*Linn.*), *Hume, S. F.* iii, p. 102.
Muscipeta affinis (*Hay*), *Hume & Dav. S. F.* vi, p. 223; *Hume, Cat.* no. 289.
Terpsiphone affinis (*Hay*), *Anders. Yunnan Exped., Aves,* p. 654; *Sharpe, Cat. B. M.* iv, p. 349; *Oates, B. B.* i, p. 264; *id. in Hume's N. & E.* 2nd ed. ii, p. 26.

Coloration. Up to the second autumn both sexes are alike. The forehead, crown, nape, and a short blunt crest are metallic black; sides of the head, chin, throat, and neck all round with the breast dark bluish ashy; remainder of the lower plumage white; back, rump, upper tail- and wing-coverts, and tail deep chestnut; wings black, broadly margined with chestnut.

The female undergoes no further change of plumage.

The male after the moult of the second autumn acquires two long median tail-feathers, but probably sheds them at the end of the breeding-season.

At the moult of the third autumn the white plumage is assumed in its entirety, and in this state resembles the white phase of *T. paradisi*, differing only in the shaft-stripes of the upper plumage being much broader, the tail more broadly edged with black, and the shafts of the median pair of tail-feathers being black for a greater length (for three quarters of their length and at the tip), and in having a short rounded crest.

The nestling is rich chestnut above, with darker tips to some of the feathers; the lower plumage white, the breast mottled with rufous.

Iris hazel-brown; eyelids plumbeous, the edges tumid and rich blue; mouth yellow; bill blue, the tip and anterior half of the margins black; legs plumbeous blue; claws dark horn-colour.

Length 8 to 18; tail 4 to 14; wing 3·6; tarsus ·7; bill from gape 1.

Distribution. The Himalayas from Sikhim to Dibrugarh in Assam, and thence south throughout the hill-tracts and Burma to the extreme southern part of Tenasserim. This species extends east of Burma and down the Malay peninsula, where it meets *T. incii*, and owing to the similarity of *T. affinis* and *T. incii* in certain stages of plumage the respective limits of these two species have not been determined with any great exactness.

T. incii resembles *T. affinis* very closely, but the large series of the former in the British Museum appears to prove beyond doubt that the male never assumes the white plumage.

Habits, &c. I found the nest of this bird in Burma at the end of April. It was cup-shaped, and composed of dry bamboo-leaves and fibres, and it was placed near the summit of a small tree. The eggs are similar to those of *T. paradisi*, and measure about ·85 by ·6.

600. **Terpsiphone nicobarica**, n. sp.　*The Nicobar Paradise-Flycatcher.*

Tchitrea affinis (*Hay*), *Walden, Ibis,* 1873, p. 309; *Hume, S. F.* ii, p. 216.

Coloration. The changes of plumage of this species cannot be followed with the same certainty as in the two preceding species, by reason of the smaller number of specimens available for examination, but certain characters are present which serve to distinguish the present form at once.

In the chestnut plumage the male has the whole head and neck all round with the chin and throat glossy black; the back ashy rufous; rump dull rufous; upper tail-coverts, tail, and the visible portions of the closed wing bright chestnut; breast dark ashy; remainder of lower plumage pale ferruginous.

The female in the chestnut plumage, which is never changed, resembles the chestnut male, but has the sides of the head, chin, throat, breast, and sides of the neck uniform dark ashy.

In the white stage the male is undistinguishable from the male of *T. affinis.*

This species differs from *T. paradisi* in all stages of plumage by having a short, round crest, by the ashy tint of the plumage of the back when in the chestnut stage; and by the same characters, when in the white stage, as those which separate *T. affinis* from *T. paradisi.*

From *T. affinis* it differs when in chestnut plumage by the male having the whole head and neck, chin and throat glossy black, *T. affinis* (male) in this stage having only the crown and crest of this colour, the other parts being constantly dark ashy. In white plumage the two species appear to be undistinguishable.

From *T. incii* of the Malay peninsula it differs in the pale coloration of the back when in chestnut plumage, and further in attaining a white stage, which *T. incii* apparently never does.

It is of the same size as *T. affinis.*

Distribution. The Andaman and Nicobar Islands, where this species appears to be a constant resident.

Genus **HYPOTHYMIS**, Boie, 1826.

The genus *Hypothymis* contains two Indian Flycatchers of a brilliant blue colour, one being of wide distribution and the other confined to the Andamans.

In *Hypothymis* the bill is very large and flattened, its length being about double its width at the nostrils. The rictal bristles are very strong and numerous. The tail is ample, being of the same length as the wing. The sexes differ, but preserve the same pattern of colour, with the exception of certain marks on the head and neck. Both the species found within the Empire are resident.

Key to the Species.

a. Abdomen, vent, and under tail-coverts white ...... *H. azurea*, p. 49.
b. Abdomen, vent, and under tail-coverts blue ...... *H. tytleri*, p. 50.

691. **Hypothymis azurea.** *The Indian Black-naped Flycatcher.*

Muscicapa azurea, *Bodd. Tabl. Pl. Enl.* p. 41 (1783).
Myiagra caerulea (*Vieill.*), *Blyth, Cat.* p. 204.
Myiagra azurea (*Bodd.*), *Horsf. & M. Cat.* i, p. 138; *Jerd. B. I.* i,
    p. 450; *Hume, N. & E.* p. 198.
Hypothymis azurea (*Bodd.*), *Anders. Yunnan Exped., Aves*, p. 655;
    Hume, Cat. no. 290; *Sharpe, Cat. B. M.* iv, p. 274; *Oates, B. B.*
    i, p. 265; *Barnes, Birds Bom.* p. 159; *Oates in Hume's N. & E.*
    2nd ed. ii, p. 27.
Hypothymis ceylonensis, *Sharpe, Cat. B. M.* iv, p. 277 (1879);
    Legge, Birds Ceyl. p. 408, pl. xviii.

The Black-naped Blue Flycatcher, Jerd.: *Kala kat-katia*, Beng.

Fig. 20.—Bill of *H. azurea.*

Coloration. Male. A patch on the nape, forehead, angle of the chin, and a crescentic bar across the fore neck black; abdomen, vent, and under tail-coverts white, or faint bluish white; remainder of lower plumage azure-blue; wings and coverts dark brown edged with blue; tail brown, suffused with blue on the median pair of feathers and the outer webs of the others; under wing-coverts and axillaries white.

Female. Head above azure-blue: sides of the head, chin, and throat duller blue, the ear-coverts almost brown; breast ashy blue; abdomen, flanks, and under tail-coverts white tinged with grey; wings, back, rump, and upper tail-coverts brown; tail darker brown, the outer edges washed with blue.

I have not been able to examine a nestling of this species.

Iris dark brown: eyelids plumbeous, the edges blue; bill dark blue, the edges and tip black: mouth yellow; legs plumbeous; claws horn-colour.

Length about 6·5: tail 3; wing 2·8: tarsus ·7; bill from gape ·75.

H. ceylonensis, from Ceylon, is said to differ in the male wanting the black bar across the throat, but I am of opinion that this alleged difference does not really hold good. Ceylonese specimens of this Flycatcher are not common in collections, but the British Museum contains six males. Of these, five have no black throat-bar, but they also have no nape-patch, which shows them to be young. The sixth bird has a small nape-patch and traces of a throat-bar. This last specimen, therefore, clearly shows that the Ceylon bird does sometimes at least exhibit a throat-patch, and this being the case there is no character by which the two supposed races can be separated.

Distribution. The whole Empire east of a line drawn approximately from Mussoorie in the Himalayas to Khandála on the Western Ghâts. This species does not ascend the Himalayas above 3000 feet or thereabouts. It occurs in Ceylon and the Nicobar Islands, but is absent from the Andamans, where it is replaced by the next species.

Habits, &c. Breeds generally from May to August, constructing a beautiful cup of fine grass coated exteriorly with cobwebs and cocoons in the fork of a branch not far from the ground. The eggs, three to five in number, are white or pinkish marked with reddish, and occasionally some purple spots and specks, and measure about ·69 by ·53.

602. Hypothymis tytleri. *The Andaman Black-naped Flycatcher.*

Myiagra tytleri, *Beavan, Ibis,* 1867, p. 324 ; *Ball, S. F.* i, p. 68 ;
   Hume, S. F. ii, p. 217 : *id. N. & E.* p. 199.
Hypothymis occipitalis (*Vig.*), *Sharpe, Cat. B. M.* iv, p. 275 (part.).
Hypothymis tytleri (*Beavan*), *Hume, Cat. no.* 290 ; *Oates in Hume's*
   N. & E. 2nd ed. ii, p. 30.

Coloration. Resembles *H. azurea.* The male differs from the male of that species in having the abdomen, vent, and under tail-coverts of the same blue as the breast ; the female similarly in having those parts dingy lilac-grey.

The differences pointed out above hold good in a considerable series of the Andamanese bird, and I think that it forms a species easily recognizable from the Continental and Nicobarese form. A richly-coloured Indian bird and a dull Andamanese bird may be difficult to separate, but such pairs of birds are not often met with and do not in my opinion affect the question. Hume recognizes the two species in his Catalogue.

In retaining the Andamanese form under Beavan's name I do so because I am not satisfied that any prior name applies to it with certainty. The forms from the Malay peninsula and the islands do not seem identical with the Andamanese bird, but rather to be referable to *H. azurea.*

Distribution. The Andaman Islands and the Great and Little Cocos.

Habits, &c. Davison found the nest of this species on 23rd April at Aberdeen, South Andaman, with three eggs. Both nest and eggs resembled those of *H. azurea.*

Genus **CHELIDORHYNX**, Hodgs., 1844.

The genus *Chelidorhynx* contains only one species of Flycatcher remarkable for the shape of its bill, which is short and pointed, and when viewed from above forms a perfect equilateral triangle.

Fig. 24.—Bill of *C. hypoxanthum.*

The rictal bristles are extremely numerous and long. The tail is of about the same length as the wing, rounded, and with the shafts thickened and white. The sexes are quite alike.

603. **Chelidorhynx hypoxanthum.** *The Yellow-bellied Flycatcher.*

Rhipidura hypoxantha, *Blyth, J. A. S. B.* xii, p. 935 (1843); *id. Cat.* p. 205.
Chelidorhynx hypoxantha (*Bl.*), *Horsf. & M. Cat.* i. p. 147; *Jerd. B. I.* i, p. 455; *Blanf. J. A. S. B.* xli, pt. ii, p. 47; *Hume, N. & E.* p. 204; *Hume, Cat.* no. 294; *Scully, S. F.* viii, p. 275; *Sharpe, Cat. B. M.* iv, p. 279; *Oates, B. B.* i, p. 269; *id. in Hume's N. & E.* 2nd ed. ii, p. 30.

The Yellow-bellied Fantail, Jerd.; *Sitte kleom,* Lepch.

Coloration. A broad band on the forehead, continued back as a broad supercilium, and the whole lower plumage bright yellow; lores, feathers round the eye, cheeks, and ear-coverts dark brown, tinged with green, the shafts of the latter part whitish; tail brown, with conspicuous white shafts and all the feathers except the middle pair tipped white; upper plumage and wing-coverts olive-brown, the greater coverts tipped with white; wings brown, edged with the colour of the back.

I have not been able to examine a young bird.

Bill black above, the lower mandible yellow; iris brown; gape orange; feet brownish (*Scully*).

Length about 4·7; tail 2·3; wing 2·1; tarsus ·6; bill from gape ·4.

Distribution. The Himalayas from Simla to Assam up to 12,000 feet; the Khási hills; Manipur; the hills east of Toungngoo in Pegu. This species appears to be a permanent resident wherever it is found.

Habits, &c. According to Blanford this Flycatcher is usually seen in small flocks hunting about trees. The nest appears to be a deep cup made of moss, hair, wool, &c., built on the branch of a tree, and the eggs white without spots.

Genus **RHIPIDURA**, Vigors & Horsf., 1826.

The genus *Rhipidura* is a very extensive one, and contains four Indian Flycatchers. In these birds the bill is large, about twice as long as broad, and the rictal bristles are very numerous and long. The tail is very ample and rounded. The sexes are alike or nearly so.

These Flycatchers are abundant everywhere and are resident. They are very lively, constantly on the move, and frequently seen with outspread tail dancing from branch to branch. They make small and very beautiful nests covered with cobwebs.

Key to the Species.

a. Forehead and sides of the crown broadly
   white.. *R. albifrontata*, p. 52.
b. Forehead black; a small white supercilium.
  a'. Abdomen black .... *R. albicollis*, p. 53.
  b'. Abdomen white or whitish.
    a''. Outer tail-feathers distinctly and
        abruptly tipped white.............. *R. javanica*, p. 54.
    b''. Outer tail-feathers merely paler towards
        the tips............. .... *R. pectoralis*, p. 55.

604. **Rhipidura albifrontata.** *The White-browed Fantail*
Flycatcher.

Rhipidura albifrontata, *Frankl. P. Z. S.* 1831, p. 116; *Horsf. & M. Cat.* i, p. 145; *Anders. Yunnan Exped., Aves,* p. 655; *Sharpe, Cat. B. M.* iv. p. 338; *Legge, Birds Ceyl.* p. 412; *Oates, B. B.* i, p. 268; *id. in Hume's N. & E.* 2nd ed. ii, p. 31.
Rhipidura aureola, *Less. Traité,* p. 380 (1831).
Leucocerca albofrontata (*Frankl.*), *Blyth, Cat.* p. 206; *Jerd. B. I.* i, p. 452; *Hume, N. & E.* p. 201.
Leucocerca aureola (*Less.*), *Hume, S. F.* i, p. 436, iii, p. 104; *id. Cat. no.* 292; *Barnes, Birds Bom.* p. 160.
Leucocerca burmanica, *Hume, S. F.* ix, p. 175, footnote (1880).

The White-browed Fantail, Jerd.; *Macharya,* Hind. in the South; *Manati,* Mal.; *Dasari-pitta,* Tel.

Coloration. Male. Crown, lores, ear-coverts, and the feathers round the eye black: forehead and a very broad supercilium to the nape white: upper plumage and wings ashy brown, the wing-coverts tipped with white; tail brown, all but the median pair of feathers tipped white, progressively more and more so to the outermost feather, which is almost entirely white; cheeks, chin, and throat black, each feather broadly terminated with white, except on the lower throat, where the white is reduced to narrow margins; sides of the breast black: remainder of lower plumage white.

Female. Very similar to the male, but browner above.

The young resemble the adult, but the back and wing-coverts are margined with rufous.

Bill, legs, and feet black; iris brown.

Length about 7; tail 3·7; wing 3·2; tarsus ·8; bill from gape ·7.

Birds of this species from Burma are characterized by a nearly total absence of white spots on the wing-coverts and by the presence of more white on the chin and throat; to this race Hume has given the name of *burmanica*. I do not propose to keep it distinct, as the specimens of this species in the British Museum from Burma are very few, and the characters pointed out above may prove to be accidental or variable.

Distribution. The whole Empire, ascending the Himalayas to 4000 or 5000 feet. This species is apparently rare from Assam down to Tenasserim, but is found in suitable localities all over the tract. It occurs in Ceylon, but not, so far as is known, in the Andamans and Nicobars.

Habits, &c. Breeds from February to August, having two or more broods. The nest, composed of fine grass and coated with cobwebs, is generally placed on a stout branch of a tree, or sometimes in a fork, and is cup-shaped. The eggs, usually three in number, are white or cream-coloured, marked with greyish brown, and measure about ·66 by ·51.

605. **Rhipidura albicollis.** *The White-throated Fantail Flycatcher.*

Platyrhynchus albicollis, *Vieill. Nouv. Dict. d'Hist. Nat.* xxvii, p. 13 (1818).

Rhipidura fuscoventris, *Frankl. P. Z. S.* 1831, p. 117; *Horsf. & M. Cat.* i, p. 144.

Leucocerca fuscoventris (*Frankl.*), *Blyth, Cat.* p. 206; *Jerd. B. I.* i, p. 451; *Hume, N. & E.* p. 209.

Rhipidura albicollis (*Vieill.*), *Anders. Yunnan Exped., Aves,* p. 656; *Sharpe, Cat. B. M.* iv, p. 317; *Oates, B. B.* i, p. 295; *id. in Hume's N. & E.* 2nd ed. ii, p. 35.

Leucocerca albicollis (*Vieill.*), *Ball, S. F.* vii, p. 211; *Hume, Cat.* no. 291; *Barnes, Birds Bom.* p. 160; *Hume, S. F.* xi, p. 104.

The *White-throated Fantail,* Jerd.; *Chak-dayal,* Beng.; *Chak-dil,* in the N.W. Provinces; *Nam-dit-nom,* Lepch.

Coloration. Crown, lores, sides of the head, and angle of the chin black; a short supercilium white; throat white, extending laterally to the sides of the neck, the bases of the feathers black, causing the white to appear dull; with these exceptions the whole plumage is dark sooty brown; tail dark brown, all but the middle pair of feathers broadly tipped with white. The female does not differ from the male.

The young have the back and wing-coverts tipped with rufous, the lower plumage fringed with rufous, and the white supercilium and white on the throat barely indicated.

Bill, legs, and feet black; mouth fleshy white; eyelids grey; iris deep brown; claws blackish horn-colour.

Fig. 22.—Bill of *R. albicollis*.

Length about 7·5; tail 4; wing 2·9; tarsus ·75; bill from gape ·7.

Distribution. The Himalayas from Murree to Dibrugarh in Assam from their base up to 5000 feet, and occasionally, as recorded by Stoliczka, up to 9000; portions of the Central Provinces and Chutia Nagpur, Ball recording this species from Manbhoom, Lohardugga, Jashpur, Sambalpur north of the Mahánadi, Raipur, and the Godávari valley, and Hume from Bod; to the west the range of this bird is somewhat undefined, Barnes expressing his belief that it does not occur in the Bombay Presidency, but Mr. B. Aitken stating that it is common both at Bombay and Poona. Sykes obtained it in the Deccan. It is found in Lower and Eastern Bengal, Assam, and all the countries between this latter province and Central Tenasserim, extending into Cochin China.

Habits, &c. Breeds in May, June, and July, constructing a nest of grass coated with cobwebs, of the shape of an inverted cone, in a small fork on a branch of a tree. The eggs, usually three in number, are fawn-colour or yellowish, marked with grey, and measure about ·65 by ·49.

606. **Rhipidura javanica.** *The Java Fantail Flycatcher.*

Muscicapa javanica, *Sparrm. Mus. Carls.* iii, pl. 75 (1788).
Leucocerca javanica (*Sparrm.*), *Blyth, Cat.* p. 206; *Hume & Dav. S. F.* vi, p. 226; *Hume, Cat.* no. 293 bis; *id. S. F.* ix, p. 175.
Rhipidura javanica (*Sparrm.*), *Horsf. & M. Cat.* i, p. 144; *Sharpe, Cat. B. M.* iv, p. 332; *Oates, B. B.* i, p. 267.
Leucocerca infumata, *Hume, S. F.* i, p. 455 (1873).

Coloration. Male. Forehead, crown, and sides of the head sooty brown; the whole upper plumage and the wings brown, suffused with rufous; tail dark brown, the four outer pairs of feathers broadly tipped with white, the pair next to these very narrowly tipped with white; a short supercilium white; chin, a band across the upper breast, and the sides of the breast blackish brown; throat white; remainder of lower plumage white tinged with buff.

Female. Resembles the male, but has the lower plumage more tinged with buff.

The young resemble the adult, but the back and the wing-coverts are broadly tipped and fringed with rufous.

Iris brown; bill black, fleshy at the base of the lower mandible (*Hume and Davison*).

Length about 7·5; tail 3·6; wing 3; tarsus ·75; bill from gape ·7.

Distribution. Tenasserim from Tavoy southwards, extending to Siam, Cochin China, the Malay peninsula and islands.

607. Rhipidura pectoralis. *The White-spotted Fantail Flycatcher.*

Leucocerca pectoralis, *Jerd. Ill. Ind. Orn.* text to plate ii (1847); *Blyth, Cat.* p. 206; *Jerd. B. I.* i, p. 454; *Hume, N. & E.* p. 203.
Muscipeta leucogaster, *Cuvier, fide Pucher. Arch. Mus.* vii, p. 335 (1854).
Rhipidura pectoralis (*Jerd.*), *Blyth, J. A. S. B.* xii, p. 935 (1843); *Sharpe, Cat. B. M.* iv, p. 335; *Oates in Hume's N. & E.* 2nd ed. ii, p. 38.
Leucocerca leucogaster (*Cuv.*), *Hume, Cat.* no. 293; *Barnes, Birds Bom.* p. 161.

The White-spotted Fantail, Jerd.

Coloration. Resembles *R. javanica.* Differs in having the breast-band ocellated with white and the outer tail-feathers merely paler towards the tips, not abruptly and broadly pure white.

Iris blackish brown; legs, feet, and bill black (*Butler*).

Length about 7; tail 3·8; wing 3; tarsus ·7; bill from gape ·6.

Distribution. Found chiefly and most abundantly in the western portion of India from Abu nearly down to Cape Comorin. To the eastward this species has been found at Raipur (Ball), Chikalda (McMaster), Goona (King), Chánda (Blanford), and a line connecting these localities probably represents its eastern limits. It appears to be found up to 6000 feet or even higher.

Habits, &c. Breeds from April to August, constructing a nest very similar to that of *R. albifrontata.* The eggs are buffy white, with a zone of brownish spots round the larger end, and measure about ·66 by ·47.

Fig. 23.—*Copsychus saularis.*

Family TURDIDÆ.

The intrinsic muscles of the syrinx fixed to the ends of the bronchial semi-rings ; the edges of both mandibles smooth, or the upper one simply notched ; hinder aspect of tarsus bilaminated, the laminæ entire and smooth ; wing with ten primaries ; tongue non-tubular ; nostrils always clear of the line of forehead, the space between the nostril and the edge of the mandible less than the space between the nostril and the culmen ; plumage of the nestling mottled or squamated to a greater or less extent : one moult in the year, but frequently supplemented by a seasonal change of plumage caused by the casting-off of the margins of the feathers in spring ; rectrices usually twelve, very seldom fourteen.

The *Turdidæ*, or Chats, Robins, Thrushes, Dippers, and Accentors, form a very large family of the *Passeres*. The only character which links all the species together is the mottled or squamated plumage of the nestling. In this character the *Turdidæ* agree with the *Muscicapidæ*, but they differ from this family in having long or moderate tarsi and in having the nostrils and base of the upper mandible quite free from all hairs. The only exception to this latter feature appears to be in *Zoothera*, in which genus the frontal hairs are developed and reach over the nostrils. The long tarsus of the birds of this genus, however, will prevent them from being confounded with any of the Flycatchers. In some few genera, especially *Ruticilla* and *Pratincola*, the shafts of the feathers of the forehead are somewhat elongated and the webs disintegrated, but these cannot be considered hairs nor do they lie over the nostrils as is always the case with the *Muscicapidæ*.

The *Turdidæ* are found over nearly the whole globe and their migratory instincts are generally very strong.

The *Turdidæ* may be divided into five subfamilies, characterized partly by habits and partly by structural characters.

Tarsus smooth ; rictal bristles present ; habits
   Muscicapine, the insect-food captured
   by sallies from a fixed perch . . . .   *Saxicolinæ*, p. 57.
Tarsus smooth, with hardly an exception * ;
   rictal bristles present : habits terrestrial,
   the insect-food captured on the ground.   *Ruticillinæ*, p. 84.
Tarsus smooth ; rictal bristles present : habits
   terrestrial and arboreal, the species being
   both insectivorous and frugivorous . . . .   *Turdinæ*, p. 120.
Tarsus smooth ; rictal bristles absent : habits
   aquatic ; eggs unspotted white . . . .   *Cinclinæ*, p. 161.
Tarsus scutellated ; rictal bristles present :
   habits terrestrial ; eggs unspotted blue. .   *Accentorinæ*, p. 165.

Subfamily SAXICOLINÆ.

The *Saxicolinæ* or Chats form a natural section of the Thrushes very nearly related to the Flycatchers and with many of their habits. The Chats feed entirely on insects, which they capture generally on the ground from a fixed perch, such as the summit of a stone, a stalk of grass, or a branch of a bush, and then return at once to their post of observation. The characteristic habit of the Chats is the frequent movement and expansion of the tail. The majority of this subfamily are migratory, and they have a very marked seasonal change of plumage caused by the abrasion of the margins of the feathers in the late autumn or early spring. The sexes usually differ very much in colour.

---

* The only exception I know of is *Thamnobia*.

In the Chats the bill is strong and the rictal bristles occasionally very numerous and strong; the wing in most is pointed; the tail, of twelve feathers, is seldom or never longer than the wing; and the tarsus and foot are of medium size and strength.

The Chats nest in holes in the ground or in walls, or among heaps of stones, and they lay eggs which, so far as is known, are always marked with brown or rufous.

Key to the Genera.

a. Bill broad at base; rictal bristles numerous and
    strong.
    a'. Tail shorter than wing; outer feathers reach-
        ing nearly to tip .... PRATINCOLA, p. 58.
    b'. Tail about as long as wing; outer feathers
        falling short of tip by about half length of
        tarsus .... OREICOLA, p. 66.
b. Bill narrow, not strikingly broad at base; rictal
    bristles few and weak.
    c'. Tail with a pattern of two colours .... SAXICOLA, p. 67.
    d'. Tail entirely of one colour .... CERCOMELA, p. 79.

Genus **PRATINCOLA**, Koch, 1816.

The genus *Pratincola* contains a considerable number of species of wide distribution, some of which are migratory and others resident. They are mostly familiar birds, displaying little fear, and one or more species are generally common in every part of the Empire except the tracts covered with forest.

In *Pratincola* the bill is rather less than half the length of the head, broad at base and well notched; the rictal bristles are very strong; the wing is rather sharp, and the first primary varies from one half to one third the length of the second; the tail is shorter than the wing and slightly rounded, and the tarsus is moderate in length. The sexes are invariably dissimilar, and the seasonal changes of plumage are very marked.

Key to the Species *.

a. Plumage entirely black and white.
    a'. Wing about 2·8; bill from nostril to tip ·3 .   *P. caprata* ♂, p. 59.
    b'. Wing about 3; bill from nostril to tip ·4 ..   *P. atrata* ♂, p. 60.

---

* I exclude the following species from the Indian list : —

PRATINCOLA ROBUSTA, Tristram.

Pratincola robusta, *Tristram, Ibis,* 1870, p. 497; *Hume, S. F.* ix, p. 133.

The type of this species, said to have been procured in Mysore by Mr. R. E. Fox more than twenty years ago, has been obligingly lent to me by Canon Tristram. It is identical with the larger Bush-Chat of Madagascar, and both Sharpe and Tristram agree with me in this identification. There are two Bush-Chats of this type in Madagascar, agreeing in coloration but differing in size. To one

b. Plumage not entirely black and white.
  c'. Chin and throat black.
    a''. No white on tail except at extreme base . *P. maura* ♂, p. 61.
    b''. Inner webs of tail-feathers white ...... *P. leucura* ♂, p. 65.
  d'. Chin, throat, and upper tail-coverts white or
    pale rufous.
    c''. Inner webs of tail-feathers white ...... *P. macrorhyncha,*
    d''. Inner webs of tail-feathers black or brown.   [p. 63.
      a'''. Under wing-coverts and axillaries black
        broadly edged with white; wing 3·6.. *P. insignis,* p. 64.
      b'''. Under wing-coverts an laxillaries rufous { *P. maura* ♀, p. 61.
        or fulvous; wing 2·6 .... { *P. leucura* ♀, p. 65.
  e'. Chin and throat brown; upper tail-coverts
    deep ferruginous.
    e''. Wing about 2·8 .... *P. caprata* ♀, p. 59.
    f''. Wing about 3 .... *P. atrata* ♀, p. 60.

608. Pratincola caprata. *The Common Pied Bush-Chat.*

Motacilla caprata, *Linn. Syst. Nat.* i, p. 335 (1766).
Saxicola bicolor, *Sykes, P. Z. S.* 1832, p. 92.
Pratincola caprata (*Linn.*), *Blyth, Cat.* p. 163; *Horsf. & M. Cat.*
  i. p. 284; *Jerd. B. I.* ii, p. 124; *Hume, N. & E.* p. 312; *Anders.*
  Yunnan Exped., Aves, p. 617; *Hume, Cat.* no. 481; *Sharpe, Cat.*
  B. M. iv, p. 195 (part.); *Oates, B. B.* i, p. 284; *Barnes, Birds*
  Bom. p. 190; *Oates in Hume's N. & E.* 2nd ed. ii, p. 41.

The White-winged Black Robin, Jerd.; *Pidha, Kala pidha,* Hind.;
Kampa nalanchi, Tel.

Coloration. Male. In the summer the whole plumage is black,
except the lower part of the rump, the upper and lower tail-coverts,
and those feathers of the wing nearest the body, which are white,
the latter forming a very conspicuous patch on the wing.

In the autumn the black feathers are more or less fringed with
rufous-brown.

---

of these, but to which will probably never be satisfactorily determined, Linnaeus
assigned the name *Motacilla sybilla* (Syst. Nat. i, p. 337). I propose, therefore,
that Linnaeus's name should be retained for the smaller bird, with wing 2·5
or 2·6, and Tristram's name for the larger, with wing 2·9.

The occurrence of *P. robusta* in India must, for the present I think, be viewed
with a certain amount of distrust, and I therefore omit it from my list. The
following is a description of the type, which appears to be a male, judging from
its fine colour:—

The breast and sides of the body are a deep cinnamon-rufous, sharply demarcated from the pure white of the abdomen, which extends to the vent and
under tail-coverts; the axillaries and under wing-coverts are white with dark
bases; the tail is black; the other parts of the plumage resemble the same
parts in *P. maura* and *P. leucura* in autumn plumage.

Length about 6; tail 2·4; wing 2·9; tarsus ·9; bill from gape ·6.

The many specimens from the Himalayas which have been identified with this
species by Brooks, Scully, and others are quite a different type of bird and are,
in my opinion, merely large specimens of *P. maura.*

Female. The upper plumage grey, with dark brown mesial streaks, the back tinged with rufous : upper tail-coverts ferruginous : tail black ; chin and throat brownish grey ; breast, upper part of the abdomen, and sides wood-brown with dark streaks ; lower part of the abdomen the same but without streaks ; under tail-coverts rufescent ; lores and feathers in front of the eye mixed with white ; quills and the larger coverts brown narrowly edged with rufescent, the other coverts brown broadly edged with light buff ; under wing-coverts bright buff with dark centres. In the winter the grey margins on the upper plumage are so ample that hardly any of the dark brown centres are visible ; otherwise there is no change.

The young are fulvous-brown, mottled all over with dusky ; in the young male the white patch on the wing makes its appearance from the earliest period.

Iris brown ; eyelids plumbeous ; bill black ; mouth dusky ; legs and claws black.

Length about 5·5 ; tail 2·2 ; wing 2·8 ; tarsus ·85 ; bill from gape ·7.

Fig. 21.—Head of *P. caprata.*

Distribution. A resident species throughout the whole of India and Burma, except the southernmost part of the peninsula of India and portions of Tenasserim. This bird ascends the Himalayas up to 8000 feet, probably in summer only. It is found in the south as far at least as Maddur in Mysore. It is more or less abundant throughout the peninsula and through Assam and the Burmese provinces to Pegu. In Tenasserim Davison observed this bird in the northern and central portions, but not in the extreme south, and Wardlaw Ramsay procured it in Karennee.

Outside Indian limits this species extends to Persia on the west, and on the east and south to the Philippines and Java.

Habits, &c. Breeds from March to June, constructing a flat, and frequently shapeless, nest in a hole in the ground, in a bank or in a well, composed of grass, roots, and hair. The eggs, usually four in number, are pale bluish green, marked in various ways with brownish red, and measure about ·67 by ·55.

609. **Pratincola atrata.** *The Southern Pied Bush-Chat.*

Pratincola atrata, *Kelaart, Blyth, J. A. S. B.* xx. p. 177 (1851) ; *Jerd. B. I.* ii. p. 124 ; *Oates in Hume's N. & E.* 2nd ed. ii. p. 46.

Pratincola bicolor (*Sykes*), *apud Hume, N. & E.* p. 314 ; *Legge, Birds Ceyl.* p. 430 ; *Hume, Cat.* no. 482 ; *Davison, S. F.* x. p. 380.

Pratincola caprata (*Linn.*), *apud Sharpe, Cat. B. M.* iv. p. 195 (part.).

Coloration. In all respects similar in plumage to *P. caprata*, sex for sex. Differs in being much larger and in having a conspicuously more massive bill.

Length about 6; tail 2·3; wing 3; tarsus ·9; bill from gape ·75.

In distinguishing between this and the preceding species, *P. caprata*, the size of the bill alone is quite sufficient. In the present species the bill, measured from the anterior margin of the nostril to the tip, is ·4; in *P. caprata* ·3 or less.

I adopt Kelaart's name for this species, as Sykes's *Saxicola bicolor* was procured in the Deccan, where, so far as I know, only *P. caprata* occurs.

Distribution. Southern India, from the Nilgiris to Cape Comorin, above 5000 feet; Ceylon. A permanent resident.

Habits, &c. Breeds from February to May, placing its nest in similar localities to those selected by *P. caprata*, and laying similar eggs, which, however, are much larger and measure about ·77 by ·6.

610. Pratincola maura. *The Indian Bush-Chat.*

Motacilla maura, *Pall. Reis. Russ. Reichs*, ii, p. 708 (1773).
Saxicola saturatior, *Hodgs. in Gray's Zool. Misc.* p. 83 (1844).
Pratincola indica, *Blyth, J. A. S. B.* xvi. p. 129 (1847); *id. Cat.* p. 170; *Jerd. B. I.* ii, p. 124; *Cabanis, Journ. f. Orn.* 1873, p. 359; *Severtz. S. F.* iii, p. 429; *Anders. Yunnan Exped., Aves,* p. 618; *Hume, Cat.* no. 483; *Barnes, Birds Bom* p. 200.
Pratincola albosuperciliaris, *Hume, S. F.* i, p. 397 (1873).
Pratincola rubicola (*Linn.*), *apud Hume, N. & E.* p. 316; *Hume & Henders. Lah. to Yark.* p. 204.
Pratincola maura (*Pall.*), *Sharpe, Cat. B. M.* iv, p. 188; *Oates, B. B.* i, p. 279; *id. in Hume's N. & E.* 2nd ed. ii, p. 48.

Adwi-kampa-nalanchi, Adaci-kampa-jitta, Tel.

Coloration. Male. After the autumn moult the forehead, crown, nape, hind neck, back, scapulars, and upper rump are black, with broad fulvous or rufous margins to the feathers; the innermost wing-coverts pure white; the remaining upper wing-coverts black, edged with rufous; primary-coverts and winglet black, edged with whitish; quills dark brown, the primaries narrowly, the other quills broadly, edged with rufous on the outer web and tip; lower rump and tail-coverts white, frequently suffused with orange-rufous; tail black, narrowly edged with pale rufous; the extreme bases of the feathers white; lores, sides of the head, chin, and throat black, most of the feathers edged with fulvous; a large patch of white on each side of the neck; breast orange-rufous; remainder of the lower plumage paler rufous; under wing-coverts and axillaries black with narrow white tips. In summer the margins of the feathers of the black portions of the plumage are almost entirely lost, and these parts become deep black.

Female. After the autumn moult the upper plumage, wings, and tail resemble those parts in the male, but the black is everywhere replaced by brown and the upper tail-coverts are uniform pale rufous; the lores, ear-coverts, and round the eye are dusky; supercilium, chin, and throat pale fulvous; remainder of lower plumage pale orange-rufous; no white on the side of the neck; under wing-coverts and axillaries fulvous. In summer the edges of the feathers are much worn down, and the plumage is paler.

The nestling has the upper plumage brown, the head and neck streaked with fulvous, the back broadly edged with fulvous; lower part of the rump and upper tail-coverts bright ferruginous; the lower plumage fulvous, with brown mottlings on the breast. After the first autumn moult the young male has the lower plumage very bright chestnut, but resembles the adult in other respects.

Bill, legs, and feet black; iris dark brown.

Length about 5; tail 1·9; wing 2·8; tarsus ·8; bill from gape ·65.

This species differs from the European *P. rubicola* in having the upper tail-coverts streakless, and the under wing-coverts and axillaries very narrowly tipped with white.

Although I have assigned Pallas's name to the Indian Bush-Chat, I am by no means satisfied that the Siberian and Indian birds are identical, nor is it certain that any of the Bush-Chats which visit the plains of India in the winter cross over to the north of the Himalayas in the summer. The Indian Bush-Chat breeds so abundantly at all moderate levels in the Himalayas that it is not improbable that the Himalayas form the northern limit of its range. Siberian specimens of Bush-Chats are not very numerous, but all I have seen are so intensely black on the head and back, so intensely rufous on the breast, and, moreover, so small, the wing not exceeding 2·6 in length, that I have not been able to match them with any breeding bird from the Himalayas, except in the case of one bird from the interior of Sikhim. This small dark race occurs also in Turkestan.

Distribution. A winter visitor to every portion of the Empire except the southern portion of the peninsula of India south of Mysore. The most southern point from which I have seen a specimen of this species is Belgaum; but Hume says (S. F. x, p.389) that it is reported common from South-west Mysore. It occurs in the Andamans.

In the summer this species is found throughout the Himalayas, from Afghanistan to Assam, up to 5000 feet. Should the Indian bird prove identical with the Siberian form, its range will extend to Japan and China on the east and to Northern Russia on the west. Specimens from Abyssinia are quite inseparable from Indian birds.

Habits, &c. Breeds in the Himalayas at all heights up to about 5000 feet, constructing a nest of grass and moss in small shrubs or in holes of walls, and laying four or five eggs, which are pale green marked with brownish red, and measure about ·7 by ·55.

611. **Pratincola leucura.** *The White-tailed Bush-Chat.*

Pratincola leucura, *Blyth, J. A. S. B.* xvi, p. 474 (1847); *id. Cat.*
p. 170; *Jerd. B. I.* ii, p. 126; *Godw.-Aust. J. A. S. B.* xxxix,
pt. ii, p. 270; *Hume, S. F.* i, p. 183, iii, p. 135, v, p. 241; *id.
Cat.* no. 484; *Brooks, S. F.* viii, p. 473; *Sharpe, Cat. B. M.* iv,
p. 194: *Oates, B. B.* i, p. 280; *Barnes, Birds Bom.* p. 200;
Hume, S. F. xi, p. 191.

Khar-pidda, Hind. at Monghyr.

Coloration. Resembles *P. maura* very closely. The male *P.
leucura* differs from the male *P. maura* in having the inner webs
and the basal half of the outer webs of all the tail-feathers, except
the middle pair, white, and the abdomen, vent, and under tail-
coverts also white.

The females, although undistinguishable from each other by mere
description, may perhaps be separable by actual comparison of specimens if the plumage be in good order. In *P. leucura* the rufous
of the upper tail-coverts and the lower plumage is much paler, in
P. maura much darker; *P. leucura* is also somewhat greyer and less
rufous on the back.

Legs and feet black; bill black; iris brown (*Hume*).

Length about 5; tail 1·9; wing 2·6; tarsus ·7; bill from
gape ·7.

Hume draws attention to the great difference in plumage
between birds of this species from Sind and from Manipur. Considering, however, how very irregular the changes and tints of the
plumage of these birds are, I do not think that there are any
grounds for separating the Manipur from the Sind race.

Distribution. Sind; the Terai and lower hills of Nepal and
Sikhim; Eastern Bengal; Dacca; Tipperah; Mymensing; Manipur; Thayetmyo and the valley of the Irrawaddy immediately
below this town; Toungngoo; Pahpoon in Tenasserim. There is
little doubt that this species is a permanent resident in those
places.

Habits, &c. This Bush-Chat is found invariably in or near
swamps where there are reeds and grass.

612. **Pratincola macrorhyncha.** *Stoliczka's Bush-Chat.*

Saxicola rubetroides, *Jameson, Jerd. B. I.* ii, *App.* p. 872 (1864,
descr. null.).
Pratincola macrorhyncha, *Stoliczka, J. A. S. B.* xli, pt. ii, p. 238
(1872); *Hume, S. F.* iv, p. 40, vii, p. 55; *id. Cat.* no. 485 bis;
Sharpe, Cat. B. M. iv, pp. 182, 473; *Barnes, Birds Bom.* p. 201.
Pratincola jamesoni, *Hume, S. F.* v, p. 239 (1877).

Coloration. Upper plumage sandy buff streaked with dark
brown; upper tail-coverts pale rufous; middle pair of tail-feathers
dark brown, narrowly margined with fulvous-white; the next pair
with the basal third of the outer, and three quarters of the inner,
web white, the remainder black; the others progressively more
white and less black; the outermost almost entirely white; wing-
coverts blackish with broad sandy margins, the last of each series

next the body almost entirely white; the earlier primary-coverts chiefly white on the outer webs, dark brown edged with sandy elsewhere; quills dark brown edged with sandy; lores and a broad supercilium pale buff; ear-coverts rufous; remainder of the sides of the head mixed brown and buff; chin and throat white; remainder of the lower plumage very pale buff, somewhat deeper on the breast; under wing-coverts white mottled with black; axillaries white, with blackish bases.

The sexes appear to be alike in the winter, but may probably differ in the summer.

The above is the plumage of adults of both sexes during the winter. I have not been able to examine birds in summer plumage; but the skins most advanced towards this plumage in the Hume Collection have a dark blackish streak from the bill down the sides of the throat and breast, expanding in width gradually and leaving the throat narrowly white. The sandy margins of the upper plumage are probably at this season much reduced in extent, leaving the upper plumage blacker.

The young resemble the adults in winter plumage, but there is no white on the tail, which is brown with fulvous margins, and the white on the wing-coverts is either absent or very much reduced.

Legs and feet black; iris brown; bill black (*Hume*).

Length nearly 6; tail 2·2; wing 3; tarsus 1; bill from gape ·7.

Distribution. A winter visitor to the Punjab, Rajputana, Northern Guzerat, Cutch, and Sind. The summer-quarters of this species are unknown. No one has met with it in Central Asia, and Hume's conjecture that it may be a resident in the above provinces of India may prove to be correct. Natives of Jodhpur assured him that these birds remained in this State and bred there during the rainy season.

Habits, &c. Hume states that this species was extremely abundant in the thin, stunted, scrub-jungle that here and there studs the sandy, semi-desert, waterless tracts which occur all round Jodhpur. It has the ordinary habits of *P. maura.*

613. Pratincola insignis. *Hodgson's Bush-Chat.*

Saxicola insignis, *Hodgs. in Gray's Zool. Misc.* p. 83 (1844, descr. null.).

Pratincola insignis, *Hodgs., Blyth, J. A. S. B.* xvi, p. 129 (1847); *Jerd. B. I.* ii, p. 127; *Hume, S. F.* v, pp. 132, 496, vii, pp. 454, 519; *id. Cat.* no. 485; *Sharpe, Cat. B. M.* iv, p. 183.

Pratincola robustior, *C. H. T. and G. F. L. Marshall, S. F.* iii, p. 380.

The Large Bush-Chat, Jerd.

Coloration. Male. In winter the lores, under the eyes, and the whole of the ear-coverts are deep black; forehead, crown, and nape black with small fulvous edges; mantle, back, and rump black with broad fulvous edges; upper tail-coverts white dashed with rusty; wing-coverts white next the body, black elsewhere; the lesser coverts near the edge of the wing fringed with fulvous ; primary-coverts with the basal half black, the terminal half white ; quills blackish, all of them broadly white at base, except the last two or three primaries and the first secondary, the primaries and secondaries narrowly, the tertiaries broadly, edged with fulvous ; tail black, with concealed white bases to the inner webs ; point of the chin and a narrow stripe along the base of the mandible black ; throat, extending laterally to the sides of the neck, white, more or less marked with rusty ; remainder of the lower plumage rusty ferruginous, the breast marked with some broad black streaks, the abdomen paler ; under tail-coverts pale fulvous-white ; under wing-coverts black edged with white ; axillaries white, with the bases of the feathers blackish.

In summer, judging from the only specimen I have seen (one collected by Hodgson at Segowlie, and figured by him), the fulvous margins on the upper plumage are cast and this part becomes black, the black streaks on the breast are absent, and there is no rusty either on the throat or the upper tail-coverts. Hodgson's bird appears to have been obtained on the 10th January, but it seems nevertheless to be in full well-worn summer plumage, and there may be some mistake about the date.

Female. In winter, and probably in summer also, the upper plumage is brown, each feather margined with dull fulvous ; upper tail-coverts rusty ; tail brown, with fulvous margins and tips and with no white at the base ; wing-coverts dark brown, margined and tipped with fulvous ; the innermost greater coverts and the last tertiary chiefly white ; quills dark brown, with small dull white bases and margined with fulvous ; sides of the head and neck, lores, and above the eye dull fulvous, the ear-coverts rufescent ; the whole lower plumage rusty brown, darker on the breast, which sometimes has a few dark-brown streaks.

A young male obtained in December has the wings, tail, and upper plumage similar to the same parts of the adult male in winter, but the lower plumage is that of the female and the ear-coverts are nearly black.

The male has the iris deep brown, the bill and legs black ; the female has the bill blackish brown, horny at base of the lower mandible (*Cleveland*).

Length about 6·5 ; tail 2·4 ; wing 3·6 ; tarsus 1·1 ; bill from gape ·85.

Distribution. A rare species, occurring on the plains of Northern India from Cawnpore to the Bhutan Doars. The Marshalls procured it near Cawnpore in February ; Mr. Cleveland in the Gorakhpur and Basti districts in October and December ; Hodgson at Segowlie, as already mentioned ; and Mandelli in the lower hills of Sikhim and the Bhutan Doars in April. The summer-quarters of this species are not known, but lie probably in the Central hills of Nepal and Sikhim.

Habits, &c. This Bush-Chat is found in flat open country thickly dotted with cane-fields, which appear to be its favourite haunts.

Genus **OREICOLA**, Bonap., 1854.

This genus differs from *Pratincola* in having a much longer tail, which is also very much more graduated.

Key to the Species.

a. Whole upper plumage, wings, and tail
black............................. *O. jerdoni* ♂, p. 66.
b. Upper plumage ashy and black; wing-
coverts largely white; tail margined
white .... *O. ferrea* ♂, p. 66.
c. Upper plumage rufous-brown or rufous-
ashy.
a'. With no supercilium .... *O. jerdoni* ♀, p. 66.
b'. With a supercilium .... *O. ferrea* ♀, p. 66.

614. Oreicola jerdoni. *Jerdon's Bush-Chat.*

Rhodophila melanoleuca, *Jerd. B. I.* ii, p. 128 (1863, *nec Vieill.*),
   App. p. 872; *Godw.-Aust. J. A. S. B.* xxxix, pt. ii, p. 270.
Oreicola jerdoni, *Blyth, Ibis,* 1867, p. 14; *Blanf. Ibis,* 1870, p. 466;
   Hume, Cat. no. 487; *Sharpe, Cat. B. M.* iv, p. 204; *Oates, B. B.*
   i, p. 282; *Salvadori, Ann. Mus. Civ. Gen.* (2) iv, p. 590; *Hume,
   S. F.* xi, p. 193.
Pratincola jerdoni (*Blyth*), *Anders. Yunnan Exped., Aves,* p. 616.

Coloration. Male. The whole upper plumage, wings, tail, and sides of the head and neck deep black; the whole lower plumage white; under wing-coverts black slightly tipped with white.

Female. The whole upper plumage brown tinged with rufous, especially on the rump and upper tail-coverts; tail brown, edged paler; wings and coverts brown edged with rufous; sides of the head mixed ashy and brown; chin and throat white; remainder of the lower plumage pale fulvous.

Bill and legs black; iris dark brown (*Jerdon*).

Length about 6; tail 2·7; wing 2·7; tarsus ·85; bill from gape ·7.

I have not been able to examine a young bird of this species.

Distribution. Purneah in Behar; Eastern Bengal; Dibrugarh in Assam; Sylhet; Cachar; Manipur; the neighbourhood of Bhámo; Bassein district; Leppadan on the Rangoon and Prome Railway, where I lately observed this species in March in thick grass on the banks of the Leppadan river. It is not known whether this Bush-Chat is migratory or not.

615. Oreicola ferrea. *The Dark-grey Bush-Chat.*

Saxicola ferrea, *Hodgs. in Gray's Cat. Mamm. &c. Nep.* pp. 71, 153
   (1846).
Pratincola ferrea (*Hodgs.*), *Blyth, Cat.* p. 170; *Horsf. & M. Cat.*
   i, p. 286; *Jerd. B. I.* ii, p. 127; *Stoliczka, J. A. S. B.* xxxvii,
   pt. i, p. 41; *Hume, N. & E.* p. 318; *Hume & Henders. Lah. to*

Yark. p. 205, pl. xii; *Anders. Yunnan Exped., Aves,* p. 617; *Hume, Cat.* no. 486; *Scully, S. F.* viii, p. 301.
Oreicola ferrea (*Hodgs.*), *Sharpe, Cat. B. M.* iv, p. 266; *Oates, B. B.* i, p. 283; *id. in Hume's N. & E.* 2nd ed. ii, p. 50.

Surruk-chak-pho, Lepch.

Coloration. Male. After the autumn moult the whole upper plumage is dark ashy grey, all the feathers except those of the rump centred with black and margined with a varying amount of rusty; coverts and quills black edged with grey, which inclines to white on the tertiaries; the inner greater coverts entirely white; tail black, the feathers increasingly margined with white, the outer web of the outermost feather being entirely white; a white super-cilium from the forehead to the nape; sides of the head black; lower plumage white, tinged with ashy across the breast and on the thighs.

The margins of the feathers of the upper plumage get worn away rapidly, and later on in the winter almost disappear, leaving the upper parts black during the summer.

Female. The whole upper plumage rufous ashy, the centres of the feathers dark, but not very distinctly visible till the spring, when the edges of the feathers are reduced in extent; upper tail-coverts chestnut; tail brown, broadly edged with chestnut; wings brown, narrowly edged with rufous; a pale grey supercilium; sides of the head reddish brown speckled with brown; chin and throat whitish; remainder of lower plumage rufous ashy.

The young are dark rufous-brown, with streaks and spots of fulvous, and broad rufous edges to the tail and wings.

Iris brown; tail black; legs dark brown.

Length nearly 6; tail 2·7; wing 2·7; tarsus ·8; bill from gape ·65.

Distribution. The Himalayas, from Murree and the Indus valley in Kashmir to the extreme east of Assam. This species is found up to 9000 feet in summer, and it descends to the valleys in the winter. It extends in the winter from Assam through the hill-ranges and Burma as far as Karennee, Central Tenasserim, and the Thoungyeen valley. This Bush-Chat is found in China.

Habits, &c. Breeds in the Himalayas from April to July, con-structing a nest of grass, moss, and hair in a hole in the ground or under the shelter of a stone or clod of earth. The eggs, four or five in number, are pale green marked with reddish brown, and measure about ·72 by ·57. It is not improbable that this species may breed in some of the hill-ranges of Burma.

Genus SAXICOLA, Bechst., 1802.

The genus *Saxicola* contains a large number of species which are essentially birds of deserts and waste lands, and they are most developed in the dry parts of South-western Asia and Northern Africa. The majority of them are migratory to a greater or less

F 2

extent, and a few appear to be resident. The sexes are usually dissimilar, and both sexes undergo a seasonal change of plumage, which in some species causes a very great alteration in their appearance. In all the species and in both sexes the tail is marked with two colours, generally black and white and occasionally black and chestnut.

In *Saxicola* the bill is about half the length of the head, slender, and not widened at the base; the rictal bristles are few and weak; the wing is sharp, the first primary being about one third the length of the second; the tail is shorter than the wing and nearly square, and the tarsus is moderate.

The young of *Saxicola* are in general like the adult female, but each feather of the plumage has a terminal dark bar and a pale centre, causing a mottled appearance. This plumage at the first autumn gives place to that of the adult.

Key to the Species.

a. Tail white or buff and brown; the lateral feathers immaculate or merely obliquely marked with black .... *S. monacha*, p. 69.
b. Tail white and black; the laterals with a broad band at the tip.
  a'. Band on lateral tail-feathers not exceeding 1 inch in breadth.
    a''. Second primary shorter than sixth.
      a'''. Sexes alike; plumage black and white; wing 4 inches or longer.. *S. albinigra*, p. 70.
      b'''. Sexes different; males black and white: female brown; wing under 3·7 inches.
        a⁴. Abdomen white, and crown black...................... *S. picata* ♂, p. 71.
        b⁴. Abdomen and crown white .... *S. capistrata* ♂, p. 72.
        c⁴. Abdomen black .... *S. opistholeuca* ♂, p. 73.
        d⁴. Throat and breast dark brown, contrasting with the pale abdomen .... *S. picata* ♀, p. 71.
        e⁴. Throat and breast buff, blending softly with the paler buff of the abdomen .... *S. capistrata* ♀, p. 72.
        f⁴. Throat, breast, and upper abdomen sooty brown........... *S. opistholeuca* ♀, p. 73.
    b''. Second primary between the fifth and sixth; sexes different *.
      c'''. Chin and throat black.
        g⁴. Back and scapulars of the same colour .... *S. pleschanka* ♂, p. 73.
        h⁴. Back buff, scapulars black .... *S. barnesi* ♂, p. 75.

---

* I am not acquainted with the females of *S. barnesi* and *S. rittata*, and consequently I do not enter either of these in the Key, nor the female of *S. pleschanka*, which comes into this same section.

616. Saxicola monacha.   *The Hooded Chat.*

Saxicola monacha, *Rüpp., Temm. Pl. Col. no. 359, fig. 1 (1825); Blanf. & Dresser, P. Z. S.* 1874, p. 227; *Seebohm, Cat. B. M.* v, p. 369; *Hume, Cat. no.* 490 bis; *Barnes, Birds Bom.* p. 203. Dromolæa monacha (*Rüpp.*), *Hume, S. F.* i, p. 186.

Coloration. Male. After the autumn moult the forehead, crown, nape, hind neck, rump, upper tail-coverts, abdomen, vent, and under tail-coverts are pure white; tail white, the terminal two-thirds of the middle pair and a dash or two on the laterals brown: remainder of plumage black, the back and breast with a few whitish fringes, the secondaries, under wing-coverts, and axillaries with white tips. In summer the white fringes on the back and breast disappear.

Female. Upper plumage buffish brown; wings and coverts brown edged with buffy white; lower rump and upper tail-coverts buff; tail as in the male, but the white replaced by buff and with more brown on the laterals; lower plumage pale buff.

Legs, feet, and bill black (*Hume Coll.*).

Length about 7; tail 2·9; wing 4·2; tarsus ·9; bill from gape 1.

Distribution. The hills dividing Sind from Khelát, extending

---

† * *Saxicola lugens*, Licht., has been recorded from Daulatpur in Sind by Mr. Murray, but under circumstances which render it doubtful whether the specimen upon which the statement is based was really obtained in Sind. This Chat resembles *S. pleschanka*, but has the under tail-coverts a deep buff, and the inner webs of the quills of the wing very largely white. (*Cf. S. F.* vii, pp. 118, 527, where this bird is referred to under the name of *S. leucomela*.) This species occurs in Persia and westwards to Palestine and Northern Africa. *S. persica*, Seebohm, is also allied to *S. pleschanka* and *S. lugens*, and differs from the latter in having the inner webs of the quills of the wing merely margined with white. In the former the whole wing is black. *S. persica* occurs in Persia in summer.

east to Schwan, whence I have seen a specimen procured by Brooks in January. This species is said to be a winter visitor to Sind, but this statement requires confirmation. It extends westwards to Baluchistan and Afghanistan, and on to Palestine, occurring also in Nubia.

617. Saxicola albinigra. *Hume's Chat.*

Saxicola alboniger, *Hume, S. F.* i, p. 2 (1873); *Blanf. & Dresser, P. Z. S.* 1874, p. 226; *Blanf. East. Pers.* ii, p. 153, pl. xi; *Seebohm, Cat. B. M.* v, p. 366; *Hume, Cat.* no. 489 bis; *Barnes, Birds Bom.* p. 202; *Biddulph, Ibis,* 1881, p. 58; *Scully, Ibis,* 1881, p. 442.
Dromolæa alboniger (*Hume*), *Hume, S. F.* i, p. 185.

Coloration. The sexes are alike. The whole head, neck, back, scapulars, sides of breast, axillaries, and under wing-coverts deep black; wings dark brown, the coverts edged with black; remainder of the plumage white, except the terminal half of the middle tail-feathers and a terminal band on the laterals, which are black.

Bill, legs, and feet black (*Hume Coll.*).

Length about 7; tail 2·8; wing 4; tarsus 1; bill from gape ·9.

Distribution. The hills dividing Sind from Khelát, ranging west to Schwan and Lárkána; Gilgit at 5000 feet; extending east to Persia.

This species is no doubt resident in Sind and Gilgit, as it probably is in the other parts of its somewhat limited range. I have seen specimens killed on the following dates:—Gilgit, January and June; Sind, November to January; Baluchistan and Mekran Coast, February, April, August, and November; Afghanistan, August and December; Persia, May.

The next two species are united by Hume and some other ornithologists, but I consider them distinct on the following grounds:— *S. picata,* a species with the crown black, visits the plains of India only in the winter, and retires for the summer to the mountains of Afghanistan and Kashmir. *S. capistrata,* a species with the crown white, is a constant resident in the plains of India and the lower parts of Afghanistan, and is never found on the mountains. The females of both species when in good plumage, from September to April, are quite distinct, and may be recognized without difficulty by their colour.

A few birds obtained in Gilgit have the crown largely white, but they were shot just before the autumn moult, when the feathers of that part are extremely worn and ragged, and this may be the result of bleaching. I do not think too much importance should be attached to the occurrence of these abnormal specimens among a very large series of typical *S. picata.* In the same way a few specimens from the plains of India exhibit some black among the white feathers of the crown. These variations are no doubt puzzling, but their cause will probably be solved hereafter without having recourse to the theory of interbreeding, which in this instance is singularly inapplicable, since the breeding-areas of the two species are totally distinct one from the other. One point is quite clear from the immense series of these Chats in the National Collection : the white or the black crown, or the intermixture of black and white, is not due to age.

618. Saxicola picata. *The Pied Chat.*

Saxicola picata, *Blyth, J. A. S. B.* xvi, p. 131 (1847); *id. Cat.* p. 167; *Horsf. & M. Cat.* i, p. 287 ; *Jerd. B. I.* ii, p. 131; *Blanf. & Dresser, P. Z. S.* 1874, p. 227 ; *Hume, S. F.* iii, p. 475; *id. Cat.* no. 489; *Seebohm, Cat. B. M.* v, p. 367 ; *Barnes, S. F.* ix, p. 217 ; *Biddulph, Ibis,* 1881, p. 56 ; *Scully, Ibis,* 1881, p. 441 ; *Biddulph, Ibis,* 1882, p. 236 ; *Barnes, Birds Bom.* p. 262; *Oates in Hume's N. & E.* 2nd ed. ii, p. 52.

Dromolæa picata (*Blyth*), *Hume, S. F.* i, p. 184; *Ball, S. F.* iii, p. 206.

The Pied Stone-Chat, Jerd.

Coloration. Male. The whole head and neck all round, back, scapulars and wings, under wing-coverts and axillaries deep black : remainder of lower plumage with the rump and upper tail-coverts white ; tail white, except the terminal half of the middle pair of feathers and a broad band at the tip of the others, which are black. There is hardly any difference between the summer and winter plumages.

Female. Upper plumage brown; rump and upper tail-coverts white ; tail as in male, but black replaced by brown ; wings brown, all the feathers broadly edged with rufous ; chin, throat, and breast dark ochraceous brown; remainder of the lower parts very pale buff or pinkish white.

The young resemble the female but are mottled below, and the crown is always of the same colour as the back.

A few adult males have sometimes a small amount of white on the crown or over the ear-coverts, and occasionally in birds about to moult nearly the whole crown is white.

Bill and legs black ; iris dark brown (*Bingham*).

Length about 7 ; tail 2·7 ; wing 3·5 ; tarsus 1 ; bill from gape ·75.

Distribution. The Pied Chat summers in Gilgit and the mountains of Afghanistan and Baluchistan, extending at this season to Persia. In winter it visits the plains of the Punjab, Sind, Guzerat, Rajputana as far east as Deesa and Sámbhar, and the Northwest Provinces down to Allahabad. At this season it is also found in the low country of Baluchistan and Afghanistan.

Habits, &c. Breeds from March to July, constructing a nest of grass, lined with feathers, in a hole of a wall or a cleft of a rock. The eggs are greenish blue, with very pale marks of rusty brown round the larger end, and measure about ·81 by ·56.

619. **Saxicola capistrata.** *The White-headed Chat.*

Saxicola leucomela (*Pall.*), *apud Jerd. B. I.* ii, p. 131.
Saxicola capistrata, *Gould, Birds Asia,* iv, pl. 28 (1865); *Hume, S. F.* iii, p. 475; *Seebohm, Cat. B. M.* v, p. 368.
Saxicola morio (*Hempr. & Ehr.*), *apud Hume, Cat.* no. 490; *Barnes, Birds Bom.* p. 203.

The White-headed Stone-Chat, Jerd.

Fig. 25.—Head of *S. capistrata.*

Coloration. Male. After the autumn moult, the forehead, crown, nape, and hind neck are greyish white, somewhat whiter over the eye and ear-coverts; sides of the head, chin, throat, neck all round, back, scapulars, wings, under wing-coverts, and axillaries black; remainder of lower plumage with rump and upper tail-coverts white; tail white, except the terminal half of the middle pair of feathers and a broad band on the tip of the others, which are black. Soon after the autumn moult the tips of the crown-feathers become reduced, and the crown is much whiter than before. When these feathers become still more worn, the crown has a tendency to exhibit patches of black. There is no other seasonal change of plumage.

Female. Resembles the female of *S. picata,* but the upper plumage is more sandy: the chin, throat, and breast are light fulvous, very little darker than the remainder of the lower plumage.

The young resemble the adult female, but are mottled below. After the first autumn the males are blackish brown with broad brown fringes, and the crown is always conspicuously paler than the back.

Bill and legs black; iris brown (*Hume*).

Length about 7; tail 2·7; wing 3·6; tarsus 1; bill from gape ·75.

Distribution. A constant resident in the plains of the Punjab, Sind, and Rajputana, extending in this latter area as far south only as Jodhpur and Sámbhar; and apparently not passing east of the Jumna river. This species extends on the west to Kandahar. It appears to be somewhat rare, but I have seen specimens killed in the above localities in every month of the year except May and July.

Seebohm records this bird from the cultivated districts of Turkestan, apparently on the authority of Severtzoff; but this gentleman states (*S. F.* iii, p. 429) that *S. lugens,* Licht., of his Turkestan list, is nothing but *S. morio* or *S. hendersoni*, and consequently there are now no grounds for stating that *S. capistrata* occurs in Turkestan.

Habits, &c. Nothing is known of the nidification of this species.

620. Saxicola opistholeuca. *Strickland's Chat.*

Saxicola opistholeuca, *Strickl., Jard. Contr. Orn.* 1849, p. 60 ; *Blyth, Cat.* p. 167 ; *Blanf. & Dresser, P. Z. S.* 1874, p. 229 ; *Butler & Hume, S. F.* iii, p. 475 ; *Hume, Cat.* no. 488 ; *Seebohm, Cat. B. M.* v, p. 376 ; *Biddulph, Ibis,* 1881, p. 55 ; *Scully, Ibis,* 1881, p. 411 ; *Biddulph, Ibis,* 1882, p. 276 ; *Barnes, Birds Bom.* p. 201.
Saxicola leucoroides, *Guérin, apud Jerd. B. I.* ii, p. 130.

The Indian White-tailed Stone-Chat. Jerd.

Coloration. Male. The whole plumage black except the rump and upper and under tail-coverts ; tail white, except the terminal half of the middle pair of feathers and a broad band at the tip of the others, which are black. There appears to be no seasonal change of plumage.

Female. Resembles the females of *S. picata* and *S. capistrata* in general colour, but is very dusky throughout and has the ear-coverts a rich brown.

Bill, legs, and feet black ; iris dark brown (*Bingham*).

Length about 6·5 ; tail 2·7 ; wing 3·6 ; tarsus 1 ; bill from gape ·75.

Distribution. A winter visitor to the plains of India from the Punjab down to Khandesh and Nágpur, and from Sind to Etáwah in the N.W. Provinces. This species extends to Afghanistan in the winter. It passes through Gilgit in the spring and autumn, and summers in Turkestan. In addition to the above localities in the plains, it has been recorded by Stoliczka from the lower hills of the Sutlej valley.

621. Saxicola pleschanka [*]. *The Siberian Chat.*

Motacilla pleschanka, *Lepech. Nov. Com. Petr.* xiv, p. 503, pl. xiv, fig. 2 (1770).
Motacilla leucomela, *Pall. Nov. Com. Petr.* xiv, p. 584, pl. xxii, fig. 3 (1770).
Saxicola morio, *Hempr. et Ehr. Symb. Phys.* fol. aa (1833) ; *Blanf. & Dresser, P. Z. S.* 1874, p. 225 (part.) ; *Wardlaw Ramsay, Ibis,* 1880, p. 55 ; *Seebohm, Cat. B. M.* v, p. 372 ; *Biddulph, Ibis,* 1881, p. 58 ; *Scully, Ibis,* 1881, p. 443 ; *Biddulph, Ibis,* 1882, p. 276.
Saxicola hendersoni, *Hume, Ibis,* 1871, p. 408 ; *Hume & Henders. Lah. to Yark.* p. 206, pl. xiii ; *Hume, S. F.* ii, p. 526 ; *Scully, S. F.* iv, p. 144 ; *Hume, Cat.* no. 492 bis ; *id. S. F.* ix, p. 326 note.
Saxicola pleschanka (*Lep.*), *Oates in Hume's N. & E.* 2nd ed. ii, p. 53.

---

[*] As the names *leucomela* and *morio* have been frequently misapplied, especially by Indian ornithologists, it is a matter for congratulation that Lepechin's name is available for this species.

Coloration. Male. After the autumn moult the feathers at the base of the upper mandible, lores, a narrow line above the eye, sides of the head, chin, and throat black, most of the feathers with fawn-coloured fringes; forehead, crown, nape, and hind neck greyish brown, with the bases of the feathers white; a pale buff supercilium; back, scapulars, and upper rump black, very broadly fringed with rufous, the black being nearly invisible; lower rump and upper tail-coverts white; tail white, except the terminal two-thirds of the middle pair of feathers and a band at the tips of the others, which are black; wings black, all the feathers margined with rufous; lower plumage from the throat downwards rufous-fawn, deepest on the breast; under wing-coverts and axillaries black, with very narrow white fringes. In spring and summer the plumage, by a course of abrasion of the tips of the feathers, becomes quite different to that of the autumn and winter. The forehead, crown, nape, and hind neck become pure white; the feathers at the base of the upper mandible, sides of the head, chin, throat, back, scapulars, wings, under wing-coverts, and axillaries uniform deep black; the remainder of the lower plumage pure white; the rump, upper tail-coverts, and tail undergo no alteration.

Female. In the autumn the forehead, crown, nape, hind neck and sides of neck, back, and scapulars are rufous-brown, with narrow paler fringes; wings dark brown, with broad rufous margins; rump and upper tail-coverts white; tail as in the male; a pale rufous supercilium; ear-coverts darker rufous; lower plumage rufous-brown, varying in different individuals, darkest on the breast, a few feathers of the breast and flanks with dark streaks; under wing-coverts and axillaries blackish, with narrow whitish fringes. In spring and summer all the margins to the feathers are lost. The chin, throat, and fore neck become dusky, and the remainder of the lower parts nearly white; the upper plumage is earthy brown, with a fulvous tinge.

Bill, legs, and feet black; iris probably brown.

Length about 6; tail 2·4; wing 3·7; tarsus ·9; bill from gape ·65; second primary longer than sixth.

A perfectly connected series of this Chat in the British Museum conclusively proves that *S. hendersoni* is merely the present species in freshly-moulted plumage.

Distribution. This Chat has a very extensive range, and is migratory, but the materials for tracing its movements are very imperfect. The only part of India in which it occurs is Gilgit and the extreme northern portions of Kashmir, where it is very common throughout the summer and breeds. At this season of the year this species is found in Turkestan and throughout Central Asia to Siberia and Western China. To the west it ranges in summer to Afghanistan, and it is said to breed in Persia and South-eastern Europe. The winter-quarters of this Chat are said to be Abyssinia and Arabia, but I have seen one specimen killed in Gilgit in December.

Habits, &c. Breeds in May and June, building its nest in a hole of a stone wall or in a pile of stones. The eggs are described as being pale blue, with small dusky red freckles, and one measured ·72 by ·56.

622. Saxicola barnesi. *Barnes's Chat.*

Saxicola erythræa, *Hemp. & Ehr., apud Blanford, East. Pers.* ii, p. 150.
Saxicola finschii, *Heugl., apud Seebohm, Cat. B. M.* v, p. 388 (part.) : *apud St. John, Ibis,* 1889, p. 163.

Coloration. Male. After the autumn moult the forehead, crown, nape, hind neck, and the middle portion of the back are buff ; the sides of the back, scapulars, wing-coverts, tertiaries, lores, sides of the head and neck, chin, and throat black, a few of the feathers with narrow pale fringes ; winglet and primary-coverts black, broadly edged with pale buffy white ; quills dark brown, more narrowly edged with pale buff ; rump, upper tail-coverts, breast, and lower plumage white ; under wing-coverts and axillaries black ; tail white, except the terminal half of the middle pair of feathers and a band at the tip of all the others, which are black. In summer the buff upper plumage wears down almost to a white colour, and the fringes on the black portions are cast.

The female is unknown to me. The female of the allied *S. finschi* is thus described by Canon Tristram :—" Back uniform cinereous ; wings brown ; rump white ; tail white, with broad black termination like the male ; throat pale ashy brown : rest of lower parts dull white."

Length nearly 7 ; tail 2·6 ; wing 3·7 ; tarsus 1·05 ; bill from gape ·2.

This species differs from *S. finschi*, the only Chat with which it can be confounded, in having only the chin and throat black, and this colour not connected with the axillaries, from which it is separated by a broad band of white. In *S. finschi* the breast is black, and connected with the black axillaries.

Distribution. Occurs in Baluchistan and Afghanistan and eastwards to Persia. This species is probably resident in all parts of its range. I have examined specimens obtained at Quetta in February, at Kandahar in March and September, and in Persia in March and June.

623. Saxicola vittata. *The Black-backed Eared Chat.*

Saxicola vittata, *Hempr. & Ehr. Symb. Phys. Aves,* fol. cc (1833) ; *Blanf. & Dresser, P. Z. S.* 1874, p. 229 ; *Seebohm, Cat. B. M.* v, p. 386 ; *Biddulph, Ibis,* 1881, p. 59 ; *Hume, S. F.* ix, p. 324 note : *Scully, Ibis,* 1881, p. 444 ; *Biddulph, Ibis,* 1882, p. 277.

Coloration. Male. In autumn the forehead, crown, nape, hind neck, and a small portion of the upper back are white, tinged with grey or brown ; lores, round eye, ear-coverts, sides of neck and of the breast, axillaries and under wing-coverts, back, scapulars, and wings black, many of the feathers fringed with greyish brown, and the wing-coverts with white ; rump, upper tail-coverts, and entire lower plumage white; tail white, except the terminal two-thirds of the middle pair of feathers and a band at the tips of the others, which are black. The summer plumage hardly differs.

Female. According to Seebohm differs from the male in having the black parts replaced by brown, and in having the head and nape suffused with brown.

Bill black, gape leaden ; iris dark brown ; legs black (*Scully*).

Length about 6 ; tail 2·4 ; wing 3·6 ; tarsus ·9 ; bill from gape ·8.

The only specimens of this Chat that I have been able to examine are two males in Seebohm's Collection, one obtained by Severtzoff in Turkestan in March, and the other by Scully in Gilgit in June. The female is unknown to me.

Distribution. Has occurred in Gilgit in May and June. This species breeds in Turkestan, and is said to winter in Arabia and the Bogos country.

624. Saxicola œnanthe. *The Wheatear Chat.*

Motacilla œnanthe, *Linn. Syst. Nat.* i, p. 332 (1766).
Saxicola œnanthe (*Linn.*), *Seebohm. Cat. B. M.* v, p. 391 ; *Biddulph, Ibis,* 1881, p. 60: *Scully, Ibis.* 1881, p. 444 ; *Hume, S. F.* ix, p. 325 note.

Coloration. Male. After the autumn moult the feathers at the base of the upper mandible, a supercilium, and a moustachial streak are white ; lores, under the eye, and ear-coverts black : upper plumage slaty grey with broad rufous fringes ; rump and upper tail-coverts white ; tail white, except the terminal two-thirds of the middle pair of feathers and the terminal quarter of the others, which are black : all the tail-feathers also tipped with pale buff ; wings black, each feather with a rufous margin ; lower plumage buff ; under wing-coverts and axillaries brown, edged with white. In summer the rufous fringes are cast, leaving the upper plumage slaty grey and the wings black.

Female. Feathers at base of upper mandible and supercilium pale rufous : lores and upper part of ear-coverts brown ; lower part of ear-coverts rich ruddy ; upper plumage brown tinged with rufous ; rump and upper tail-coverts white : tail as in male : wing as in male, but not so black ; lower plumage rich buff ; axillaries and under wing-coverts brown, edged with white. In summer the colours are less brilliant, and the fringes to the wing-feathers are reduced in extent.

Bill and claws black ; legs and toes brownish black (*Scully*); iris dark brown (*Seebohm*).

Length about 6 ; tail 2·3 ; wing 3·8 ; tarsus 1·05 ; bill from gape ·8.

Distribution. Has been noticed in Gilgit during the spring migration in March and April. It is highly improbable that Jerdon should have procured this Chat in Central India, and there can be little doubt that the species recorded by him under the name of *S. œnanthe* was *S. isabellina.*

The Wheatear Chat has an immense range and migrates great distances. According to season it is found over a great part of Asia, Europe, Africa, and North America.

625. Saxicola isabellina. *The Isabelline Chat.*

Saxicola isabellina, *Cretzschm. Rüpp. Atlas,* p. 52 (1826); *Stoliczka, J. A. S. B.* xli, pt. ii, p. 239; *Blanf. & Dresser, P. Z. S.* 1874, p. 229; *Scully, S. F.* iv, p. 142; *Hume, Cat.* no. 491; *Seebohm, Cat. B. M.* v, p. 399; *Scully, Ibis,* 1881, p. 411; *Barnes, Birds Bom.* p. 203.
Saxicola œnanthe (*Linn.*), *apud Jerd. B. I.* ii, p. 132.

Coloration. Male. After the autumn moult the upper plumage is sandy brown, the longer feathers of the rump and the upper tail-coverts white; wings dark brown, every feather with a fulvous margin and tip, the margins broader on the secondaries and greater coverts; middle pair of tail-feathers with basal third white, remaining two-thirds black; the other tail-feathers with rather more than the basal half white, and remainder black, and all of them tipped narrowly with white; a white supercilium from the nostrils to the end of the ear-coverts; lores black; ear-coverts fulvous-brown; chin whitish; whole lower plumage buff; under wing-coverts and axillaries fulvous. In summer the margins of the wing-feathers are much reduced in breadth.

Female. Hardly differs from the male, but has the lores generally paler.

Legs and feet black; bill black; iris brown (*Hume*).

Length about 7; tail 2·3 to 2·6; wing 3·6 to 4; tarsus 1·15 to 1·25; bill from gape ·8 to ·9.

Distribution. A winter visitor to the plains of India from the Punjab south to Ahmednagar and east to Chunar and Benares, which are the extreme limits of this species as indicated by the specimens I have examined. This Chat breeds and passes the summer in Turkestan, Afghanistan, Baluchistan, and Persia, and passes through Gilgit in the spring and autumn. It has a very wide range, extending to South-east Europe and North-east Africa on the one side, and to the east of Asia on the other.

Habits, &c. Breeds in Afghanistan, according to Barnes, in March, and in Turkestan, according to Scully, in April and May, but neither the nest nor eggs have yet been taken in these countries.

626. Saxicola deserti. *The Desert Chat.*

Saxicola deserti, *Temm. Pl. Col.* pl. 359, fig. 2 (1825); *Jerd. B. I.* ii, p. 132; *Stoliczka, J. A. S. B.* xxxvii, pt. ii, p. 42; *Hume, S. F.* i, p. 188; *Blanf. & Dresser, P. Z. S.* 1874, p. 224 (part.); *Scully, S. F.* iv, p. 143; *Hume, Cat.* no. 492; *Seebohm, Cat. B. M.* v, p. 383; *Barnes, Birds Bom.* p. 205.
Saxicola atrogularis, *Blyth, J. A. S. B.* xvi, p. 131 (1847); *Blyth, Cat.* p. 167; *Horsf. & M. Cat.* i, p. 287; *Hume & Henders. Lah. to Yark.* p. 205.

The Black-throated Wheatear, Jerd.

Coloration. Male. After the autumn moult the upper plumage is rich buff, turning to pale fulvous-white on the rump and upper tail-coverts; basal third of tail-feathers white, remainder black; wing-coverts and quills black, all the feathers more or less margined with white; the tertiaries broadly margined with buff; the inner coverts white; feathers at base of upper mandible and a supercilium pale buff; sides of head and neck, chin, and throat black, fringed with pale buff; remainder of lower plumage buff, brightest on the breast; under wing-coverts and axillaries black, tipped with white; the inner webs of quills, when viewed from below, narrowly margined with white. In spring and summer the supercilium becomes more distinct, the fringes of all the black feathers disappear, and the mantle is marked with dusky; the under wing-coverts and axillaries frequently become entirely black.

Female. Resembles the male in general appearance, but has the colours duller and the supercilium paler; the chin, throat, sides of head and of neck pale brown, not black; ear-coverts rich brown; all the wing-coverts and quills brown, broadly margined with buff, and the inner coverts not white as in the male; under wing-coverts brown, tipped white.

Bill, legs, and feet black; iris brown (*Hume*).

Length about 6·5; tail 2·7; wing 3·7; tarsus 1; bill from gape ·85.

Distribution. A winter visitor to the plains of India, where the limits of this species are almost identical with those of *S. isabellina*, being the latitude of Bombay on the south and Nágpur on the east. This Chat has been obtained at Sámbhar in June, and it is not improbable that this and other Chats, which are considered winter visitors, may remain in small numbers to breed in some of the less-frequented parts of the deserts of Rájputana and Sind. The Desert-Chat breeds in Turkestan, and at the seasons of passage to and fro must occur in Kashmir and Gilgit. It is common, according to Stoliczka, in Western Tibet. It ranges west as far as Algeria.

627. Saxicola montana. *Gould's Chat.*

Saxicola montana, *Gould, Birds Asia.* iv, pl. 30 (1865); *Seebohm, Cat. B. M.* v, p. 384; *St. John, Ibis,* 1889, p. 164.

Coloration. Male. Resembles the male of *S. deserti*, but has

almost the whole of the inner webs of the quills of the wing pure
white quite up to the shaft; is also much larger, the wing measuring 3·9 or 4 inches.

Female. Resembles the female of *S. deserti*, but has a large
amount of white on the inner webs of the quills, whereas *S. deserti*
has none or hardly any; is also considerably larger.

Distribution. I have examined specimens of this Chat killed by
Biddulph at Skárdo on the Indus river in Kashmir in October;
by Blanford at Gwádar, in Baluchistan, in December, and near
Dizak in March.

Outside our limits this species occurs in Afghanistan in September, and in Turkestan and Tibet during the summer. It ranges
to the west as far as the island of Socotra, whence I have seen a
specimen killed in February.

628. Saxicola chrysopygia. *The Red-tailed Chat.*

Dromolæa chrysopygia, *De Filippi, Arch. Zool. Genova*, ii, p. 381
(1863).
Saxicola kingi, *Hume, Ibis*, 1871, p. 29; *Stoliczka, J. A. S. B.* xli,
pt. ii, p. 239; *Hume, S. F.* i, p. 187, iii, p. 476; *id. Cat. no.* 491 bis;
Barnes, Birds Bom. p. 204.
Saxicola chrysopygia (*De Fil.*), *Blanf. & Dresser, P. Z. S.* 1874,
p. 230; *Blanf. East. Pers.* ii, p. 151, pl. x, fig. 1; *Hume, S. F.*
vii, p. 57; *Seebohm, Cat. B. M.* v, p. 389.

Coloration. Sexes alike. Forehead, crown, nape, back, and scapulars brown; rump and upper tail-coverts pale chestnut; tail
brighter chestnut, the terminal half of the middle pair of feathers
and a band at the tip of the others black: lesser, median, and
primary coverts and winglet brown, edged with grey; greater
coverts and quills brown, edged with rufous; lores dark brown;
ear-coverts rich hair-brown; a supercilium greyish white; chin,
throat, and breast white tinged with ashy; remainder of lower
plumage pale brown with a vinaceous tinge, and turning to buff
on the vent and under tail-coverts; axillaries and under wing-
coverts greyish white.

Legs, feet, and bill black; iris dark brown (*G. King*).

Length about 6·5; tail 2·5; wing 3·8; tarsus 1·05; bill from
gape ·85.

Distribution. Apparently a winter visitor to the plains of
North-western India, being found in the Punjab west of the
Jhelum river, Sind, Cutch, Northern Guzerat, and Rajputana as
far east as Jodhpur. This species extends to the west as far as
Persia, where it is certainly found in the summer.

Genus CERCOMELA, Bonap., 1856.

The genus *Cercomela* contains one Indian species, the position
of which is somewhat doubtful. The only young bird I have been
able to examine resembles the adult, but, on the other hand, the

habits and colour of the egg of this species ally it to the *Saxico-linæ*. It is probable that the position of *Cercomela* is among the *Brachypteryginæ*.

The Rock-Chat seems to have the habits of *Saxicola*, frequenting stony tracts of land and breeding in holes of rocks and old buildings. The sexes are alike, the plumage is very dull, and there is little or no seasonal change of plumage.

In *Cercomela* the bill and rictal bristles resemble those of *Saxicola*; the wing is blunt and the first primary is about half the length of the second; the tail is entirely of one colour and much shorter than the wing, and the tarsus is short.

Cercomela melanura (Temm.) occurs in Palestine, Arabia, and North-east Africa, and has been met with at Aden. This species was included among the birds of India by Jerdon, on the authority of Blyth, who identified it by a drawing in the possession of Sir A. Burnes. The bird, from which the drawing was taken, is stated to have been killed in Sind. I do not propose to include this species in my list, as I do not consider its occurrence in India sufficiently well authenticated. The general colour of the plumage is grey, paler beneath, and the tail and upper tail-coverts are black : tail 2·3; wing 3·1.

629. **Cercomela fusca.** *The Brown Rock-Chat.*

Saxicola fusca, *Blyth, J. A. S. B.* xx, p. 523 (1851) ; *id. Cat.* p. xi.
Cercomela fusca (*Blyth*), *Jerd. B. I.* ii, p. 134; *Stoliczka, J. A. S. B.*
    xli, pt. ii, p. 240 ; *Hume, N. & E.* p. 319; *Adam, S. F.* i, p. 380;
    Butler, S. F. iii, p. 477 : *Hume, Cat.* no. 494; *Barnes, Birds Bom.*
    p. 206 ; *Oates in Hume's N. & E.* 2nd ed. ii, p. 54.
Myrmecocichla fusca (*Blyth*), *Seebohm, Cat. B. M.* v, p. 360.

Shama, Cent. Prov.

Coloration. Upper plumage dull rufous-brown, the feathers of the upper tail-coverts darker : wings brown, every feather edged with rufous-brown ; sides of the head and lower plumage dull ferruginous ; tail very dark brown.

Legs and feet black ; bill black ; iris dark brown (*Hume*).

Length about 6·5 ; tail 2·6 ; wing 3·5 ; tarsus 1 ; bill from gape ·85.

Distribution. A resident in a considerable portion of the central parts of the Indian peninsula. The western limits of this species appear to be a line drawn from Cutch through Jodhpur to Hardwar. Thence it extends to Chunar, near Benares, on the east, and to Jubbulpur on the south, and I have not been able to trace its distribution more accurately than this.

Habits, &c. Breeds from March to July, constructing a nest of grass and roots, lined with hair and wool, in holes of walls, quarries, banks, and cliffs, and laying three or four eggs, which are blue marked with rufous, and measure about ·82 by ·62.

Subfamily RUTICILLINÆ.

The *Ruticillinæ*, or Redstarts and Robins, connect the Chats with the Thrushes. They feed principally on the ground, their tarsi are lengthened, and their feet are well adapted for running. They are almost entirely insectivorous, and they are seldom or never gregarious like the Thrushes.

Many of the *Ruticillinæ* are migratory; others are resident. The seasonal change of plumage in the majority of the species, caused by the abrasion of the margins of the feathers, is considerable, especially in the Redstarts.

The Redstarts and Robins have the habit of frequently moving the tail and drooping the wings; they mostly build their nests in holes of trees and rocks, and their eggs are of various colours, generally spotted, but in the case of *Ruticilla* plain blue.

Key to the Genera.

a. Tail forked.
  a'. Tail much longer than wing; middle
    rectrices one third the length of tail . . HENICURUS, p. 82.
  b'. Tail about equal to wing; middle rec-
    trices half the length of tail . . . . HYDROCICHLA, p. 86.
  c'. Tail much shorter than wing; middle
    rectrices reaching nearly to tip of tail. MICROCICHLA, p. 88.
b. Tail rounded or square.
  d'. Tail in both sexes largely chestnut *.
    a''. Tail considerably longer than twice
      tarsus.
      a'''. Tail much rounded; sexes alike. . CHIMARRHORNIS, p. 89.
      b'''. Tail nearly square; sexes different. RUTICILLA, p. 90.
    b''. Tail about twice tarsus.
      c'''. Rictal bristles very long and
        strong . . . . RHYACORNIS, p. 97.
      d'''. Rictal bristles weak or obsolete. . CYANECULA, p. 99.
  e'. Tail without any chestnut.
    e''. First primary shorter than one-third
      of second.
      e'''. Difference between wing and tail
        less than tarsus . . . . DAULIAS, p. 100.
      f'''. Difference between wing and tail
        twice tarsus . . . . GRANDALA, p. 110.
    d''. First primary longer than one-third
      of second.
      g'''. Tail equal to or shorter than wing.
        a'. Outer tail-feathers falling short
          of tip of tail by a distance less
          than half length of middle toe.

* The only exception is in the female of *Rhyacornis fuliginosus*.

a^5. Bill straight and Thrush-like;
rictal bristles well developed.
  a^4. Tail about twice tarsus.
    a^3. Throat of male brilliantly
coloured .... CALLIOPE, p. 101.
    b^3. Throat of male coloured
like remainder of lower
parts .... TARSIGER, p. 104.
  b^4. Tail considerably more than
twice tarsus.
    c^3. Tail uniformly of one
colour.
      a^2. Tips of tail-feathers
mucronate.......... IANTHIA, p. 105.
      b^2. Tips of tail-feathers
rounded........... ADELURA, p. 108.
    d^3. Tail largely white...... NOTODELA, p. 111.
  b^5. Bill slender and curved; rictal
bristles obsolete .... THAMNOBIA, p. 113.
b^4. Outer tail-feathers falling short
of tip of tail by a distance
quite equal to length of middle
toe.
  c^3. Tail of one colour .... CALLENE, p. 113.
  d^3. Tail black and white ...... COPSYCHUS, p. 116.
b'''. Tail much longer than wing .... CITTOCINCLA, p. 118.

Genus HENICURUS, Temm., 1823.

The genus *Henicurus* comprises certain birds with the general appearance of Pied Wagtails, but differing from them in having a forked tail and ten primaries, together with a coarse bill.

The Forktails are found in mountain-streams flitting from pool to pool and feeding on insects which are found on the edge of the water. They are solitary and not very shy when disturbed, flying some distance further on, and on being disturbed a second time frequently disappearing into the jungle to return to the stream shortly afterwards. They wag their tails incessantly, and seldom perch except on rocks and bare branches near the ground. They build nests of moss in the banks of streams or under rocks and snags, and lay spotted eggs.

In *Henicurus* the bill is nearly as long as the head, stout and straight, and the lower mandible is much bulged out in the middle; the rictal bristles are well developed; the wing is large, the first primary being about half the length of the second; the tail is much longer than the wing, deeply forked, and the median feathers of about one third the length of the outer ones; the tarsus is long and of a very pale colour. The sexes are alike. None of the species are known to migrate.

Key to the Species.

a. Back spotted.
  a'. Spots on back lunate ; feathers on lower
    breast fringed with white .... *H. maculatus*, p. 83.
  b'. Spots on back round ; feathers on lower
    breast entirely black .... *H. guttatus*, p. 84.
b. Back plain.
  c'. Chin and throat black.
    a''. Back slate-coloured .... *H. schistaceus*, p. 84.
    b''. Back black...................... *H. immaculatus*, p. 85.
  d'. Chin, throat, and breast black........ *H. leschenaulti*, p. 86.

630. **Henicurus maculatus.** *The Western Spotted Forktail.*

Enicurus maculatus, *Vigors, P. Z. S.* 1830, p. 9 ; *Gould, Cent.*
pl. xxvii ; *Blyth, Cat.* p. 159 ; *Horsf. & M. Cat.* i, p. 346 ; *Jerd.
B. I.* ii, p. 212 ; *Hume, N. & E.* p. 374.
Henicurus maculatus, *Vigors, Blyth, Ibis,* 1867, p. 29 ; *Stoliczka,
J. A. S. B.* xxxvii, pt. ii, p. 47 ; *Elwes, Ibis,* 1872, p. 260 ; *Hume
& Henders. Lah. to York.* p. 222 ; *Hume, Cat.* no. 584 ; *Scully,
S. F.* viii, p. 310 ; *Sharpe, Cat. B. M.* vii, p. 317 ; *Oates in Hume's
N. & E.* 2nd ed. ii, p. 57.

The Spotted Fork-tail, Jerd. ; *Khanjan*, N.W. Him.

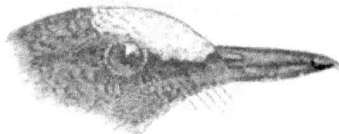

Fig. 26.—Head of *H. maculatus.*

Coloration. Forehead and anterior half of crown white ; remainder of the head, the whole neck, breast, back, lesser and median wing-coverts, primary-coverts, and winglet black ; the feathers of the hind neck thickly marked with large round white spots, those of the back and scapulars with large ovate transverse bars ; greater wing-coverts black, tipped white ; the earlier primaries black ; the other quills broadly white at base and also tipped white ; rump, upper tail-coverts, abdomen, and under tail-coverts white ; the lower part of the black of the breast fringed with white ; the two outer pairs of tail-feathers white ; the others black, broadly white at base and tipped white ; under wing-coverts black ; axillaries white.

In many specimens the crown and nape are rufous-ashy, with black margins to the feathers.

The young have the head, back, and lower parts down to the breast rufous-ashy, the breast with whitish shaft-streaks.

Bill black : iris dark brown ; legs, feet, and claws white, with a pink tinge at junction of toes with tarsi and on all the toe-joints (*Hume*).

G 2

Length about 11; tail up to 6·2; wing 4·5; tarsus 1·15; bill
from gape 1·05.

Distribution. The Himalayas from Murree to Nepal at all elevations up to 12,000 feet.

Habits, &c. Breeds from April to June, constructing a cup-
shaped nest, chiefly of moss, on a ledge of rock or at the root of a
tree, or on a bank near water, and laying four or five eggs, which
are greenish white marked with yellowish or reddish brown, and
measure about ·96 by ·72.

631. Henicurus guttatus. *The Eastern Spotted Forktail.*

Enicurus maculatus, *Vig., Jerd. B. I.* ii, p. 212 (part.).
Enicurus guttatus, *Gould, P. Z. S.* 1865, p. 664; *Hume, N. & E.*
    p. 375; *Oates, S. F.* iii, p. 342.
Henicurus guttatus, *Gould, Blyth, Ibis,* 1867, p. 29; *Elwes, Ibis,*
    1872, p. 261; *Hume, S. F.* vii, p. 349; *id. Cat. no.* 584 bis; *Scully,*
    S. F. viii, p. 311; *Sharpe, Cat. B. M.* vii, p. 316; *Oates, B. B.*
    i, p. 26; *Hume, S. F.* xi, p. 227; *Oates in Hume's N. & E.* 2nd
    ed. ii, p. 58.

Oong-sam ching-pho, Lepch.; *Chubia leka,* Bhut.

Coloration. Resembles *H. maculatus.* Differs in having the
white marks on the back fewer in number, small, and perfectly
round in shape, and in having the breast deep black without any
white fringes; smaller in size.

Length about 10; tail 5·3; wing 4; tarsus 1·2; bill from
gape 1.

Distribution. Nepal; Sikhim; North Khási hills; Manipur;
Arrakan. I have seen a typical specimen of this species from
Mussoorie, but it is seldom that this bird occurs so far west.

Habits, &c. Breeds in Sikhim from 2000 feet upwards in May
and June. The nest and eggs do not appear to differ from those
of *H. maculatus.* The eggs measure about ·93 by ·68.

632. Henicurus schistaceus. *The Slaty-backed Forktail.*

Enicurus schistaceus, *Hodgs. As. Res.* xix, p. 189 (1836); *Blyth,*
    Cat. p. 159; *Horsf. & M. Cat.* i, p. 346; *Jerd. B. I.* ii, p. 214;
    Hume, N. & E. p. 376.
Henicurus schistaceus, *Hodgs., Elwes, Ibis,* 1872, p. 253; *Godw.-*
    Aust. J. A. S. B. xlv, pt. ii, p. 80; *Hume & Dav. S. F.* vi, p. 361;
    Hume, Cat. no. 586; *Scully, S. F.* viii, p. 311; *Sharpe, Cat. B. M.*
    vii, p. 315; *Oates, B. B.* i, p. 27; *Hume, S. F.* xi, p. 229; *Oates*
    in Hume's N. & E. 2nd ed. ii, p. 60.

Coloration. A frontal band and the feathers immediately over
and behind the eye white; lores, cheeks, ear-coverts, chin, upper
part of the throat, and the sides of the lower part black; crown,
nape, sides of neck and back slaty blue; lesser wing-coverts black,
margined with slaty black; median coverts black; greater coverts
black, tipped with white; scapulars slaty blue, tipped with white;
quills black, broadly white at base; the secondaries and tertiaries

with white tips; rump, upper tail-coverts, and the two outer pairs of tail-feathers white; the other tail-feathers black, with white bases and tips; lower part of throat and the whole lower plumage white, the breast with narrow indistinct black cross-bars.

The young have the head and back ashy brown; chin and lower plumage white, the throat and breast mottled with brown.

Bill black; iris blackish brown; feet fleshy white; the tarsi livid in front; claws whitish.

Length up to 10; tail up to 5; wing 3·8; tarsus 1·05; bill from gape 1.

Distribution. The Himalayas from Kumaun * to the Daphla hills in Assam; the Nága hills; Manipur; Arrakan; the whole of Tenasserim. This species extends into Southern China.

Habits, &c. Breeds in Sikhim and Tenasserim from March to July. The nest and eggs do not differ in any respect from those of the preceding species. The eggs measure about ·85 by ·65.

633. Henicurus immaculatus. *The Black-backed Forktail.*

Enicurus immaculatus, *Hodgs. As. Res.* xix, p. 190 (1836); *Blyth, Cat.* p. 159; *Horsf. & M. Cat.* i, p. 346; *Jerd. B. I.* ii, p. 213; *Hume, S. F.* iii, p. 141.
Henicurus immaculatus, *Hodgs., Elwes, Ibis,* 1872, p. 254; *Anders. Yunnan Exped., Aves,* p. 610; *Hume, Cat.* no. 585; *Sharpe, Cat. B. M.* vii, p. 314; *Oates, B. B.* i, p. 25; *Hume, S. F.* xi, p. 228; *Oates in Hume's N. & E.* 2nd ed. ii, p. 62.

Coloration. Forehead and the feathers above and behind the eye white; the plumes at the base of the bill, the lores, sides of the head, sides of the neck, chin, throat, crown, nape, and back black; lesser and median wing-coverts black; primary-coverts and winglet black; scapulars and greater coverts black, tipped with white; primaries black; secondaries and tertiaries black, the bases broadly white and tipped white; rump, upper tail-coverts, and the lower plumage white, the feathers at the side of the breast tipped with black; axillaries white; the two outer pairs of tail-feathers white, the others black, with broad white bases and narrower tips.

The young have the head, neck, back, and breast sooty black, and there is no white on the forehead and about the eyes. The nestling is probably spotted.

Bill and inside of mouth black; iris brown; feet and claws pale yellowish white.

Length nearly 10; tail 5·3; wing 4; tarsus 1·2; bill from gape ·95.

Distribution. The Himalayas from Garhwál to Dibrugarh in Assam; Cachar; Sylhet; the Gáro and Khási hills; Manipur;

---

* Hume assures us that he procured the nest of this species in Kumaun (N. & E. p. 376), and I accordingly give the range as extending westwards to this State, but I have seen no specimen from any locality west of Nepal.

the neighbourhood of Bhámo; Arrakan; Pegu, probably not extending to the east of the Sittoung river. This species is found at low elevations chiefly.

Habits, &c. I found the nest of this Forktail placed on a bank at the side of a nullah in the Pegu hills on the 20th April, and containing three fresh eggs.

634. Henicurus leschenaulti. *Leschenault's Forktail.*

Turdus leschenaulti, *Vieill. Nouv. Dict. d'Hist. Nat.* xx, p. 269 (1818).
Enicurus leschenaulti (*V.*), *Horsf. & M. Cat.* i, p. 315; *Godw.-Aust. J. A. S. B.* xliii, pt. ii, p. 168.
Henicurus leschenaulti (*V.*), *Elwes, Ibis,* 1872, p. 258; *Hume, S. F.* v, p. 249; *Tweedd. Ibis,* 1877, p. 310; *Hume & Dav. S. F.* vi, p. 360; *Hume, Cat.* no. 584 ter; *Sharpe, Cat. B. M.* vii, p. 313; *Oates, B. B.* i, p. 27; *Hume, S. F.* xi, p. 228.
Henicurus sinensis, *Gould, apud Godw.-Aust. J. A. S. B.* xlv, pt. ii, p. 80; *Salvad. Ann. Mus. Civ. Gen.* (2) vii, p. 412.

Coloration. Forehead and two thirds of the crown white; the remainder of the head, neck, breast, back, scapulars, and wings black, the scapulars and the greater coverts broadly tipped white, the secondaries and tertiaries white at their bases, and the later ones tipped white; rump, upper tail-coverts, abdomen, vent, under tail-coverts, under wing-coverts, and axillaries white; the two outer pairs of rectrices entirely white; the next black, with white base and tip; the others black, tipped white.

The nestling has the head, neck, and back chocolate-brown, the sides of the head with pale shafts; chin and throat grey; breast brown, with yellowish streaks; the other parts as in the adult; the white of the crown is assumed at a late period.

Length about 11; tail up to 5·9; wing up to 4·5; tarsus 1·3; bill from gape 1·1.

This species is slightly smaller than *H. sinensis,* Gould, from China, and is inseparable from it in coloration. The two species may, however, be always separated by the structure of the tail. In *H. leschenaulti* the outermost tail-feathers are about as long as those next to them, whereas in *H. sinensis* these feathers are shorter than the penultimate pair by about two inches.

Distribution. Sikhim; the Bhutan Doars; Upper Assam; the Daphla hills; the E. Nága hills; the Khási hills; Manipur; the northern and central portions of Tenasserim below 2500 feet. This species is not occur Java.

Genus **HYDROCICHLA**, Sharpe, 1883.

This genus differs from *Henicurus* in having the tail about equal in length to the wing, and the middle pair of tail-feathers about half the length of the tail. There are two species of this genus found in India, in one of which the sexes are alike and in the other dissimilar.

Key to the Species.

a. Crown of head white; nape black...... *H. frontalis*, p. 87.
b. Crown of head and nape chestnut ...... *H. ruficapilla*, p. 87.

635. **Hydrocichla frontalis.** *The White-crowned Forktail.*

Enicurus frontalis, *Blyth, J. A. S. B.* xvi, p. 156 (1847); *id. Cat.*
p. 159; *Horsf. & M. Cat.* i, p. 346.
Henicurus frontalis, *Blyth, Elwes, Ibis,* 1872, p. 259, pl. ix; *Oates,*
S. F. v, p. 248; *Tweedd. Ibis,* 1877, p. 310; *Hume & Dav. S. F.* vi,
p. 360; *Hume, Cat.* no. 584 quat.
Hydrocichla frontalis (*Blyth*), *Sharpe, Cat. B. M.* vii, p. 321; *Oates,*
B. B. i, p. 29.

Coloration. Forehead and front of crown white; head, neck,
breast, back, lesser and median wing-coverts, primary-coverts and
winglet black; scapulars and greater wing-coverts black, tipped with
white; primaries black; secondaries and tertiaries black, with broad
white bases; rump and upper tail-coverts, abdomen, and under
tail-coverts white; the two outer pairs of tail-feathers white, the
others black with white bases and tips.

The youngest bird I have been able to examine has the whole
head, neck, back, and breast dusky brown, with no trace of white
on the forehead or crown. Nestling birds will, no doubt, prove to
be spotted as in the other species.

Bill black; legs flesh-colour.

Length nearly 8; tail 3·6; wing 3·6; tarsus 1·4; bill from
gape ·95.

Distribution. The extreme south of Tenasserim, extending down
the Malay peninsula to Sumatra and Borneo.

636. **Hydrocichla ruficapilla.** *The Chestnut-backed Forktail.*

Enicurus ruficapillus, *Temm. Pl. Col.* iii, pl. 534 (1832); *Blyth, Cat.*
p. 159.
Henicurus ruficapillus, *Temm., Elwes, Ibis,* 1872, p. 257; *Hume &*
Dav. S. F. vi, p. 361; *Hume, Cat.* no. 588 bis.
Hydrocichla ruficapilla (*Temm.*), *Sharpe, Cat. B. M.* vii, p. 319;
Oates, B. B. i, p. 28.

Coloration. Male. Forehead white; a frontal band, the lores,
cheeks, ear-coverts, chin, and throat black; the forehead, crown, nape,
upper part of back, and the sides of the neck chestnut; back, lesser
and median coverts, primary-coverts, and winglet black; greater
coverts and scapulars black, tipped with white; primaries black;
secondaries and tertiaries black, with broad white bases and narrower
tips; rump, upper tail-coverts, and the lower plumage white; the
feathers of the breast and the upper part of the abdomen margined
with black; the two outer pairs of tail-feathers white; the others
black, with broad white bases and tips.

Female. Resembles the male, except that the whole of the back
is chestnut, tinged with olivaceous on the lower portion.

The young bird is probably spotted in its first stage of plumage.

Legs, feet, and claws pale pinky- or fleshy-white; bill black; iris dark brown (*Hume & Davison*).

Length about 8; tail 3·3; wing 3·7; tarsus 1·15; bill from gape 1·1.

Distribution. Tenasserim, from Nwalabo mountain southwards. This species extends down the Malay peninsula and is found in Borneo.

Genus MICROCICHLA, Sharpe, 1883.

This genus differs from the two preceding genera in its very short tail, which is very much shorter than the wing and very slightly forked, the middle feathers reaching nearly to the tip of the outermost ones. There is only one species known, and the sexes are alike.

637. Microcichla scouleri. *The Little Forktail.*

Enicurus scouleri, *Vigors, P. Z. S.* 1831, p. 174; *Gould, Cent.* pl. xxviii; *Blyth, Cat.* p. 159; *Horsf. & M. Cat.* i, p. 347; *Jerd. B. I.* ii, p. 214; *Hume, N. & E.* p. 377.

Enicurus nigrifrons, *Hodgs., Gray, P. Z. S.* 1859, p. 102; *Jerd. B. I.* ii, p. 215; *Hume, Cat.* no. 588.

Henicurus scouleri, *Vigors, Elwes, Ibis,* 1872, p. 255; *Hume, Cat.* no. 587; *Scully, S. F.* viii, p. 311; *Biddulph, Ibis,* 1882, p. 270.

Microcichla scouleri (*Vigors*), *Sharpe, Cat. B. M.* vii, p. 322; *Oates in Hume's N. & E.* 2nd ed. ii, p. 62.

The Short-tailed Forktail, The Black-fronted Forktail, Jerd.; *Oomy-sumbrek-pho,* Lepch.

Coloration. Forehead and anterior half of crown white; the whole head, neck, upper part of breast, back, lesser and median wing-coverts, primary-coverts, and winglet black; scapulars and greater coverts black tipped white; primaries black; secondaries and tertiaries black with broad white bases, the later quills also with white margins; rump and upper tail-coverts white, with a broad black band across the former; the two outer pairs of tail-feathers white; third pair white, with a black tip; the others progressively with less white and broader black tips; lower plumage white, the breast and the sides of the body mottled with black.

The young have the forehead and anterior half of the crown black like the rest of the upper plumage; the chin and throat white and the breast more mottled than in the adult.

Bill black; iris dark brown; feet and claws pure fleshy white (*Scully*).

Length 5; tail 1·9; wing 2·9; tarsus ·95; bill from gape ·6.

Distribution. The whole of the Himalayas, from Gilgit to the Daphla hills in Assam; the E. Nága hills. This species is found up to 11,000 feet in summer. It extends into Western China.

Habits, &c. This Forktail appears to differ from all the other Forktails in its habit of plunging into the water. The nest does not appear to have been found yet.

Genus CHIMARRHORNIS, Hodgs., 1844.

The genus *Chimarrhornis* is closely allied to the Forktails in habits, and connects them with the Redstarts. The sole species of this genus is found on the Himalayas and on certain hill-ranges of Assam and Burma.

Fig. 27.— Head of *C. leucocephalus*.

In *Chimarrhornis* the sexes are alike, and the tail, which is chestnut tipped with black, is much rounded; the wing is large and rounded, the first primary being about half the length of the second; the tarsus is long.

638. Chimarrhornis leucocephalus. *The White-capped Redstart.*

Phœnicura leucocephala, *Vigors*, P. Z. S. 1830, p. 35; *Gould, Cent.* pl. xxvi, fig. 1.
Ruticilla leucocephala (*Vig.*), *Blyth, Cat.* p. 169; *Horsf. & M. Cat.* i, p. 309; *Blanf. J. A. S. B.* xli, pt. ii, p. 51.
Chæmorrornis leucocephala (*Vig.*), *Jerd. B. I.* ii, p. 143; *Stoliczka, J. A. S. B.* xxxvii, pt. ii, p. 44; *A. Anders. S. F.* iii, p. 355; *Hume & Henders. Lah. to Yark.* p. 214; *Hume, Cat.* no. 505; *Scully, S. F.* viii, p. 305; *Oates, B. B.* i, p. 24.
Chimarrhornis leucocephala (*Vig.*), *Anders. Yunnan Exped., Aves,* p. 613; *Sharpe, Cat. B. M.* vii, p. 47; *Hume, S. F.* xi, p. 197; *Oates in Hume's N. & E.* 2nd ed. ii, p. 63.

Gir-chandia, Hind.; *Kali-pholia* at Mohun Ghât; *Mati-tap-pho,* Lepch.; *Chubia-muti,* Bhut.

Coloration. Crown and nape white; with this exception, the whole head and neck, with the breast, back, and the whole of the wings black; rump, upper tail-coverts, abdomen, flanks, and under tail-coverts chestnut; tail chestnut, broadly tipped with black.

The young have the plumage blackish brown, the feathers of the back, rump, and of the whole lower plumage fringed with rufous; crown and nape white with black edges; tail and wings as in the adult. The rufous and black tips and fringes are soon cast off, and the adult plumage assumed in October or November.

Bill black; gape fleshy white; iris deep brown; feet blackish brown; claws black (*Scully*).

Length about 7·5; tail 3·2; wing 3·9; tarsus 1·2; bill from gape ·8.

Distribution. The Himalayas from Afghanistan and Gilgit to Assam; the Khási hills; Manipur; the second defile, Irrawaddy river; Arrakan.

This species is found at high elevations on the Himalayas in summer, Stoliczka recording it from about 20,000 feet above the sea. In winter it descends to below 3000 feet, or probably to the plains at the foot of the hills.

This Redstart extends to China.

Habits, &c. Frequents the banks of rivers and nullahs, feeding at the edge of the water on insects, and constantly moving its tail up and down and expanding the feathers. Breeds in May, constructing a cup-shaped nest of green moss and fibres, lined with hair, in the hollow of a bank on the side of a stream. The eggs are greenish white covered with rufous spots.

Genus **RUTICILLA**, Brehm, 1828.

The genus *Ruticilla* contains the true Redstarts, which may be recognized by their rather long tail, which is more than twice the length of the tarsus and nearly or quite square at the tip; by the large amount of chestnut in the tail; and by the sexes being differently coloured. The bill is short, slender, and black, and the rictal bristles moderate or short. The wing is sharply pointed, and the first primary less than half the second. The tarsus is of moderate length.

The Redstarts feed on the ground largely, but they also catch insects on the wing and perch freely. They constantly vibrate their tail. Nearly all the species migrate to a greater or less extent, and those that inhabit the Himalayas move vertically according to season. They breed in holes of trees and rocks, and lay unspotted blue eggs.

The nestlings of the Redstarts are streaked with fulvous above and have the feathers margined with brown below. In each species the nestling has the same pattern of tail as the adult, rendering specific recognition comparatively easy.

The seasonal change of plumage in the Redstarts, due to the wearing away of the edges of the feathers in the winter and spring, is very great.

Key to the Species[*].

a. All tail-feathers except the middle pair
 abruptly tipped with black .... *R. frontalis*, p. 91.
b. None of the tail-feathers tipped.
 a'. A large white patch on throat .... *R. schisticeps*, p. 92.

---

[*] The following species are reported to have occurred in India, but either by error or on insufficient evidence :—

R. PHŒNICURA (Linn.).—Two skins of this species now in the British Museum, originally deposited in the Indian Museum, as noticed by Horsfield and Moore (Cat. i, p. 304), are said to have been procured at Sahâranpur by Dr. Jameson. The two specimens in question, which have been at one time stuffed and mounted, are typical *R. phœnicura*. The occurrence of this species in India requires confirmation. It resembles *R. rufiventris*, but has the anterior part of the crown

b'. No white patch on throat.
  a''. Middle tail-feathers distinctly different
    to others; wing under 3·5.
    a'''. Secondaries with white on both
      webs........................ *R. aurorea*, p. 93.
    b'''. No white on inner webs of secon-
      daries.
      a⁴. Terminal portion of shafts of all
        lateral tail-feathers black ...... *R. erythronota*, p. 94.
      b⁴. Shafts of lateral tail-feathers uni-
        formly chestnut.
        c⁵. Throat and breast black.
          a⁶. Portion of outer webs of secon-
            daries broadly margined with
            white .... *R. hodgsoni* ♂, p. 95.
          b⁶. No white margins to secon-
            daries .... *R. rufiventris* ♂, p. 95.
        c⁵. Throat and breast buff or ashy
          brown.
          c⁶. Lower plumage in general
            ashy brown............. *R. hodgsoni* ♀, p. 95.
          d⁶. Lower plumage in general
            buff, frequently suffused
            with orange .... *R. rufiventris* ♀, p. 95.
  b''. Middle tail-feathers of much the same
    colour as others; wing over 4 ...... *R. erythrogaster*, p. 97.

639. Ruticilla frontalis. *The Blue-fronted Redstart.*

Phœnicura frontalis, *Vigors, P. Z. S.* 1831, p. 172; *Gould, Cent.*
pl. xxxvi, fig. 1.
Ruticilla frontalis (*Vig.*), *Blyth, Cat.* p. 168; *Horsf. & M. Cat.* i.
p. 308; *Jerd. B. I.* ii, p. 141; *Blanf. J. A. S. B.* xli, pt. ii, p. 50;
Hume & Henders. Lah. to Yark. p. 211; *Hume, Cat.* no. 503;
Scully, S. F. viii, p. 302; *Biddulph, Ibis,* 1881, p. 63; *Scully, Ibis,*
1881, p. 446; *Seebohm, Cat. B. M.* v, p. 349; *Hume, S. F.* xi,
p. 195; *Oates in Hume's N. & E.* 2nd ed. ii, p. 64.

Tak-tirriri-pho, Lepch.

Coloration. Male. After the autumn moult the forehead and

white with a distinct greyish-white supercilium, and the back is grey without a
trace of black at any season.

R. ERYTHROPROCTA, Gould. — This bird has been recorded from India under a
misapprehension as to what the species really is. The only specimens known
are two from the Gould collection, now in the British Museum. They were
obtained in Asia Minor. This Redstart resembles *R. rufiventris*, but has the
under wing-coverts and axillaries black tipped with ashy, and the black of the
lower parts produced further down on the abdomen.

R. MESOLEUCA (Hempr. & Ehr.). — This species is said to have been procured
at Dankapur in Sind by Mr. Murray's native collector. The specimen obtained
reached Mr. Hume's hands without a label. Both Hume (S. F. vii, p. 115)
and Blanford (S. F. vii, p. 527) entertain doubts of the occurrence of this species
in India, and I think the matter requires confirmation. For the present it is
perhaps advisable to omit it from my list. *R. mesoleuca* resembles *R. rufiventris*,
but has the anterior third of the crown pure white and the supercilium whitish;
the outer webs of the secondaries and later primaries are, moreover, margined
with white.

supercilium are bright blue; remainder of head, neck, back, and lesser and median wing-coverts dull blue, the feathers with brown fringes; greater coverts dark brown, suffused with blue on the outer webs and tipped with rufous; quills dark brown, margined with pale rufous; rump, upper tail-coverts, breast, and lower

Fig. 28.—Head of *R. frontalis*.

plumage chestnut; tail chestnut, except the middle pair of feathers and broad tips to the others, which are black; under wing-coverts and axillaries chestnut with blue bases. In summer the fringes to the feathers of the blue parts of the plumage get worn away, leaving those parts pure blue.

Female. Forehead, crown, nape, scapulars, and back rich brown with a tinge of fulvous; rump, upper tail-coverts, and tail as in the male; quills and coverts brown, edged with fulvous; a ring of pale feathers round the eye; chin fulvous; throat and breast fulvous-brown; remainder of lower parts orange-brown, becoming brighter and purer orange on the abdomen and under tail-coverts.

The young nestling has the tail similar to that of the adult.

Bill, legs, feet, and claws black; iris deep brown (*Hume*).

Length about 6; tail 2·6; wing 3·4; tarsus ·9; bill from gape ·7.

Distribution. The Himalayas from Gilgit and Kashmir to Assam; the Khási hills; North Cachar; Manipur. This species is seldom found below 5000 feet in summer, and it occurs up to 14,000 feet or even higher at that season. It extends into Tibet and Western China.

Habits, &c. Little is known of the nidification of this Redstart; eggs said to belong to this species, procured in Sikhim in June, measured about ·82 by ·59.

640. Ruticilla schisticeps. *The White-throated Redstart.*

Ruticilla schisticeps, *Hodgs. Cat. Mamm. etc. Nep. Coll.* p. 69, App. p. 153 (1846); *Jerd. B. I.* ii, p. 140; *Blanf. J. A. S. B.* xlvii, pt. 2, p. 1, pl. i; *Hume, Cat.* no. 501; *Seebohm, Cat. B. M.* v, p. 351.
Ruticilla nigrogularis, *Moore, P. Z. S.* 1854, p. 29, pl. xli; *Horsf. & M. Cat.* i, p. 307; *Jerd. B. I.* ii, p. 140; *Hume, S. F.* iv, p. 497; *id. Cat.* no. 502.

The Slaty headed Redstart, Jerd.

Coloration. Male. Forehead and crown cobalt-blue, turning to duller blue on the nape; feathers at base of upper mandible, sides

of the head and neck, chin, throat, back, and scapulars black, the
longer scapulars tipped with chestnut, and all the feathers of these
black portions of the plumage, and also of the crown, fringed with
fulvous after the autumn moult ; a well-defined white patch on the
throat ; rump, upper tail-coverts, and lower plumage rich chestnut,
the middle of the abdomen albescent : tail black, all but the middle
pair of feathers chestnut on the basal third of their length : the
median wing-coverts, the innermost greater coverts, and the outer
margins of tertiaries and later secondaries white : remainder of
wings black : under wing-coverts and axillaries black, with broad
white tips.

Female. Forehead, crown, nape, back, upper part of rump,
scapulars, and sides of neck rich brown ; lower part of rump and
upper tail-coverts chestnut ; tail dark brown, the basal half of all
but the middle pair of feathers dull rufous ; wings as in the male,
but the black replaced by brown and the white margins narrower ;
lower plumage pinkish ashy, albescent on the abdomen, and with
a large white patch on the throat ; under wing-coverts and axillaries brown, tipped white.

Bill and legs black.

Length about 6 ; tail 2·9 ; wing 3·3 ; tarsus ·9 ; bill from
gape ·55.

Distribution. Confined to the higher portions of Nepal and
Sikhim and the adjoining parts of Tibet and Central Asia, such as
Kansu.

644. Ruticilla aurorea. *The Daurian Redstart.*

Motacilla aurorea, *Pall. Reis. Russ. Reichs*, iii, p. 695 (1776).
Phœnicura leucoptera, *Blyth, J. A. S. B.* xii, p. 962 (1843)
Ruticilla leucoptera (*Bl.*), *Blyth, Cat.* p. 168.
Ruticilla aurorea (*Pall.*), *Horsf. & M. Cat.* i, p. 305 ; *Jerd. B. I.* ii,
   p. 139 ; *Hume, Cat.* no. 500 ; *Scebohm, Cat. B. M.* v, p. 345 ;
   Oates, B. B. i, p. 46 ; *Hume, S. F.* xi, p. 195.

Reeves' Redstart, Jerd.

Coloration. Male. After the autumn moult the forehead, crown,
nape, and mantle are slaty grey, the feathers fringed with slaty
brown, the portion over the eye and along the sides of the neck
purer grey : lower back, scapulars, and wing-coverts black with
fulvous margins ; rump, upper tail-coverts, and tail chestnut,
except the middle pair of feathers of the latter, which are black :
quills black, each secondary with a patch of white, forming a large
wing-spot ; feathers at base of bill, chin, throat, and sides of head
black, with whitish margins : remainder of lower plumage chestnut. In the spring and summer the fringes of the feathers are
much reduced in extent, but seldom entirely lost except on the
throat.

Female. Everywhere brown, paler beneath and albescent on the
abdomen ; a circle of pale feathers round the eye ; rump, upper
tail-coverts, and tail as in the male : wings brown edged with

fulvous, and the secondaries with white patches as in the male,
but reduced in extent and slightly fulvous.

Legs and feet black; bill blackish brown to black, yellow or
yellowish at gape in the male and sometimes on base of lower
mandible also; iris deep brown (*Hume*).

Length about 6; tail 2·5; wing 2·9; tarsus ·9; bill from gape ·65.

Distribution. A winter visitor to Bhutan; Assam; the Khási
hills; Cachar; Sylhet; the Nága hills; Manipur; Thayetmyo.
In winter this species extends to Java and Timor, and in summer
it is found in Siberia, Mongolia, Japan, and China.

642. Ruticilla erythronota. *Eversmann's Redstart.*

Sylvia erythronota, *Eversm. Add. Pall. Zoogr. Ross.-Asiat.* fasc. ii,
    p. 11 (1841).
Ruticilla rufogularis, *Moore, P. Z. S.* 1854, p. 27, pl. lix; *Horsf. &
    M. Cat.* i, p. 306.
Ruticilla erythronota (*Eversm.*), *Hume, S. F.* vii, p. 389; *id. Cat.*
    no. 498 bis; *Scully, Ibis,* 1881, p. 445; *Biddulph, Ibis,* 1882, p. 277;
    Seebohm, Cat. B. M. v, p. 348.

Colouration. *Male.* After the autumn moult the forehead, crown,
nape, and hind neck are pale blue, nearly concealed by broad slaty-
grey fringes; lores, cheeks, point of chin, sides of the head and of
the neck, produced round the upper back as a collar, black with
narrower slaty-grey fringes; back, scapulars, throat, breast, and
flanks chestnut fringed with grey; rump and tail chestnut, the
middle pair of feathers black, as also the tip of the outer web of
the outermost feather, and the terminal portion of the shaft of all
the feathers; lesser wing-coverts black, tipped with white; median
coverts and the inner greater coverts pure white; remaining
coverts and the quills brown edged with pale fulvous, the primary-
coverts very largely white; abdomen and under tail-coverts pale
fulvous; under wing-coverts and axillaries white with black bases.
In summer the fringes are all dropped.

Female. Forehead, crown, nape, back, scapulars, and upper part
of rump brown; lower part of rump and upper tail-coverts chestnut; tail as in the male; wing-coverts and quills brown, broadly
edged with fulvous white; no white on wing; a ring of pale
feathers round the eye; lower plumage greyish brown, tinged with
dull orange in places, and paler on the abdomen.

In the dry state the legs and bill are black.

Length about 6·5; tail 2·9; wing 3·4; tarsus ·9; bill from
gape ·6.

Distribution. A winter visitor to every portion of Kashmir, extending on the west to Hazára and Afghanistan and on to Asia
Minor. The most easterly locality from which I have seen a
specimen of this bird is Kotokhai in the Himalayas. In summer
this Redstart is found in Turkestan, and even in Mongolia and
Siberia, if *R. alaschanica*, Prjev., should prove to be the same
species, as is probable.

613. Ruticilla hodgsoni. *Hodgson's Redstart.*

Ruticilla erythrogastra (*Güld.*), *apud Blyth, Cat.* p. 168 (part.).
Ruticilla hodgsoni, *Moore, P. Z. S.* 1854, p. 26, pl. lviii; *Horsf. & M. Cat.* i, p. 303; *Jerd. B. I.* ii, p. 138; *Godw.-Aust. J. A. S. B.* xlv, pt. ii, p. 199; *Hume, Cat.* no. 498; *Scully, S. F.* viii, p. 302; *Seebohm, Cat. B. M.* v, p. 344; *Hume, S. F.* xi, p. 195.

Thar-capni, Nep.

Coloration. Male. After the autumn moult the forehead, lores, sides of head and neck, chin, throat, and upper breast are deep black with a few grey fringes; crown, nape, and back ashy, the portion of crown above the forehead and at the sides nearly white; lower rump, upper tail-coverts, and tail chestnut, except the terminal two thirds of the middle pair of feathers of the latter, which are black; wing-coverts black, edged with ashy; quills dark brown, a few of the later secondaries margined with white about their middle portion, forming a patch in the closed wing; lower plumage chestnut. Males in summer are unknown to me, but probably differ in wanting the grey fringes on the throat and breast.

Female. Upper plumage and wings brown tinged with ashy, the feathers of the wings edged paler; lower rump, upper tail-coverts, and tail chestnut except the middle pair of tail-feathers, which are blackish; a ring of whitish feathers round the eye; lower plumage ashy brown, albescent on the abdomen and turning to pale rufous on the flanks, vent, and under tail-coverts.

Bill black; gape fleshy yellow; iris dark brown; feet black or brownish black, soles yellow; claws black (*Scully*).

Length about 6; tail 2·8; wing 3·4; tarsus ·9; bill from gape ·7.

Distribution. Nepal; Sikhim; Bhutan; the Daphla hills in Assam; the Nága hills; Manipur. This species is only a winter visitor to the above localities. It summers in Western China and probably in Central Asia. This Redstart has been erroneously recorded from Afghanistan and Gilgit.

614. Ruticilla rufiventris. *The Indian Redstart.*

Œnanthe rufiventris, *Vieill. Nouv. Dict. d'Hist. Nat.* xxi, p. 431 (1818).
Ruticilla indica, *Blyth, Cat.* p. 168 (1849).
Ruticilla phœnicuroides, *Moore, P. Z. S.* 1854, p. 25, pl. lvii; *Horsf. & M. Cat.* i, p. 301; *Jerd. B. I.* ii, p. 136.
Ruticilla nipalensis (*Hodgs.*), *Moore, P. Z. S.* 1854, p. 26; *Horsf. & M. Cat.* i, p. 302.
Ruticilla rufiventris (*Vieill.*), *Jerd. B. Ind.* ii, p. 137; *Blanf. J. A. S. B.* xli, pt. ii, p. 50; *Hume, N. & E.* p. 321; *id. S. F.* v, p. 36; *id. Cat.* no. 497; *Seebohm, Cat. B. M.* v, p. 342; *Hume, S. F.* xi, p. 194; *Oates in Hume's N. & E.* 2nd ed. ii, p. 64.

Thir-thira, Thirtir-kampa, Hind.; *Phir-ira, Lal-girdi,* Beng.; *Nuni-budi-gadu,* Tel.

Coloration. Male. In typical autumn plumage the forehead, sides of the head, chin, throat, breast, and sides of neck are black with grey fringes, the black more or less concealed ; crown, nape, hind neck, back, and scapulars ashy grey, this grey appearance caused by broad fringes which generally quite conceal the black bases of the feathers ; lesser and median wing-coverts black, edged with ashy ; the other coverts and the quills brown, edged with rufous ; rump and upper tail-coverts bright chestnut ; tail chestnut except the middle pair of feathers, which are brown : abdomen, vent, under tail- and wing-coverts, and axillaries deep orange-brown.

In typical summer plumage the whole head, neck, back, scapulars, lesser and median wing-coverts, and the breast are deep black, with an ashy supercilium and some ashy on the crown just behind the forehead. The rufous margins to the greater coverts and quills are reduced or disappear.

Between these two stages every intermediate form occurs regardless of season, the deep black plumage sometimes making its appearance immediately after the moult, and some birds even at midsummer retaining the broad ashy-grey fringes in varying degrees. Some males are said to breed in female plumage.

Female. Upper plumage brown tinged with fulvous ; the wings broadly edged with fulvous : rump, upper tail-coverts, and tail chestnut, except the middle pair of feathers, which are brown : a circle of pale feathers round the eye ; lower plumage buffy brown, suffused with orange on the abdomen, flanks, vent, and under tail-coverts.

Bill, legs, feet, and iris black ; base of bill yellow (*Bingham*).

Length about 6 ; tail 2·6 ; wing 3·3 ; tarsus ·9 ; bill from gape ·7.

Distribution. A common winter visitor to a great portion of the Empire, this species occurs from the Himalayas down to Bangalore and the Nilgiris, and from Sind to Assam, thence ranging down to Manipur. It appears to be common from September to April. Some few birds are found in the plains in summer, but do not apparently breed. In the Hume Collection there are specimens shot at Sámbhar in July and at Ahmednagar in June.

This Redstart extends on the west to Persia and on the east to China, and large numbers appear to summer in Turkestan and Mongolia. Within our limits it breeds on the higher mountains of Kashmir above 10,000 feet. It also breeds in Afghanistan. Mandelli procured a specimen in Native Sikhim in June, and probably it may be found to breed throughout the Himalayas at great heights.

Habits, &c. The nest of this species has seldom been found, and little is known of its nidification. Wardlaw Ramsay found the nest in Afghanistan on the 1st July in an old tree-stump, but the young had apparently left it some time before.

645. **Ruticilla erythrogaster.** *Güldenstädt's Redstart.*

Motacilla erythrogastra, *Güld. Nov. Com. Petrop.* xix, p. 461, pls. 16, 17 (1775).

Ruticilla erythrogastra (*Güld.*), *Blyth, Cat.* p. 168 ; *Horsf. & M. Cat.* i, p. 304 ; *Jerd. B. I.* ii, p. 139 ; *Blanf. J. A. S. B.* xli, pt. ii, p. 54 ; *Hume & Henders. Lah. to Yark.* p. 210 ; *Scully, N. F.* iv, p. 141 ; *Hume, Cat.* no. 480 ; *Biddulph, Ibis,* 1881, p. 65 ; *Scully, Ibis,* 1881, p. 445 ; *Seebohm, Cat. B. M.* v, p. 347.

Ruticilla vigorsi, *Moore, P. Z. S.* 1854, p. 27, pl. lx ; *Horsf. & M. Cat.* i, p. 304.

The White-winged Redstart, Jerd.

Coloration. *Male.* After the autumn moult the crown and nape are white with a few ashy margins ; forehead, sides of head and neck, back, scapulars, upper wing-coverts, chin, throat, and upper breast deep black, a few of the feathers margined with grey ; wings black, the middle portion of all the quills except the tertiaries white ; remainder of the plumage with the tail deep chestnut. Soon after the autumn, the few margins present on the black portions of the plumage drop off, and the crown becomes pure white.

Female. Upper plumage brown tinged with ashy : the lower portion of rump, upper tail-coverts, and tail ferruginous, the middle tail-feathers and the tips of the others dusky ; wings brown, edged with pale fulvous ; sides of head and whole lower plumage uniform fulvous-grey. The female has no seasonal change of plumage.

Bill black, yellow at gape ; iris brown ; legs, feet, and claws black (*Hume Coll.*).

Length about 7 ; tail 3 ; wing 4·2 ; tarsus 1·05 ; bill from gape ·7.

Distribution. The Himalayas from Kashmir and Gilgit to Sikhim. In summer this species is found at very high altitudes, from 10,000 to 14,000 feet or even higher. In winter it descends to 5000 feet. This Redstart extends on the west to the Caucasus ; on the north, through Turkestan and Tibet, to Lake Baikal in Southern Siberia ; and on the east into China.

Habits, &c. This species, like *Chimarrhornis leucocephalus,* affects streams and lakes, but is more frequently seen, according to Blanford, on rocky hill-sides. Its nest has not yet been found by any naturalist.

Genus RHYACORNIS, Blanford, 1872.

The genus *Rhyacornis* contains one species, which is closely allied to both *Chimarrhornis* and *Ruticilla.* It differs from both these, however, in the shortness of its tail, which is about twice the length of the tarsus, and in its strong rictal bristles. The female, moreover, has no chestnut on the tail.

The only member of this genus inhabits mountain-streams, and is always found near water, especially where this forms a rapid or a cascade. It has the habit of expanding its tail frequently.

646. **Rhyacornis fuliginosus.** *The Plumbeous Redstart.*

Phœnicura fuliginosa, *Vigors, P. Z. S.* 1831, p. 35.
Ruticilla fuliginosa (*Vig.*), *Blyth, Cat.* p. 169; *Horsf. & M. Cat.* i,
　p. 348; *Jerd. B. Ind.* ii, p. 142; *Hume & Henders. Lah. to Yark.*
　p. 212, pl. xv.
Nymphæus fuliginosus (*Vig.*), *Hume, N. & E.* p. 322.
Rhyacornis fuliginosa (*Vig.*), *Blanf. J. A. S. B.* xli, pt. ii, p. 50;
　Hume, Cat. no. 505; *Scully, S. F.* viii, p. 304; *Hume, S. F.* xi,
　p. 196; *Oates in Hume's N. & E.* 2nd ed. ii, p. 65.
Xanthopygia fuliginosa (*Vig.*), *Sharpe, Cat. B. M.* iv, p. 253; *Oates,*
　B. B. i, p. 281.
Chimarrhornis fuliginosa (*Vig.*), *Stoliczka, J. A. S. B.* xxxvii, pt. ii,
　p. 43.

The Plumbeous Water-Robin, Jerd.; *Suradam parbo-pho,* Lepch.;
Chubbia nakki, Bhut.

Coloration. Male. The whole plumage dull cyaneous; tail-
coverts, both upper and lower, the vent, and tail bright chestnut;
wing black with bluish margins.

Female. The whole upper plumage dull bluish brown; upper
and under tail-coverts white; base of tail white, the amount on
the outer four pairs of feathers increasing towards the outside,
the outermost feather being white with a narrow dusky margin;
lores and ear-coverts dusky mottled with white: the whole lower
plumage ashy brown, each feather with a whitish centre and a
paler ashy margin: upper wing-coverts and tertiaries brown, edged
rufous and tipped with whitish; quills brown, narrowly edged
with rufescent.

The nestlings of both sexes resemble the female, and have the
same amount of white in the tail, but the whole upper plumage is
closely spotted and streaked with dull white or pale fulvous; the
lower plumage is mottled and cross-barred with brown.

Bill black; gape fleshy white; iris dark brown; feet dark
horny brown; claws black (*Scully*).

Length about 5·5; tail 2·1; wing 3; tarsus ·9; bill from
gape ·65.

Distribution. The Himalayas from Eastern Kashmir to Assam;
the Khási hills; Cachar; Manipur; Arrakan. Blyth records this
species from Thayetmyo, where, however, I failed to meet with it.
This Redstart is found on the Himalayas from low elevations up
to 13,000 feet, according to season. It extends into China and
Mongolia.

Habits, &c. Breeds apparently in every portion of its extensive
range. The nest is made of moss, lined with hair, wool, or soft
fibres, and placed on a shelf of a rock or in the hollow of a bank
by the side of a stream. The nesting-season is May and June.
The eggs are greenish white, thickly mottled with yellowish or
reddish brown, and measure about ·76 by ·6

Genus CYANECULA, Brehm, 1828.

The genus *Cyanecula* contains the Blue-throats, birds which are very closely allied to the English Robin. The Blue-throats may be recognized by their very short tail, which is only twice the length of the tarsus, and by the chestnut colour of the basal half of the tail. The males, moreover, have the chin and throat a brilliant blue. The females are of a dull colour, but have the tail chestnut as in the male.

The Blue-throats feed on the ground, and are generally found in India in thick grass-jungle, and more rarely in open country. They prefer swampy ground. They run well, elevating the tail on arriving at the end of each short course of running, and sometimes expanding it. They are said to be good songsters. They breed in holes on the ground, and lay blue eggs spotted with reddish brown. The only two species of this genus are highly migratory.

Key to the Species.

a. Throat blue, with a chestnut spot in the centre .... *C. suecica* ♂, p. 99.
b. Throat blue, either entirely or with a white spot in the centre .... *C. wolfi* ♂, p. 100.
c. Throat buffish white .... } *C. suecica* ♀, p.99.
} *C. wolfi* ♀, p. 100.

647. Cyanecula suecica. *The Indian Blue-throat.*

Motacilla suecica, *Linn. Syst. Nat.* i, p. 336, part. (1766).
Motacilla cærulecula, *Pall. Zoogr. Ross.-Asiat.* i, p. 480 (1811).
Cyanecula suecica (*Linn.*), *Blyth, Cat.* p. 167; *Horsf. & M. Cat.* i, p. 311; *Jerd. B. I.* ii, p. 152; *Hume & Henders. Lah. to Yark.* p. 214; *Anders. Yunnan Exped.* p. 614; *Legge, Birds Ceyl.* p. 443; *Hume, Cat.* no. 514; *Scully, S. F.* viii, p. 304; *Barnes, Birds Bom.* p. 200.
Erithacus cærulecula (*Pall.*), *Seebohm, Cat. B. M.* v, p. 308; *Oates, B. B.* i, p. 15.

Huseni-pedda, Hind.; *Nil kanthi*, Hind. in the N.; *Gunpigera, Gurpedra,* Beng.; *Dumbak*, Sind.

Coloration. Male. Whole upper plumage with wings brown, the feathers of the head and back with darker centres; chin and throat bright blue, with a chestnut spot in the centre of the throat; below the blue a band of black and below this a broader band of chestnut; lores black; a stripe from the nostrils to the eye fulvous; cheeks and ear-coverts mixed fulvous and black; belly, flanks, vent, and under tail-coverts buffish white; middle tail-feathers brown, the others chestnut on the basal half and brown on the terminal half.

Female. The whole lower plumage buffish white, with a broad brown-spotted gorget across the breast.

H 2

It is seldom that the male is in the full plumage described above. The amount of blue and chestnut on the throat varies much; and sometimes only the presence of a few blue feathers serves to indicate that the bird is a male.

The nestling is blackish above streaked with fulvous, and fulvous below, each feather edged with black.

Bill black, the base flesh-colour; iris brown; eyelids plumbeous; inside of mouth yellowish; legs dusky fleshy; claws brown.

Length 5·9; tail 2·3; wing 2·9; tarsus 1·1; bill from gape ·75.

Distribution. A winter visitor to almost every portion of the Empire and Ceylon. The only parts from which this species has not yet been recorded are the Nicobar Islands and the portion of Tenasserim south of Tavoy, but even in these it probably occurs.

In summer this species is found immediately north of the Himalayas and thence through Asia to the Arctic Circle, extending west throughout Europe and east to the Pacific. In winter it is found not only in India but in North Africa on the one hand and in Southern China on the other.

648. Cyanecula wolfi. *The White-spotted Blue-throat.*

Motacilla suecica, *Linn. Syst. Nat.* i, p. 336, part. (1766).
Sylvia cyanecula, *Wolf, Taschenb.* i, p. 240 (1810).
Sylvia wolfii, *Brehm, Beitr. zur Vogelk.* ii, p. 173 (1822).
Cyanecula leucocyana, *Brehm, Vög. Deutschl.* p. 353 (1831).
Cyanecula wolfii (*Brehm*), *Hume, S. F.* vii, p. 391; *id. Cat.* no. 514 bis.
Cyanecula leucocyanea, *Brehm, Biddulph, Ibis,* 1881, p. 65; *Scully, Ibis,* 1881, p. 447; *Biddulph, Ibis,* 1882, p. 278.
Erithacus cyanecuins (*Wolf*), *Seebohm, Cat. B. M.* v, p. 311.

Coloration. Resembles *C. suecica,* the male differing from the male of that species in having the patch on the throat white instead of chestnut, or in wanting a spot altogether. The females and young of the two species appear to be inseparable.

Distribution. A rare visitor to the extreme north of Kashmir, occasionally straggling even to the plains. Biddulph secured a specimen in Gilgit in April, and he records this species as very common on both sides of the Digar pass, between the Nubra and Indus valleys. In the Hume Collection there is a specimen which was obtained in Tirhoot in April, and Hume states that he has seen some half-dozen specimens from various parts of India.

The headquarters of this Blue-throat are Europe in the summer, and North Africa and Palestine in the winter.

Genus DAULIAS, Boie, 1831.

The genus *Daulias* contains the Nightingales, birds of plain plumage but of great powers of song. The one species that has been known to occur in India is of extreme rarity in that country, only two instances of its occurrence being known.

In *Daulias* the whole plumage is brown, somewhat ruddy on the tail, but making no approach to the chestnut exhibited in the preceding genera. The sexes are quite alike. The first primary is much smaller than in any other genus of this subfamily, being considerably less than a third of the length of the second. The tail is long and rounded, and the tarsus is also long.

The Nightingales frequent dense brushwood and are shy birds. They feed principally on the ground like Robins, and they nest near the ground in dense underwood.

649. Daulias golzi. *The Persian Nightingale.*

Luscinia golzii, *Cabanis, Journ. für Orn.* 1873, p. 79.
Luscinia hafizi, *Severtz. Turkest. Jevotn.* p. 120 (1873).
Daulias golzii (*Cab.*), *Hume, S. F.* iv, p. 500; *id. Cat.* no. 511 ter.
Erithacus golzii (*Cab.*), *Seebohm, Cat. B. M.* v, p. 297.

Coloration. The whole upper plumage and the margins of the wing-feathers russet-brown, brighter on the upper tail-coverts and tail; wings brown; lores, cheeks, and the whole lower plumage pale buff.

Length 7·5; tail 3·4; wing 3·6; tarsus 1·1; bill from gape ·85.

This species may be separated from *D. luscinia*, Linn., which occurs in England and Europe, and from *D. philomela*, Bechst., of Eastern Europe and South-western Asia, by its long tail and by its first primary, which is equal to the primary-coverts. In both the above species the tail is less than three inches long; in the first the first primary is considerably longer than the primary-coverts; in the second it is considerably shorter.

Distribution. Two specimens of this rare Nightingale have been procured in Oudh, one in October and the other in November. They are both in the Hume Collection. No other instance of the occurrence of this species in India has been recorded. It extends to Turkestan and to the Caucasus.

Genus CALLIOPE, Gould, 1836.

The members of the genus *Calliope* are characterized by the absence of chestnut in the tail, a comparatively long first primary, a short tail, and by the males having a brilliant red throat. In habits *Calliope* agrees closely with *Cyanecula*. All the species of this genus are migratory, and the sexes are very different in coloration. The tarsus is very long, and these birds spend most of their time on the ground in thick cover.

Key to the Species.

a. No white in the tail .... *C. camtschatkensis,* p. 102
b. Base or tip of tail or both white.
  a'. Chin and throat red.
    a''. Cheeks black .... *C. pectoralis* ♂, p. 103.
    b''. Cheeks white .... *C. tschebaiewi* ♂, p. 104.
  b'. Chin and throat white............... { *C. pectoralis* ♀, p. 103.
                                     { *C. tschebaiewi* ♀, p. 104.

650. **Calliope camtschatkensis.** *The Common Ruby-throat.*

Motacilla calliope, *Pall. Reise Russ. Reichs,* iii, p. 697 (1776).
Turdus camtschatkensis, *Gmel. Syst. Nat.* i, p. 817 (1788).
Calliope camtschatkensis (*Gmel.*), *Blyth, Cat.* p. 169; *Horsf. & M.
　　Cat.* i, p. 313; *Jerd. B. I.* ii, p. 150; *Hume & Dav. S. F.* vi, p. 337;
　　Anders. Yunnan Exped., Aves, p. 615; *Hume, Cat.* no. 512; *Barnes,
　　Birds Bom.* p. 209; *Hume, S. F.* xi, p. 193.
Calliope yeatmani, *Tristram, Ibis,* 1870, p. 444.
Erithacus calliope (*Pall.*), *Seebohm, Cat. B. M.* v, p. 305; *Oates, B. B.*
　　i, p. 14.

Gumpigora, Beng.; *Gangula,* Nep.

Fig. 29.—Head of *C. camtschatkensis.*

Coloration. Male. The whole upper plumage olive-brown, the
head darker, and all the feathers indistinctly edged paler; a line
from the forehead over the eye white; lores and under the eye
black; a broad moustachial streak white; throat and fore neck
scarlet, each feather margined at the tip with white, and the whole
patch bordered by black; upper breast brownish grey, paling and
becoming buffy grey on the lower breast and sides of the body;
abdomen and under tail-coverts white; tail brown, edged on the
outer webs with olive-brown; wing-coverts and quills brown, edged
with bright olive-brown; axillaries buff.

Female. Superciliary streak buffy white; lores and in front of
the eye dusky brown; the bright scarlet of the throat and the
surrounding black line absent, and replaced by dull white; moustachial streak olive-brown; other parts as in the male.

The *young* are mottled, and moult into the plumage of the adult
female at the first autumn, and the crimson throat-patch is assumed
in the first winter without a moult.

Bill light brown, white at the gape; mouth flesh-colour; iris
brown; legs pale plumbeous; claws horn-colour.

Length 6; tail 2·4; wing 2·9; tarsus 1·15; bill from gape ·8.

Distribution. A winter visitor to Nepal and Sikhim, extending
through the plains of the Eastern portion of India proper as far
south as the latitude of Raipur in the Central Provinces. This
species is common in Bengal, Bhutan, and Assam, and extends down
to Arrakan, Pegu, Karennee, and the northern portion of Tenasserim. As an accidental visitor this bird may be expected to occur in
almost every part of India, and Jerdon records an instance of its
being found near Bombay.

In winter the Common Ruby-throat extends its migration as
far as the Philippines, and in summer it is found throughout
Northern Asia up to the Arctic Circle.

651. **Calliope pectoralis.** *The Himalayan Ruby-throat.*

Calliope pectoralis, *Gould, Icon. Av.* pt. i, pl. iv (1837); *Blyth, Cat.*
p. 169; *Horsf. & M. Cat.* i, p. 313; *Jerd. B. I.* ii, p. 150; *Stoliczka,
J. A. S. B.* xxxvii, pt. ii, p. 45; *Blanf. J. A. S. B.* xli, pt. ii p. 52;
Hume, N. & E. p. 325; *id. Cat.* no. 513; *Scully, S. F.* viii, p. 304;
Biddulph, Ibis, 1881, p. 64; *Oates in Hume's N. & E.* 2nd ed. ii,
p. 67.

Erithacus pectoralis (*Gould*), *Seebohm, Cat. B. M.* v, p. 305.

The White-tailed Ruby-throat, Jerd.

Coloration. Male. After the autumn moult the whole upper
plumage, wing-coverts, and sides of the neck are dark slaty,
blacker on the crown; forehead and a short supercilium white;
middle of chin and of throat bright crimson; lores, sides of head,
sides of chin and of throat, and the whole breast deep black, every
feather fringed with ashy; abdomen, vent, and under tail-coverts
white; wings brown, edged with olivaceous; middle tail-feathers
black, the others with the basal half white and the terminal half
black tipped with white. In summer the ashy fringes are cast and
the upper plumage is tinged with olivaceous.

Female. Olive-brown, the outer webs of the quills suffused with
fulvous; lores, edge of forehead, and a short supercilium dingy
white; chin and middle of throat white, contrasting with the ashy-
brown of the sides of the throat and the breast; abdomen pale
fulvous; middle tail-feathers olive-brown, the others blackish brown
tipped with white.

The nestling has the upper plumage fulvous-brown, much darker
on the crown, all the feathers with fulvous streaks; lower plumage
fulvous, all the feathers margined with dark brown; the tail-feathers
at first tipped with fulvous instead of white, and the male from
the earliest age has the base of the tail white. At the first autumn
moult the young male assumes the dark upper plumage of the adult
male, but retains the lower plumage of the adult female; traces of
the black breast are assumed during the first summer, but the
breast does not become fully plumaged till the moult of the second
autumn. The female becomes adult at the first autumn moult.

Bill black, brownish at tip and base of lower mandible; iris
brown; feet brown; the tarsi rather livid; claws dusky (*Scully*).

Length about 6; tail 2·4; wing 3; tarsus 1·2; bill from gape ·75.

Distribution. A constant resident on, or a summer visitor to, the
higher portions of the Himalayas from Gilgit to Sikhim and Bhutan.
A winter visitor to the intermediate and lower ranges of the same
mountains, being occasionally found in the plains at the foot as in
the Bhutan doars and at Sultanpur in Oudh. In summer this species
is also found in Turkestan.

Habits, &c. Breeds in Kashmir and Sikhim at 10,000 feet and
upwards. A nest, said to belong to this species and found in Sikhim,
is described as being a saucer-shaped pad of fine moss and roots
placed in a deep crevice of a rock. The eggs are described as being
pale salmon-buff and as measuring about ·9 by ·66.

652. Calliope tschebaiewi. *The Tibet Ruby-throat.*

Calliope pectoralis, *Gould, apud Godw.-Aust. J. A. S. B.* xxxix,
pt. ii, p. 270, xlv, pt. ii, p. 79; *Anders. Yunnan Exped., Aves,*
p. 645; *Hume, S. F.* xi, p. 199.
Calliope tschebaiewi, *Prjev., Rowley's Orn. Misc.* ii, p. 180, pl. liv,
fig. 1 (1877).
Erithacus tschebaiewi (*Prjev.*), *Seebohm, Cat. B. M.* v, p. 308.

Coloration. Male. Differs from the male of *C. pectoralis,* when
adult, in being olive-brown, tinged with russet above, and in
having the cheeks white, not black.

The females and immature birds of both species are inseparable.

Bill and legs black; iris brown (*Cockburn*).

Length about 6; tail 2·3; wing 3; tarsus 1·2; bill from
gape ·75.

Distribution. A winter visitor to the Himalayas from Sikhim to
the Dikrang valley in Assam, extending to the Khási hills, where
it is very common at Shillong; Godwin-Austen procured this
species at Mymensing and Anderson near Bhámo. In summer
this bird is found in Tibet and Kansu.

Genus **TARSIGER**, Hodgs., 1844.

The genus *Tarsiger* contains one species, in which the sexes
resemble each other somewhat closely, and have the whole lower
plumage yellow. In structure this genus differs in no respect from
Calliope.

Tarsiger chrysæus is a constant resident at moderate heights on
the Himalayas.

653. Tarsiger chrysæus. *The Golden Bush-Robin.*

Tarsiger chrysæus, *Hodgs. P. Z. S.* 1845, p. 28; *Blyth, Cat.* p. 169;
Horsf. & M. Cat. i, p. 310; *Jerd. B. I.* ii, p. 149; *Stoliczka, J. A.
S. B.* xxxvii, pt. ii, p. 45; *Hume, N. & E.* p. 325; *Sharpe, Cat. B.
M.* iv, p. 260; *Hume, Cat. no.* 511; *Oates in Hume's N. & E.*
2nd ed. ii, p. 67.

The Golden Bush-Chat, Jerd.; *Manshil-pho,* Lepch.

Coloration. Male. Forehead, crown, nape, hind neck, and the
middle portion of the back olive-green; a superciliary streak
reaching to the nape, the lesser wing-coverts, scapulars, sides of
the back, the rump, the upper tail-coverts, and the whole lower
plumage bright orange-yellow, many of the feathers more or less
fringed very narrowly with brown; median and greater coverts
and the quills black, margined with olive-green; tail orange-
yellow, broadly tipped with black, the median pair of feathers
black, with a yellow margin on the outer webs; lores, round the
eye, and the ear-coverts black.

Female. The whole upper plumage and the exposed parts of the

wings olive-green; median pair of tail-feathers olive-green, the
others golden yellow, broadly tipped, and margined on the outer
webs, with olive-green; a yellowish-white ring round the eye; ear-
coverts olive-brown, with pale shafts; lores and an indistinct
supercilium olive-yellow; the whole lower plumage ochraceous
yellow, most of the feathers with tiny dusky fringes and the flanks
washed with olivaceous.

The young have the whole plumage dark olive-brown, the
feathers streaked with fulvous and tipped with black.

Lower mandible and edge of the upper along the commissure
yellow; rest of the bill black; iris very dark brown; legs, feet,
and claws fleshy, tinged with brown (*Hume*).

Length about 6; tail 2·3; wing 2·7; tarsus 1·2; bill from
gape ·7.

Distribution. The Himalayas from Chamba to Sikhim, apparently
up to 5000 feet; the Khási hills; the Nága hills; Manipur. This
species extends into Western China.

Habits, &c. Nests on the ground from May to August, in holes
of rocks and banks, and lays three or four eggs, which are pale
blue, and measure about ·8 by ·58.

Genus **IANTHIA**, Blyth, 1847.

The genus *Ianthia* contains three species of Indian birds, in
which the males are very brightly coloured and the females are
dull. They inhabit the Himalayas, and migrate locally according
to season.

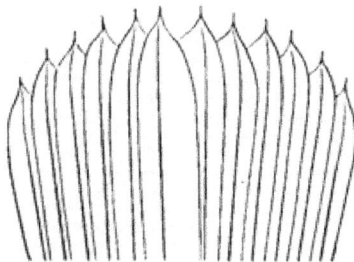

Fig. 30.—Tail of *I. indica*.

This genus differs from *Tarsiger* and *Calliope* in having a much
longer tail, the feathers of which are moreover pointed at the tips.
Very little is on record about the habits of the members of this
genus, but they probably do not differ in any important particular
from those of the Blue-throats and Ruby-throats.

Key to the Species.

a. Sides of body orange-chestnut, contrasting
   with remainder of lower plumage ....   *I. rufilata*, p. 106.
b. Sides of body of same colour as remainder of
   lower plumage.
   a'. A white supercilium ....   *I. indica*, p. 107.
   b'. No white supercilium.................   *I. hyperythra*, p. 108.

654. Ianthia rufilata.   *The Red-flanked Bush-Robin.*

Nemura rufilatus, *Hodgs. P. Z. S.* 1845, p. 27; *Horsf. & M. Cat.* i,
    p. 280; *Hume, N. & E.* p. 324; *id. S. F.* xi. p. 198.
Ianthia rufilata (*Hodgs.*), *Blyth, Cat.* p. 170; *Blanf. J. A. S. B.* xli,
    pt. ii, p. 52; *Brooks, J. A. S. B.* xli, pt. ii, p. 77; *id. S. F.* iii,
    p. 240; *Oates in Hume's N. & E.* 2nd ed. ii, p. 68.
Ianthia cyanura (*Pall.*), *apud Jerd. B. I.* ii, p. 146; *Stoliczka, J. A.
    S. B.* xxxvii, pt. ii, p. 44.
Nemura cyanura (*Pall.*), *apud Hume, Cat.* no. 508; *Scully, S. F.* viii,
    p. 304; *Biddulph, Ibis,* 1881, p. 64.
Tarsiger rufilatus (*Hodgs.*), *Sharpe, Cat. B. M.* iv, p. 256; *Scully,
    Ibis,* 1881, p. 446.

The *White-breasted Blue Wood-Chat,* Jerd.; *Mangzhil-pho,* Lepch.

Fig. 34.—Head of *I. rufilata.*

Coloration. Male. Forehead and a broad eyebrow, rump, upper
tail-coverts, and the median wing-coverts bright ultramarine-blue;
ear-coverts, lores, and round the eye black; upper plumage, edges
of wing-coverts and quills, sides of the head, throat, and neck
extending down to the sides of the breast deep purplish blue; tail
black, the outer webs suffused with deep blue; chin, throat, middle
of breast, and remaining lower plumage white, sullied with ashy
brown on the breast; a very large and conspicuous patch of
orange-chestnut on each side of the body; under wing-coverts and
axillaries white.

Males from Sikhim are very bright; those from other parts have
the upper plumage a greenish blue.

Female. Upper plumage olive-brown, the coverts and wings
edged with rufous; rump greenish blue; upper tail-coverts deep
blue; tail dark brown, the outer webs suffused with deep blue;
chin and throat narrowly white; sides of the head and neck and
the whole breast ochraceous; a large patch of orange-chestnut on
each side of the body; remainder of the sides ochraceous; middle
of abdomen and under tail-coverts white; under wing-coverts and
axillaries pale yellowish buff.

Some females have a well-marked bluish-grey supercilium, and others are without it.

The young resemble the female, and have the orange patch on each side of the body and also a blue tail and rump, but the upper plumage is everywhere streaked with fulvous as well as the sides of the head and the throat.

Bill black; legs and feet deep brown; iris brown (*Hume*).

Length nearly 6; tail 2·6; wing 3·3; tarsus 1·05; bill from gape ·6.

I. cyanura is an allied species from Northern Asia.

Distribution. The Himalayas from Gilgit and Kashmir generally to Sikhim; the Khási hills; Tipperah; Manipur. This species is found up to 11,000 feet in the Himalayas in summer, and descends to lower levels in winter.

Habits, &c. Breeds in May and June, constructing a nest of moss and grass in holes in banks and under tree-roots, and laying four eggs, which are white with a green tinge, spotted sparingly round the larger end with minute specks of reddish brown, and measuring about ·71 by ·56.

655. Ianthia indica. *The White-browed Bush-Robin.*

Sylvia indica, *Vieill. N. Dict. d'Hist. Nat.* xi, p. 267 (1817).
Nemura flavo-olivacea, *Hodgs. P. Z. S.* 1845, p. 27.
Tarsiger superciliaris, *Hodgs., Moore, P. Z. S.* 1854, p. 76; *Horsf. & M. Cat.* i, p. 311; *Hume, Cat. no.* 510.
Erythaca flavo-olivacea, *Blyth, Cat.* p. 171.
Ianthia superciliaris (*Hodgs.*), *Jerd. B. I.* ii, p. 148; *Blanf. J. A. S. B.* xli, pt. ii, p. 161.
Tarsiger indicus (*Vieill.*), *Sharpe, Cat. B. M.* iv, p. 259.

The Rufous-bellied Bush-Chat, Jerd.

Coloration. Male. The whole upper plumage dull slaty blue; a very well-defined supercilium from the point of the forehead to the nape white; lores and in front and under the eye black; sides of the head blackish blue; coverts and quills dark brown, edged with olive-yellow, the coverts next the body more or less suffused with blue; tail black, suffused with blue on the outer webs; lower plumage orange-rufous, the sides of the throat mottled with white and the middle of the abdomen whitish.

Female. The whole upper plumage olive-brown, tinged with fulvous on the rump; a partially-concealed white supercilium extending to the nape; sides of the head and a ring round the eye ochraceous, mottled with whitish; wings and tail brown, edged with the colour of the back; entire lower plumage ochraceous, tinged with rufous on the breast and paler on the abdomen.

I have not been able to examine a young bird of this species.

Bill black; legs pale horny-brown; iris brown (*Jerdon*).

Length about 6; tail 2·9; wing 3·2; tarsus 1·15; bill from gape ·7.

Distribution. Nepal and Sikhim, extending into Western China. There are few birds about which so little is known as this species.

656. Ianthia hyperythra. *The Rufous-bellied Bush-Robin.*

Ianthia hyperythra, *Blyth, J. A. S. B.* xvi, p. 132 (1847); *id. Cat.*
p. 170; *Jerd. B. I.* ii, p. 147; *Godw.-Aust. J. A. S. B.* xxxix,
pt. ii, p. 106.

Nemura hyperythra (*Blyth*), *Horsf. & M. Cat.* i, p. 280; *Hume, Cat.*
no. 509.

Tarsiger hyperythrus (*Blyth*), *Sharpe, Cat. B. M.* iv, p. 257.

The Rusty-throated Blue Wood-Chat, Jerd.

Coloration. Male. Forehead continued back as a supercilium,
the upper tail-coverts, and a patch on the lesser wing-coverts near
the edge of the wing bright ultramarine-blue; ear-coverts, lores,
and in front of the eye black; upper plumage and the sides of the
head and neck deep purplish blue; wing-coverts and quills black,
edged with purplish blue; tail black, suffused with purplish blue
on the outer webs; chin, throat, breast, and abdomen chestnut;
vent and under tail-coverts white; under wing-coverts and axillaries pale chestnut.

Female. Upper plumage and the visible portion of the closed
wing olive-brown tinged with rufous; rump slaty blue; upper
tail-coverts deep blue; tail black, the outer webs suffused with
deep blue; sides of the head fulvous olive-brown; lower plumage,
under wing-coverts, and axillaries rich ochraceous, becoming white
on the vent and under tail-coverts.

I have not been able to examine a young bird, but it will prove,
without doubt, to be spotted.

Bill black in the dry skin; legs and feet brown.

Length about 5·5; tail 2·3; wing 3·2; tarsus 1; bill from
gape ·6.

Distribution. Sikhim and the Khási hills. In the former tract
this species is a resident, probably moving vertically according to
season. This bird is figured by Hodgson, and there are likewise
some specimens collected by him in the British Museum, probably
from Nepal, but there is no certainty on this point.

Genus ADELURA, Bonap., 1854.

The sole member of this genus is frequently associated with the
true Redstarts, but the total absence of the chestnut in the tail,
which forms so conspicuous a feature in all the Redstarts, induces
me to place the present type apart from them. In habits this
species appears to be a Redstart, and structurally it does not
differ from *Ruticilla*. From *Ianthia*, which it somewhat resembles
in coloration, this genus differs in having the tips of the tail-
feathers rounded.

657. Adelura cæruleicephala. *The Blue-headed Robin.*

Phœnicura cæruleocephala, *Vigors, P. Z. S.* 1830, p. 35; *Gould,
Cent.* pl. xxx, fig. 2.

Ruticilla cœruleocephala (*Vig.*), *Blyth, Cat.* p. 168; *Horsf. & M. Cat.* i, p. 307; *Jerd. B. I.* ii, p. 141; *Stoliczka, J. A. S. B.* xxxvii, pt. ii, p. 42; *Hume & Henders. Lah. to Yark.* p. 214, pl. xiv; *Hume, N. & E.* p. 322; *id. S. F.* vii, p. 391; *id. Cat.* no. 504; *Seebohm, Cat. B. M.* v, p. 353; *Oates in Hume's N. & E.* 2nd ed. ii, p. 69.

Adelura (Ruticilla) cœruleocephala (*Vig.*), *Brooks, S. F.* iii, p. 240.

The Blue-headed Redstart, Jerd.

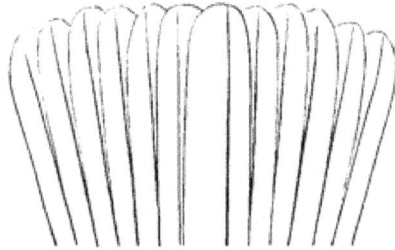

Fig. 32.—Tail of *A. cœruleocephala.*

Coloration. Male. After the autumn moult the forehead, crown, and nape are pale blue, each feather with a broad brown fringe; back, scapulars, upper rump, sides of neck, chin, throat, and breast black, the feathers broadly fringed with brown; lores, cheeks, and ear-coverts plain black; lower rump and upper tail-coverts black, with very narrow brown fringes; tail entirely black; wings dark brown or black, the median coverts, the inner greater coverts, and broad margins to tertiaries and later secondaries white; abdomen, vent, and under tail-coverts white; under wing-coverts and axillaries black, with white tips. In spring all the fringes of the feathers are lost, the forehead, crown, and nape become uniform blue, and the black parts of the plumage deep black.

Female. Upper plumage rich brown, the lower part of the rump tinged with rufous, and the upper tail-coverts ferruginous; tail brown, narrowly edged with ferruginous; wings brown, the coverts and tertiaries broadly edged and tipped with fulvous-white; the other quills narrowly margined paler; lower plumage ashy brown tinged with fulvous, becoming almost pure white on the abdomen and upper tail-coverts; a ring of pale feathers round the eye.

The nestling is mottled all over, and the young male may be known at all ages by the white margins on the wings.

Bill dark brown; legs, feet, and claws black; iris dark brown (*Hume*).

Length about 6; tail 2·7; wing 3·2; tarsus ·8; bill from gape ·65.

Distribution. The Himalayas from Afghanistan and Gilgit to Bhutan. I have seen no specimen of this bird from Sikhim, and only one from Bhutan, collected by Pemberton. This species is found from 10,000 feet upwards in summer, but at much lower levels in winter. In summer it extends into Turkestan.

Habits, &c. Breeds in Gilgit and Afghanistan in May and June, and generally throughout the higher parts of the Himalayas. According to Wardlaw Ramsay the nest is composed of small twigs and grass, lined with hair, and is placed in a crevice or hole in the face of a cliff. The eggs, five in number, are of a dull cream-colour, with a darker zone of the same round the thicker end, and measure about ·84 by ·62. According to Hume this species lays a blue unspotted egg, but it appears from his account that in the single instance in which he found the nest he did not secure the bird, and consequently there may have been some mistake about it. I am also disposed to believe that Wardlaw Ramsay's identification of the eggs is correct, as the bird is not a Redstart according to my views.

Genus GRANDALA, Hodgs., 1843.

The genus *Grandala* contains one bird of remarkable structure, the position of which it is somewhat difficult to determine. It is placed by Seebohm among the Thrushes, and by Jerdon among the *Saxicolinæ*, and I place it here in an intermediate position, considering it more allied to the Robins than to the Thrushes or Chats. The proper position of this species may probably be among the *Brachypteryginæ*.

The plumage of the nestling of this species is streaked, and so far it resembles that of the adult female; but the streaks are more numerous and less distinctly defined, giving it a decided, though not typical, Thrush-like appearance.

Fig. 33.—Head of *G. cœlicolor*.

In *Grandala* the bill is about half the length of the head and slender; the nasal membrane is clothed with plumelets to its middle portion, and the rictal bristles are rather long; the wing is excessively long, the first primary very minute, and the second reaching to the tip of the wing; the tail is rather longer than half the wing and square; the tarsus is slender and smooth and fairly long. The sexes are coloured differently, and the plumage is soft and copious.

Only one species of this genus is known. Seebohm unites it with *Sialia*, a genus of American birds, with which, however, it has, in my opinion, no affinities.

658. **Grandala cœlicolor**. *Hodgson's Grandala.*

Grandala cœlicolor, *Hodgs. J. A. S. B.* xii, p. 447 (1843); *Blyth,
Cat.* p. 166; *Horsf. & M. Cat.* i, p. 281; *Jerd. B. I.* ii, p. 119;
Blanf. J. A. S. B. xli, pt. ii, p. 49; *Hume, Cat. no.* 478.
Grandala schistacea, *Hodgs. J. A. S. B.* xii, plate illustrating p. 447
(1843).
Sialia cœlicolor (*Hodgs.*), *Seebohm, Cat. B. M.* v, p. 328.

The Long-winged Blue Chat, Jerd.

Coloration. Male. Wing, tail, greater wing-coverts, primary-
coverts, and winglet black; remainder of plumage bright smalt-
blue, most brilliant on the rump and upper tail-coverts.

Female. The whole plumage brown with a bluish tinge, the
rump and upper tail-coverts decidedly blue; the head, back, sides
of head and neck, and the whole lower plumage except the flanks
streaked with fulvous-white; wings and tail brown; the quills
with a patch of white near the base, forming a wing-spot; some
of the secondaries tipped white; under tail-coverts broadly edged
with white.

The young resemble the female closely, but they have the streaks
broader and extending on to the flanks. The young male probably assumes the adult plumage at the first autumn moult; before
finally acquiring it some of the feathers of the head and neck are
fringed with brown.

Bill and feet jet-black; iris dark brown (*Jerdon*).

Length about 9; tail 3·6; wing 5·8; tarsus 1·15; bill from
gape ·9.

Distribution. The highest parts of the Himalayas from Garhwál
to Sikhim. Blanford did not meet with this species below 15,000
feet in Sikhim, and he observed it as high as 17,000 feet. It extends into the mountains of Tibet and Western China.

Habits, &c. Probably found in pairs in the summer, but in flocks
in the winter; described as having the flight of a Starling, and
feeding on the ground on insects.

Genus **NOTODELA**, Lesson, 1831.

The genus *Notodela* contains one species, which is largely distributed from Nepal to Tenasserim. The sexes are very different,

Fig. 31.—Head of *N. leucura.*

the male being blue and the female rufous, but both sexes have a
large amount of white on the tail, which is considerably longer
than twice the tarsus.

The White-tailed Blue Robin frequents the ground, flying up into trees when disturbed, and expanding and closing its tail frequently. It does not appear to be shy, and it is said to be very silent.

659. Notodela leucura. *The White-tailed Blue Robin.*

Musci-sylvia leucura, *Hodgs. P. Z. S.* 1845, p. 27.
Myiomela leucura (*Hodgs.*), *Horsf. & M. Cat.* i, p. 280; *Jerd. B. I.* ii, p. 118; *Blanf. J. A. S. B.* xli, pt. ii, p. 161; *Blyth & Wald. Birds Burm.* p. 100; *Hume & Dav. S. F.* vi, p. 331; *Hume, Cat.* no. 477; *Oates, B. B.* i, p. 23; *Hume, S. F.* xi, p. 190.
Notodela leucura (*Hodgs.*), *Blyth, Cat.* p. 166; *Hume, N. & E.* p. 306; *Sharpe, Cat. B. M.* vii, p. 23; *Oates in Hume's N. & E.* 2nd ed. ii, p. 70.

The White-tailed Blue Chat, Jerd.; *Mangshia,* Lepch.

Coloration. Male. Forehead, eyebrow, and the smaller upper wing-coverts near the bend of the wing bright cobalt-blue; the whole upper plumage black suffused with blue; lores, sides of the head and neck, and lower plumage deep black, with a few of the feathers of the abdomen fringed with blue; a concealed patch of white on the side of the neck; wings black with bluish edges; tail black, all the feathers except the outermost and the middle pair with a large patch of white on the outer web, increasing in size towards the middle of the tail; under tail-coverts fringed with white.

Female. The whole plumage rufescent brown, and the exposed parts of the closed wings and tail bright ferruginous: no concealed white spot on the side of the neck; tail brown with white patches, as in the male, but smaller in extent; the feathers of the chin, throat, lores, and sides of the head with paler shafts; a ferruginous ring round the eye.

The young are reddish brown, with bright shaft-streaks and with ferruginous tips to the feathers of the upper wing-coverts; the white patches on the tail-feathers are present from the earliest age; in the young male the tail and wings are black, in the female rufous; the adult plumage appears to be acquired by a moult when the young are about a year old.

Bill, legs, feet, and claws black; iris deep brown (*Hume & Davison*).

Length about 7; tail 3; wing 3·8; tarsus 1·1; bill from gape ·9.

Distribution. Nepal, Sikhim, the Daphla hills in Assam, the Khási hills, Cachar, Manipur, Karennee, Muleyit mountain in Tenasserim. Blyth, in his catalogue, recorded a specimen obtained by Hutton at Mussoorie, but no specimen is contained in the British Museum from any locality west of Nepal. This bird is found from about 4000 to 9000 feet, and appears to be a constant resident in the parts it affects.

Habits, &c. Breeds in April and May, constructing a cup-shaped nest of roots and leaves, sometimes hooded, on the ground under the shelter of a rock or on the face of a bank. The eggs are salmon-pink, very faintly freckled with grey all over, and measure about ·91 by ·65.

Genus CALLENE, Blyth, 1847.

The genus *Callene* differs from *Notodela* in having a very much longer tail, and one the feathers of which are greatly graduated and without any white pattern. The tarsus is extremely long. Little is recorded of the habits of the sole Indian member of the genus, which, however, are not likely to differ materially from those of *Notodela*.

660. Callene frontalis. *The Blue-fronted Callene.*

Cinclidium frontale, *Blyth, J. A. S. B.* xi, p. 181 (1842).
Callene frontalis (*Blyth*), *Blyth, Cat.* p. 178; *Horsf. & M. Cat.* i, p. 356; *Jerd. B. I.* i, p. 496; *Hume, N. & E.* p. 220; *id. Cat.* no. 340; *Sharpe, Cat. B. M.* vii, p. 15; *Oates in Hume's N. & E.* 2nd ed. ii, p. 71.

The Blue-fronted Short-wing, Jerd.

Coloration. Male. Lores and a frontal band black; forehead and a short eyebrow cobalt-blue; with the exception of the abdomen, which is dark brown, and the under tail-coverts brown fringed with white, the whole plumage is slaty-blue with the edges of the feathers brighter; wings and tail dark brown, the outer webs suffused with blue; a portion of the under wing-coverts white; the lesser upper wing-coverts bright cobalt-blue.

Female. The whole plumage rufescent brown, and the visible portions of the closed wings and tail bright ferruginous; tail brown; the feathers of the chin, throat, lores, and sides of the head with paler shafts; a ferruginous ring round the eye; a portion of the under wing-coverts white.

The young are dusky brown, with pale mesial streaks on the feathers of the scapulars and the lower plumage. The young male assumes the adult plumage at the commencement of its first year.

Bill black: legs brown.

Length about 8; tail 3·7; wing 3·6; tarsus 1·5; bill from gape ·8.

This bird, though structurally very different from *Notodela leucura*, resembles it closely in coloration.

Distribution. Sikhim only. There is no evidence that Hodgson met with this bird in Nepal; on the contrary, his drawing appears to have been taken from a Sikhim specimen. There is nothing known of the habits of this species.

Genus THAMNOBIA, Swains., 1831.

The genus *Thamnobia* contains two species, one or other of which is found over a considerable portion of India.

I have much hesitation in placing this genus here. It is the only genus of the *Turdidæ*, with the exception of the *Accentorinæ*, in which the tarsus is strongly or at all scutellated; the bill is, moreover, quite of a different character to that of any of the Thrush tribe, and the rictal bristles are reduced to a minimum. The young are mottled to a slight extent only. A better place may possibly hereafter be found for it.

In *Thamnobia* the bill is slender and curved downwards, the wing is very rounded, and the tarsus is strongly scutellated in front.

The two species of this genus appear to run into each other at the common point of meeting in about the latitude of Bombay; but with reference to this, two points should be carefully regarded. They are both subject to two kinds of change of plumage. One change is caused by the ordinary wearing away of the margins of the feathers during the winter, and the other, coincident in time with this, is caused by the further abrasion of the feathers after the margins are worn off. In consequence of these changes it is difficult for nine months of the year to be quite certain to which species any particular specimen may belong if the abrasion of the feathers has been at all normal. I have had no difficulty, however, in separating autumnal freshly-moulted birds, and they can be ranged into two series, each of which is found to occupy a different geographical area. In a certain zone, from Ahmednagar to the mouth of the Godávari valley, both species occur, but they are to be separated even here if birds in good plumage be examined.

The Indian Robins, as they are termed by residents in India, are familiar birds, being found in compounds, &c., and nesting in houses, or in their immediate vicinity. These birds feed a good deal on the ground, and have the habit of erecting the tail after the fashion of Robins. Both species are resident. The sexes are different, and while the males of the two species are not difficult to discriminate, the females are very close to each other.

Key to the Species.

a. With white on the wing-coverts.
  a'. Upper plumage sandy brown........ *T. cambaiensis* ♂, p. 114.
  b'. Upper plumage black .... *T. fulicata* ♂, p. 115.
b. With no white on the wing-coverts .... { *T. cambaiensis* ♀, p. 114. / *T. fulicata* ♀, p. 115.

661. Thamnobia cambaiensis. *The Brown-backed Indian Robin.*

Sylvia cambaiensis, *Lath. Ind. Orn.* ii, p. 554 (1790).
Thamnobia cambaiensis (*Lath.*), *Blyth, Cat.* p. 165; *Horsf. & M. Cat.* i, p. 283; *Jerd. B. I.* ii, p. 122; *Stoliczka, J. A. S. B.* xxxvii, pt. ii, p. 40, xli, pt. ii, p. 237; *Hume, S. F.* i, p. 182; *id. N. & E.* p. 300; *Ball, S. F.* vii, p. 216; *Hume, Cat.* no. 480; *Sharpe, Cat. B. M.* vii, p. 55; *Barnes, Birds Bom.* p. 198; *Oates in Hume's N. & E.* 2nd ed. ii, p. 71.

Coloration. Male. When freshly moulted in September, the whole upper plumage is sandy brown; upper tail-coverts and tail black; wings dark brown, the lesser coverts and a portion of the median ones white, the remaining coverts with bluish edges; lores, sides of the head and neck, chin, throat, breast, upper part of abdomen, and the sides of the body glossy black with a few sandy edges; lower part of abdomen and the under tail-coverts deep chestnut.

The male continues in this plumage up to February, when the feathers of the upper plumage become much darker in colour, owing apparently to the wearing away or casting off of the tips. During the summer the plumage resembles that of *T. fulicata* in many respects, but is seldom or never so dark.

Female. Ear-coverts and round the eye rufous, the former with pale shafts; chin, cheeks, and a frontal band over the lores pale fulvous; with these exceptions, the whole plumage is sandy brown, tinged with ashy below; tail and wings dark brown; middle of the abdomen, vent, and under tail-coverts chestnut.

The young are rufous, the feathers of the back obsoletely barred and the wing-coverts and quills broadly edged with brighter rufous; upper tail-coverts smoky brown; tail very dark brown; lower plumage ashy brown tinged with rufous and slightly mottled; under tail-coverts, vent, and middle of abdomen pale chestnut.

Iris dark brown; legs, feet, and bill black (*Hume Coll.*).

Length about 6·5; tail 2·7; wing 3; tarsus 1·05; bill from gape ·75.

Distribution. A resident in a very large portion of India proper. On the west this species extends to Sind and the Punjab; on the north to the lower ranges of the Himalayas, ascending them at times up to 5000 or 6000 feet; on the east to the Rajmehal hills and Midnapur, and on the south to Ahmednagar and the Godávari valley.

Habits, &c. Breeds from March to August, constructing a flimsy nest of miscellaneous materials in holes of walls, banks, &c., and laying four to six eggs, which are greenish white mottled with reddish brown, and measure about ·79 by ·59.

662. **Thamnobia fulicata.** *The Black-backed Indian Robin.*

Motacilla fulicata, *Linn. Syst. Nat.* i, p. 336 (1766).
Thamnobia fulicata (*Linn.*), *Blyth, Cat.* p. 165; *Horsf. & M. Cat.* i, p. 281; *Jerd. B. I.* ii, p. 121; *Hume, N. & E.* p. 307; *id. Cat. no.* 479; *Legge, Birds Ceyl.* p. 440; *Sharpe, Cat. B. M.* xii, p. 54; *Davison, S. F.* x, p. 358; *Barnes, Birds Bom.* p. 198; *Oates in Hume's N. & E.* 2nd ed. ii, p. 76.

The Indian Black Robin, Jerd.; *Kalchuri,* Hind.; *Nalanchi,* Tel.; *Wannati-kuruvi,* Tam.

Coloration. Male. When freshly moulted in September, the whole plumage is glossy black except the lesser wing-coverts and a portion of the median, which are white, and the under tail-coverts and a portion of the lower part of the abdomen, which are chestnut; wings brown or dull black. This plumage is retained till February, when the coloration becomes paler either by bleaching or abrasion of the feathers, and in this state the birds resemble *T. cambaiensis*, but there is always some black left to indicate the species.

Fig. 35. -Head of *T. fulvata*.

Female. Lores, forehead, and chin rufous-ashy; ear-coverts rufous, with pale shafts; the whole upper plumage brown with a rufous tinge, the wing-coverts edged paler; quills brown, edged with the colour of the back; lower plumage ashy grey, varying in different individuals; middle of the lower abdomen and the under tail-coverts chestnut.

The young are rufous-brown above, obsoletely barred or tipped brighter; wing-coverts and quills broadly edged with rufous; upper tail-coverts dusky; tail blackish; lower plumage brown mottled with rufous, the chin paler; middle of the abdomen and the under tail-coverts pale chestnut.

The young of this species are more distinctly spotted than are those of *T. cambaiensis*.

Iris dark brown; legs, feet, and bill black (*Butler*).

Length about 6·5; tail 2·5; wing 2·9; tarsus 1·05; bill from gape ·75.

Distribution. Ceylon; the southern portion of India up to Ahmednagar on the west and the Godávari valley on the east.

Habits, &c. Precisely those of the last species in all respects. Eggs of the same type and size.

Genus COPSYCHUS, Wagler, 1827.

The genus *Copsychus* contains the well-known Magpie-Robin of India and some other allied species. It differs from all the other genera of this subfamily in having a tail which is about equal to the wing in length, considerably graduated, and coloured black and white. The sexes are different, although both possess the same pattern of colour.

663. Copsychus saularis. *The Magpie-Robin.*

Gracula saularis, *Linn. Syst. Nat.* i, p. 165 (1766).
Copsychus saularis (*Linn.*), *Blyth, Cat.* p. 166; *Horsf. & M. Cat.* i, p. 275; *Jerd. B. I.* ii, p. 114; *Hume & Henders. Lah. to Yark.* p. 202; *Hume, N. & E.* p. 303; *id. S. F.* ii, p. 230; *Hume &*

Dac. S. F. vi, p. 332; *Anders. Yunnan Exped., Aves,* p. 613;
Legge, Birds Ceyl. p. 433; *Hume, Cat.* no. 475; *Oates, B. B.* i,
p. 20; *Sharpe, Cat. B. M.* vii, p. 61; *Barnes, Birds Bom.* p. 197;
Oates in Hume's N. & E. 2nd ed. ii, p. 80.

Copsychus musicus (*Raffl.*), *apud Hume & Dac. S. F.* vi, p. 333;
Hume, Cat. no. 475 bis; *Oates, B. B.* i, p. 21.

Doyar or *Dayal,* Hind. and Beng.; *Pedda malanchi, Sarela-ga la,* Tel.
Zununid-pho, Lepch.; *Thapate-beay,* Burm.

Fig. 36.—Head of *C. saularis.*

Coloration. Male. Head, neck, breast, and upper plumage glossy
black; abdomen, sides of the body, and under tail-coverts white;
wing black, the last two secondaries with a considerable amount of
white on the outer webs, the lesser and median coverts and the
outer webs of the later greater coverts also white; the median
two pairs of tail-feathers black, the others white, the fourth pair,
however, varying from white with a small black tip to white with
a greater or less amount of black in combination; under wing-
coverts and axillaries white, with ashy bases varying in extent.

Female. Wings and tail dark brown, with white distributed as
in the male; chin, throat, breast, and sides of the neck dark grey;
forehead, lores, and cheeks mottled with white and grey; the
whole upper plumage uniform dark brown glossed with bluish;
sides of the body, vent, and under tail-coverts pale fulvescent;
middle of the abdomen whitish; under wing-coverts white.

The young have the crown and nape ashy brown; upper plumage
dark brown, streaked or barred with rufous; wings dark brown,
with rufous tips to the lesser coverts and broad rufous margins to
the quills; the white in the wing disposed as in the adult, and
the tail brown, with the white portions similarly disposed; throat
and breast greyish brown tipped with rufous; remainder of the
lower plumage white. The adult plumage is assumed almost as
soon as the young bird is fully fledged.

Bill black; mouth flesh-colour; eyelids plumbeous; iris hazel-
brown; legs dark plumbeous; claws horn-colour.

Length about 8; tail 3·6; wing 3·7; tarsus 1·15; bill from
gape 1.

Throughout its great range *C. saularis* is very constant in its
type of plumage, the only variation noticeable being in the coloration of the tail and the under wing-coverts and axillaries.

Throughout Continental India and Burma to about Moulmein
most of the birds have the fourth pair of tail-feathers, from the
outside, white with a small black tip. South of Moulmein and

118   TURDIDÆ.

throughout Tenasserim these feathers contain more black and less white, and in this respect approach *C. musicus* of Java. It is not, however, difficult to find birds in Ceylon and parts of India with these feathers almost entirely black, and consequently I do not think that this character can be utilized for the separation of the two species.

With regard to the other point, Indian and Burmese birds have the under wing-coverts and axillaries almost entirely white, but in the southern parts of Tenasserim some birds are found with these parts more black than white; but still they cannot be considered to be *C. musicus*, in which these parts are almost entirely black, and consequently I do not admit this latter species into the Indian list.

Distribution. Occurs in almost every part of the Empire and Ceylon, ascending the Himalayas up to about 5000 feet; rare in the extreme North-west and Sind, and probably absent from the Nicobars.

Habits, &c. A common and familiar bird wherever it occurs. This species has many of the habits of the common English Robin, being equally confiding and entering verandahs of houses without fear. It is a fairly good songster, feeds on the ground on insects, and has the habit of raising its tail perpendicularly at the end of its run. This bird breeds from March to July, constructing a rough nest in holes of trees, or walls, or in houses, and laying five eggs, which are greenish marked with reddish brown, and measure about ·87 by ·66.

Genus **CITTOCINCLA**, Gould, 1836.

The genus *Cittocincla* contains the Shamas, of which two species inhabit India, one being universally distributed and one confined to the Andamans.

This genus differs from *Copsychus* in its longer tail, which considerably exceeds the wing in length. The Shamas frequent thick woods and tree-jungle, and feed on the ground. One species at least sings very well, but *C. albiventris* is said to have no voice.

Key to the Species.

a. Abdomen rufous .... *C. macrura*, p. 118.
b. Abdomen white .... *C. albiventris*, p. 120.

661. **Cittocincla macrura.** *The Shama.*

Turdus macrourus, *Gmel. Syst. Nat.* i, p. 820 (1788).
Kittacincla macroura *(Gm.), Blyth, Cat.* p. 165; *Horsf. & M. Cat.* i, p. 279; *Jerd. B. I.* ii, p. 116; *Barnes, Birds Bom.* p. 197.
Cittocincla macrura *(Gm.), Legge, Birds Cyl.* p. 457; *Oates, B. B.* i, p. 22; *id. in Hume's N. & E.* 2nd ed. ii, p. 86.

Cercotrichas macrurus (*Gm.*), *Hume, N. & E.* p. 306; *id. & Dav. S. F.* vi, p. 333; *Hume, Cat.* no. 476.
Cittocincla tricolor (*Vieill.*), *apud Sharpe, Cat. B. M.* vii, p. 85.

Shama, Hind.; *Pula nalanchi, Tonka nalanchi*, Tel.

Coloration. Male. Head and neck all round, breast, back, scapulars, and wing-coverts glossy black; rump and upper tail-coverts white; the median four tail-feathers entirely black, the others black at the base and then white; quills, winglet, and primary-coverts dull black, with a slight gloss on the outer webs only; abdomen, vent, and under tail-coverts bright chestnut; thighs white.

Female. In the female, which resembles the male in the distribution of colours, all those parts are slaty brown which are black in the male; the bright chestnut parts of the male are pale rufous in the female, and the quills and wing-coverts are narrowly edged with rufous. In other respects the sexes are alike.

The young vary a good deal. The general colour of the upper plumage is dark brown, the wing-coverts and some of the feathers of the back tipped with rufous; the quills margined with rufous; the lower plumage is chiefly pale rufous, mottled with brown on the throat and breast. Young birds assume the adult plumage very soon after they are fully fledged.

The coloration of this bird is very constant to type throughout the large area of its distribution, the only variation apparent being in the darker coloration of some of the females in Tenasserim.

Bill black; legs pale flesh-colour; claws light horn-colour; mouth flesh-colour; eyelids plumbeous; iris dark brown.

Length about 11; tail about 6; wing 3·7; tarsus 1·1; bill from gape ·95; the female has the tail about one inch shorter than the male.

Distribution. Ceylon; the hills along the western coast of India from Cape Comorin to Khandála; the Eastern Ghats, according to Jerdon; Orissa; the Central Provinces; Chutia Nagpur; Western Bengal; the lower ranges of the Himalayas, from Nepal to Dibrugarh in Assam; the Khási hills; Cachar; Tipperah; Manipur; universally distributed over the whole of Burma and Karennee. The range of this common species is not well exhibited by the series in the Hume Collection, and it may be more widely spread over India proper than the above localities indicate. This bird is a permanent resident, and does not ascend the hills to any great height.

Habits, &c. Frequents thick jungle and is very shy; a most excellent songster; breeds from April to June, constructing a nest of leaves and grass &c. in a hole of a tree at no great height from the ground, and laying four eggs, which are very similar in colour to those of *Copsychus saularis*, and measure about ·85 by .62.

665. Cittocincla albiventris. *The Andaman Shama.*

Kittacincla albiventris, *Blyth, J. A. S. B.* xxvii, p. 269 (1858); *Ball, S. F.* i, p. 73; *Hume, S. F.* ii, p. 232; *Walden, Ibis,* 1873, p. 307, pl. xii, fig. 1.
Cercotrichas albiventris (*Blyth*), *Hume, Cat.* no. 476 bis.
Cittocincla albiventris, *Blyth, Sharpe, Cat. B. M.* vii, p. 90.

Coloration. Both sexes are alike, or nearly so, the female merely differing from the male in having the chin and throat less glossy and they both resemble the male of *C. macrura,* from which they differ in the colour of the abdomen and vent, which is white instead of chestnut. The under tail-coverts and flanks are pale ferruginous. The tail is much shorter.

The nestling bird is dark brown, spotted with ferruginous; the wings are margined with the same, and the coverts spotted.

Legs and feet pale fleshy; bill black (*Hume*).

Length about 9; tail about 5; wing 3·6; tarsus 1; bill from gape 1. The female has usually a shorter tail.

Distribution. The Andamans.

Subfamily TURDINÆ.

The *Turdinæ* comprise the true Thrushes. These differ chiefly from the *Saxicolinæ* and *Ruticillinæ* in being of larger size, in having a greater tendency to be gregarious, and in being less dependent on insects for their food—berries forming a considerable portion of their diet during winter.

The Thrushes are mostly migratory; some few are resident, and when this is the case they are generally confined to limited areas. The majority undergo a seasonal change of plumage through the margins of the feathers dropping off; but these changes are never very striking, and frequently hardly appreciable. The Thrushes feed a great deal on the ground, and their long tarsi enable them to hop with great facility; they are good songsters; they mostly build cup-shaped nests in trees, and they lay spotted eggs.

The *Turdinæ* resemble each other closely in structure, and it is by no means easy to divide them into genera. I have had recourse to the type of coloration in subdividing them, and I have found the colour of the under wing-coverts and axillaries of considerable importance in classification.

The young of the Thrushes are greatly spotted, and they acquire the adult plumage at the first autumn moult. I have not attempted to describe the young of each species, as, from the nature of the coloration, the descriptions, to be of any utility, must of necessity be somewhat lengthy, and space does not permit of this; and it may be doubted if any description of young Thrushes, however elaborate, would enable the student to identify the species.

Key to the Genera.

a. Bill narrow; breadth at forehead not more
    than half length of culmen; rictal bristles
    well developed.
    a'. Sexes different in coloration.
        a''. Axillaries and under wing-coverts in
            both sexes uniformly of one colour or
            very nearly so; lower plumage never
            blue nor chestnut combined with black
            or blue .... MERULA, p. 121.
        b''. Axillaries and under wing-coverts in
            both sexes of two colours in strong
            contrast; arrangement of colours in
            axillaries transposed in under wing-
            coverts .... GEOCICHLA, p. 135.
        c''. Axillaries and under wing-coverts in
            males of one colour, in females more
            or less barred with two colours; lower
            plumage of males wholly blue, or chest-
            nut combined with blue or black, in
            females squamated.
            a'''. Tail very much longer than half
                wing .... PETROPHILA, p. 142.
            b'''. Tail about equal to half wing ...... MONTICOLA, p. 147.
    b'. Sexes alike in coloration.
        d'. Axillaries and under wing-coverts en-
            tirely of one colour .... TURDUS, p. 148.
        e. Axillaries and under wing-coverts of
            two colours; arrangement of colours
            in axillaries transposed in under wing-
            coverts.
            e'''. Lower plumage distinctly barred or
                spotted; rictal bristles few and
                lateral .... OREOCINCLA, p. 151.
            d'''. Lower plumage squamated; rictal
                bristles numerous, and anterior ones
                projecting forwards over nostrils .... ZOOTHERA, p. 156.
b. Bill broad; breadth at forehead more than half
    length of culmen; rictal bristles obsolete .. COCHOA, p. 158.

Genus MERULA, Leach, 1816.

I restrict this genus to those Thrushes in which the sexes are
different in coloration and in which the under wing-coverts and
axillaries in both sexes are uniformly of one colour or nearly so.
The lower plumage of the Thrushes of this genus is, moreover,
never blue, nor chestnut combined with black or blue.

In *Merula* the bill is about half the length of the head; the
rictal bristles moderate; the wing long and sharp, the first primary
being small; the tail rather ample, and the tarsus long. The underside of the wing has no pattern.

Key to the Species.

a. General colour of plumage black or brown,
   unrelieved by any distinctive marks.
   a'. Legs black or brown................. *M. maxima,* p. 123.
   b'. Legs yellow or orange.
     a''. Wing quite 5 inches and generally
        more.
        a'''. Lower plumage uniformly dark
           brown; crown of male not much
           darker than back............... *M. simillima,* p. 124.
        b'''. Lower plumage albescent on abdo-
           men and under tail-coverts; crown
           of male black, forming a cap con-
           spicuously darker than back...... *M. nigripileus,* p. 126.
     b''. Wing about 4·5 inches, rarely reach-
        ing 5.
        c'''. Upper plumage with all feathers
           margined .... *M. kinnisi,* p. 124.
        d'''. Upper plumage uniform.
           a⁴. Sides of head of much the same
              colour as other parts of head .... *M. bourdilloni,* p. 125.
           b⁴. Sides of head rufous * .... *M. erythrotis,* p. 126.
b. Plumage variegated.
   c'. Hind neck of different colour from back.
     e''. Crown and back of same colour ...... *M. albicincta,* p. 127.
     d''. Crown and back of different colours .. *M. castanea,* p. 128.
   d'. Hind neck of same colour as back.
     e''. Feathers of upper plumage variegated
        with dark central marks .... *M. fuscata,* p. 129.
     f''. Feathers of upper plumage not varie-
        gated.
        e'''. Tail, throat, and upper breast chiefly
           chestnut .... *M. ruficollis,* p. 130.
        f'''. No chestnut on tail, throat, or upper
           breast.
           c⁴. Wings boldly marked with a large
              patch of grey or rufous .... *M. boulboul,* p. 130.
           d⁴. Wings uniform.
              a⁵. Under wing-coverts and axil-
                 laries wholly or in part chest-
                 nut or orange-brown.
                 a⁶. Sides of breast and abdomen
                    grey or brown.
                    a⁷. Throat and breast uni-
                       formly of one colour.
                       a⁸. Throat and breast black. *M. atrigularis* ♂, p. 131.
                       b⁸. Throat and breast slaty
                          grey .... *M. unicolor* ♂, p. 132.
                    b⁷. Throat and breast streaked.
                       c⁸. Under wing - coverts
                          orange-brown; axillaries
                          rufous-grey .... *M. atrigularis* ♀, p. 131.

──────────────────

* Of *M. kinnisi, M. bourdilloni,* and *M. erythrotis* the series to which I
have access is so very small and unsatisfactory that the characters for these
three species given here may not prove to hold good in all cases.

d². Under wing - coverts
and axillaries uniformly
chestnut-brown ......   *M. unicolor* ♀, p. 132.
b². Sides of breast and abdomen
orange-ferruginous ....   *M. protomomelæna,*
b³. Under wing-coverts and axil-   [p. 133.
laries slaty grey.
c⁶. Breast and sides of abdomen
chestnut-brown.
c⁷. Wing 4·8; second primary
longer than fifth ....   *M. obscura,* p. 134.
d⁷. Wing 5·4; second primary
shorter than fifth ......   *M. subobscura,* p. 135.
d⁶. Breast and sides of abdomen
slaty grey* ....   *M. feæ,* p. 135.

666. **Merula maxima.**   *The Central-Asian Blackbird.*

Merula vulgaris ?, *Jerdon, Ibis,* 1872, p. 137.
Merula vulgaris, *Ray, Scully, S. F.* iv, p. 139.
Merula vulgaris, *Leach, Hume, Cat.* no. 359 bis.
Merula maxima, *Seebohm, Cat. B. M.* v, p. 405 (1881) ; *St. John, Ibis,*
1889, p. 161.

Coloration. Male. Entirely black throughout.

Female. Upper plumage dark slaty brown with an olive tinge ;
tail black ; wings dark brown, all the feathers edged with olivaceous ; lores dark brown, with a whitish line over them ; sides of
the head ashy brown, the lower portion of the ear-coverts with white
shafts ; lower plumage slaty grey, the chin, throat, and breast
streaked with blackish ; axillaries and under wing-coverts uniform
slaty brown.

A young bird procured by Jerdon in Kashmir is black ; but the
abdomen, vent, thighs, under tail-coverts, and under wing-coverts
are barred with buff, and the feathers of the rump and upper tail-
coverts are tipped with the same.

The male has the bill yellow, tip of upper mandible blackish ;
legs and feet dark brown ; claws black (*Scully*).

The female has the bill brownish black ; legs and feet blackish
brown ; claws black (*Scully*).

Length about 11 ; tail 4·8 to 5·15 ; wing 5·4 to 5·9 ; tarsus 1·45 ;
bill from gape 1·2. Scully gives the length of the tail of a female
bird of this species as 5·6 ; but this is probably a misprint, as the
tail of a male, as given by the same author, is only 5·15.

This species differs from its European ally in being much larger,
the wing in *Merula vulgaris* being seldom more than 5 inches and
the tail 4·5 inches.

_____

* *M. kessleri,* Prjev., was obtained by Mandelli in Tibet (S. F. v, p. 484), not
far from the Sikhim frontier. In the male the abdomen and flanks are deep
chestnut, in the female dull chestnut-brown ; in both sexes the rump is dull
rufous, the wings and tail nearly black ; the head and breast in the male are
black or dark brown, in the female paler and streaked on the throat. Wing
5·7 ; tail 4·7.

Distribution. I have examined specimens of this Blackbird from Kashmir, Kandahar, Bala Murghab, Tashkend, and Yárkand. It meets *M. vulgaris* in Persia, and both species occur in that country.

Habits, &c. According to Scully, this bird is said not to be uncommon during the winter near Káshgarh and Yárkand. It seemed to keep principally among *Eleagnus* trees and thorn-bushes in the vicinity of unfrozen bits of water. It migrated northwards in spring. St. John states that it is common about Kandahar.

667. **Merula simillima.** *The Nilgiri Blackbird.*

Turdus simillimus, *Jerd. Madr. Journ. L. S.* x, p. 253 (1839).
Merula simillima (*Jerd.*), *Blyth, Cat.* p. 162 ; *Horsf. & M. Cat.* i, p. 401 ; *Jerd. B. I.* i, p. 524 ; *Hume, N. & E.* p. 232 ; *Fairbank, S. F.* v, p. 403 ; *Seebohm, Cat. B. M.* v, p. 251 ; *Hume, Cat.* no. 360 ; *Oates in Hume's N. & E.* 2nd ed. ii, p. 88.

Coloration. Male. Forehead, crown, and nape black ; the whole upper plumage and the outer webs of the feathers of the wings and tail dark ashy brown ; the whole lower plumage, axillaries, and under wing-coverts dark brown, the feathers indistinctly edged paler.

Female. The whole upper plumage, including the forehead, crown, and nape, dark ashy brown ; the whole lower plumage, including the axillaries and under wing-coverts, brownish grey, some of the feathers of the abdomen with whitish shafts ; chin and throat streaked with dark brown.

Iris brown ; bill reddish orange ; orbital skin and eyelids yellow ; legs orange-yellow (*Wardlaw Ramsay*).

Length about 10·5 ; tail 4·2 ; wing 5 to 5·2 ; tarsus 1·25 ; bill from gape 1·2.

Distribution. A resident on the higher portions of the Nilgiri hills, the Brahmagiris in Coorg, and the Palni hills.

Habits, &c. Frequents dense woods, occasionally entering gardens. Breeds from March to May, constructing a massive cup-shaped nest of ferns, grass, moss, and roots, more or less plastered together with mud, in a branch of a tree up to 20 feet from the ground. The eggs, usually four or five in number, are greenish marked with brownish red, and measure about 1·17 by ·86.

668. **Merula kinnisi.** *The Ceylon Blackbird.*

Merula kinnisii, *Kelaart, Blyth, J. A. S. B.* xx, p. 177 (1851) ; *id. Ibis,* 1867, p. 304 ; *Hume, Cat.* no. 360 bis ; *Seebohm, Cat. B. M.* v, p. 252 ; *Oates in Hume's N. & E.* 2nd ed. ii, p. 90.
Turdus kinnisi (*Blyth*), *Legge, Birds Ceyl.* p. 449.

Coloration. Male. Upper plumage black, each feather with a bluish-grey margin at all seasons apparently ; quills and wing-coverts black, with similar margins ; tail black, the feathers with narrow and less distinct bluish-grey margins ; lower plumage slaty brown, each feather with a pale margin.

Female. The small series of this bird in the British Museum appears to consist entirely of males. Legge thus describes the female :—Above dark bluish slate, pervaded with brownish on the head, the margins of all the feathers black ; outer webs of primaries and secondaries washed with brownish slaty ; tail blackish brown ; beneath slaty washed with earthy brown, the feathers of the abdomen sometimes with light shaft-streaks ; under wing-coverts edged with earthy brown.

In the male the iris is pale brown ; eyelid and bill orange-yellow ; legs and feet paler yellow than the bill ; claws yellowish horny. In the female the bill is yellowish orange ; eyelid yellow ; legs and feet pale yellow (*Legge*).

Length about 9·5 ; tail 3·6 to 4 ; wing 4·3 to 4·5 ; tarsus 1·3 ; bill from gape 1·2.

This species differs from *M. maxima* in being very much smaller and in having yellow feet ; from *M. simillima* in being smaller and blacker ; from *M. bourdilloni* also in being smaller, and in having the feathers of the upper plumage margined with bluish grey ; and from *M. erythrotis* in having the whole head black or brown.

Distribution. A resident in the forests of Ceylon above 2500 feet elevation. Breeds from April to June, constructing a cup-shaped nest in trees, and laying four eggs, which are pale green marked with reddish-brown and umber, and measure about 1·05 by ·82.

669. **Merula bourdilloni.** *Bourdillon's Blackbird.*

Merula kinnisi (*Kelaart*), *apud Hume, S. F.* vii, p. 35 ; *Terry, S. F.* x, p. 474.
Merula bourdilloni, *Seebohm, Cat. B. M.* v, p. 254, pl. xv (1881) ; Oates in *Hume's N. & E.* 2nd ed. ii, p. 91.

Coloration. Male. The whole upper plumage deep black ; tail black ; wings black, the outer webs of the feathers suffused with slaty grey ; the whole lower plumage blackish brown.

Female. The only bird of this sex that I have seen has the whole upper plumage dark brown tinged with olivaceous, the chin and throat white brown, and the lower plumage fulvous ashy.

In the male the bill, legs, feet, and claws bright orange-red ; iris dark brown (*Hume Coll.*).

Length about 9·5 ; tail 3·6 to 4 ; wing 4·6 to 5 ; tarsus 1·25 ; bill from gape 1·2.

This species, long accepted as *M. kinnisi,* differs from that species in being larger and in having the upper plumage (in the males) deep black without slaty margins. The legs would also appear to be of a different colour, judging from the recorded colours above. I have not been able to compare females of the two species together.

Distribution. The hills of Southern Travancore, extending north to the Palnis. This species does not appear to be found below 3000 feet.

Habits, &c. Breeds on the Palnis in May and June. The nest and eggs resemble those of *M. simillima*, but the size of the eggs has not been recorded.

670. **Merula erythrotis.** *Davison's Blackbird.*

Merula erythrotis, *Davison, Ibis*, 1886, p. 205.

Coloration. The type of this species, the only specimen I have seen, is in the British Museum. It appears to be a female, and resembles the only female of *M. bourdilloni* contained in the same collection in general appearance. It differs, however, in having a supercilium, the lores, the sides of the head, cheeks, and chin rufous-brown.

Davison discovered this specimen, with another similar to it, in the Museum of Trivandrum in Travancore, and they are thought to have been obtained in the Pálghat hills in Travancore. Davison remarks on the triangular bare patch of skin behind the eye in this species, but this patch is present, in a more or less distinct form, in all Thrushes, and forms no character of any value.

Length about 10; tail 4·1; wing 4·8; tarsus 1·2; bill from gape 1·2.

Distribution. Probably the Pálghat hills in Travancore.

671. **Merula nigripileus.** *The Black-capped Blackbird.*

Turdus (Merula) nigripileus, *Lafresnaye, Delessert, Voy. de l'Inde*, pt. ii, p. 27 (1843); *Blyth, Cat.* p. 162; *Jerd. B. I.* i, p. 525; *Butler, S. F.* iii, p. 470; *Hume, Cat.* no. 359; *Vidal, S. F.* ix, p. 63; *Seebohm, Cat. B. M.* v, p. 250; *Barnes, Birds Bom.* p. 173; *Oates in Hume's N. & E.* 2nd ed. ii, p. 91.

Kasturi, Hind.; *Pada palisa,* Tel.

Coloration. Male. Forehead, crown, nape, and sides of head black, paling on the lower portion of the ear-coverts; hind neck, mantle, sides of the neck, and the whole lower plumage brownish grey, infuscated on the throat and suffused with ashy on the flanks; the under tail-coverts whitish along the shafts; lower back, scapulars, rump, the wings, and tail dark ashy; axillaries and under wing-coverts ashy.

Female. Whole upper plumage brown, tinged with ashy on the rump; wings and tail with the outer webs of the feathers suffused with ashy; ear-coverts with pale shafts; chin and throat dull greyish white, streaked with brown; breast, upper abdomen, and flanks brownish grey; lower abdomen dull whitish; under tail-coverts whitish, broadly edged with ashy brown; axillaries and under wing-coverts brownish grey like the breast.

In the male the legs and feet are dirty straw-colour; iris dark brownish red; bill dirty orange: in the female the legs and feet are dark straw-colour; bill dark orange (*Hume*).

Length about 10·5; tail 4; wing 5·1; tarsus 1·2; bill from gape 1·2.

The male of this species cannot be confounded with any other Blackbird, but the female resembles the females of the other species somewhat in colour and *M. simillima* in size. From this latter she may generally be recognized by the whitish abdomen.

Distribution. The western parts of India from the Nilgiris and Mysore up to Mount Abu. This species extends into the interior of the peninsula, and has been recorded from Chikalda, Raipur, and Sambalpur. It appears to be a summer visitor only to the extreme northern portions of its range, but to be resident elsewhere.

Habits, &c. Breeds throughout its limits from May to July, constructing a nest of twigs and grass mixed with earth and moss, and laying three eggs, which are greenish marked with brownish red and purple, and measure about 1·03 by ·82.

672. **Merula albicincta.** *The White-collared Ouzel.*

Turdus albociuctus, *Royle, Ill. Bot. Himal. &c.* pp. lxxvii, lxxviii (1839).
Turdus albicollis, *op. cit.* pl. viii, fig. 3.
Merula nivicollis, *Hodgs. Gray's Zool. Misc.* p. 83 (1844).
Merula albicincta (*Royle*), *Blyth, Cat.* p. 162; *Horsf. & M. Cat.* i, p. 197; *Jerd. B. I.* i, p. 526; *Stoliczka, J. A. S. B.* xxxvii, pt. ii, p. 35; *Blanf. J. A. S. B.* xli, pt. ii, p. 49; *Hume, Cat.* no. 362; *Scully, S. F.* viii, p. 285; *Seebohm, Cat. B. M.* v, p. 245; *Oates in Hume's N. & E.* 2nd ed. ii, p. 92.

Kundoo kastura, Hind.

Coloration. Male. Entire plumage black, except the hind neck, upper back, sides of neck, chin, throat, and upper breast, which are white and form a broad collar; the chin and throat frequently streaked with brown; under tail-coverts with white shafts.

Female. Forehead, crown, nape, and sides of the head rufous-brown; the white parts in the male replaced by dull ashy; lesser wing-coverts, scapulars, back, rump, and upper tail-coverts rufous-brown; wings and tail dark brown tinged with olivaceous; breast, under wing-coverts, and axillaries rufous, gradually becoming darker and blacker on the remainder of the lower plumage, many of the feathers of which are fringed with light rufous; under tail-coverts with white shafts.

The nestling exhibits no indication of the white collar till the autumn moult.

Bill yellow, dusky at extreme tip; iris deep brown; tarsi and toes buffy yellow; claws brown-horny (*Scully*).

Length about 11; tail 4·4; wing 5·5; tarsus 1·3; bill from gape 1·3.

Distribution. The Himalayas from Eastern Kashmir to Sikhim. In the summer this species is found up to 13,000 feet, and in winter down to 7000 feet. Godwin-Austen procured this Ouzel at Remta in Manipur.

Habits, &c. Found singly or in pairs. Breeds in May. The

nest has not been described, but the eggs are said to be greyish white marked with reddish brown, and to measure about 1·2 by ·85.

673. **Merula castanea.** *The Grey-headed Ouzel.*

Merula castanea, *Gould, P. Z. S.* 1835, p. 185; *Blyth, Cat.* p. 162; *Horsf. & M. Cat.* i, p. 197; *Jerd. B. I.* i, p. 526; *Cock & Marshall, S. F.* i, p. 354; *Godw.-Aust. J. A. S. B.* xxxix, pt. ii, p. 268; *Hume, N. & E.* p. 235; *id. Cat. no.* 363; *Seebohm, Cat. B. M.* v, p. 259; *Oates in Hume's N. & E.* 2nd ed. ii, p. 93.

Lal kastura, Hind.

Coloration. Male. Forehead, crown, nape, and sides of the head dark grey; chin, throat, and neck all round pale greyish white; upper back dark chestnut; lower back, scapulars, rump, and upper tail-coverts bright chestnut; wings and tail black; lower plumage chestnut, the middle of the abdomen whitish; under tail-coverts black, with mesial white streaks and fulvous margins near the tip of the feathers; axillaries and under wing-coverts chestnut-brown.

Female. Resembles the male in general pattern of colour, but the head and neck are a darker grey, the chestnut of the upper and lower parts is paler, and the wings and tail are brown; the under tail-coverts are brown instead of black, but marked in the same manner as in the male.

Bill, orbits, and legs yellow; iris brown (*Jerdon*); legs dull yellow, iris dark brown (*Godw.-Aust.*).

Length about 11; tail 4·3; wing 5·3; tarsus 1·3; bill from gape 1·2.

This species and *M. albicincta* were at one time thought to be the same, but no one now doubts their distinctness from each other.

Distribution. The Himalayas from Murree to Sikhim. Godwin-Austen procured this species in the Tura range, Gáro hills. Griffith appears to have obtained it in Assam. This Ouzel, according to Stoliczka (J. A. S. B. xxxvii, pt. ii, p. 35), comes to Kotgarh in the winter, and probably lives during the summer months in Central Asia and Eastern Tibet. This distribution has not, however, been confirmed, and the nest of this species has been found at Kotgarh and Murree, showing that some birds at least remain in the Himalayas during the summer and at comparatively low levels.

Habits, &c. Associates in flocks (during the winter?) according to Hutton. This species appears to construct its nest in banks, making it of moss and fern-leaves with a little earth, and lining it with grass. The eggs are pale green marked with brownish red and pinkish purple, and measure about 1·2 by ·85.

674. **Merula fuscata.** *The Dusky Ouzel.*

Turdus fuscatus, *Pall. Zoogr. Rosso-Asiat.* i, p. 451, pl. xii (1811).
Planesticus fuscatus (*Pall.*), *Jerd. B. I.* i, p. 530; *Godw.-Aust. J. A.
S. B.* xliii, pt. ii, p. 159, xlv, pt. ii, p. 72.
Turdus dubius, *Bechst. apud Hume, Cat.* no. 366.
Merula fuscata (*Pall.*), *Seebohm, Cat. B. M.* v, p. 262.

The Dusky Thrush, Jerd.

Coloration. Male. After the autumn moult the forehead, crown,
and nape are black, with narrow grey margins; the remaining
upper plumage black, with broad rufous-grey margins, the rufous
increasing in intensity towards the tail: wing-coverts and quills
blackish, each feather margined exteriorly with dull chestnut; tail
black, very narrowly edged with rufous; a distinct pale buff supercilium from the nostril to the nape; lores and ear-coverts black;
chin, throat, upper breast, and sides of the head and neck pale buff,
with a few brown marks; lower breast and sides of the body black,
the former with narrow, the latter with broad, white margins;
abdomen white; under tail-coverts brown, broadly edged with
white; axillaries and under wing-coverts dull chestnut.

Female. Differs from the male in having the dark portions of
the upper plumage brown, in having the chin and throat much
spotted with black, and in having the black on the lower breast
much less in extent.

In the spring the margins on the upper plumage disappear, and
these parts become nearly uniform black or brown. Some birds
from Siberia, however, exhibit a large amount of rufous on the
upper plumage even in the height of summer, the black or brown
parts becoming very worn and faded.

Young birds after the first autumn moult have the black centres
to the feathers of the upper plumage smaller than in adults, the
chin and throat very much streaked and spotted and less black on
the lower parts.

Iris dark brown; bill horny brown, yellowish towards the base
of the lower mandible; legs light brown (*Wardlaw-Ramsay*): iris
dark brown, bill black above, dull yellow below, legs dull brown
(*Godw.-Aust.*).

Length about 9·5; tail 3·5; wing 5; tarsus 1·25; bill from
gape 1.

Distribution. A rare winter visitor to the north-eastern portion
of the Empire. Hodgson procured this species in Nepal; Godwin-
Austen at Harmutti in the Daphla hills in Assam and on the peak
of Japvo, the highest point of the Burrail range, at 10,000 feet;
Hume at Shillong and Dibrugarh in Assam, and Wardlaw-Ramsay
at Tonngngoo in Burma. A specimen in the Hume Collection
from the Bhutan Doars, referred to this species, appears to me to
be *M. atrigularis* in immature plumage.

This Ouzel summers in the eastern portion of Siberia, and is
found in winter in Japan and China. Occasionally it wanders
into Europe.

675. Merula ruficollis. *The Red-throated Ouzel.*

Turdus ruficollis, *Pall. Reis. Russ. Reichs.* iii, p. 694 (1776); *Blyth, Cat.* p. 161; *Horsf. & M. Cat.* i, p. 194; *Hume, Cat.* no. 364; *Biddulph, Ibis,* 1881, p. 53; *Hume, S. F.* ix, p. 318, xi, p. 120.
Planesticus ruficollis (*Pall.*), *Jerd. B. I.* i, p. 528; *Godw.-Aust. J. A. S. B.* xxxix, pt. ii, p. 102.
Merula ruficollis (*Pall.*), *Seebohm, Cat. B. M.* v, p. 269.
Turdus hyemalis (*Dybowski*), *apud Biddulph, Ibis,* 1882, p. 271.

The Red-tailed Thrush, Jerd.

Coloration. Male. The whole upper plumage ashy brown, the shafts of the feathers of the crown dark; wings dark brown, the outer webs suffused with ashy brown; tail chestnut, the terminal half or third of the middle pair of feathers brown, the others successively with less black at the tip; a narrow pale chestnut supercilium; lores and ear-coverts ashy brown; cheeks, chin, throat, breast, and sides of the neck chestnut, the feathers of these parts immediately after the autumn moult very narrowly margined with white, and all but the very oldest birds with a row of black spots down each side of the throat; remaining lower plumage white, the sides of the body mottled with brown; under tail-coverts chestnut at base; axillaries and under wing-coverts orange-brown.

Female. Resembles the male, but has the chestnut of the lower parts much paler, and the breast spotted with black. In very old females, however, these spots disappear, and the sexes are then very closely alike.

Tarsi greyish fleshy; feet fleshy brown; upper mandible and tip of lower brown, rest of lower mandible, gape, and margins of upper mandible, except at tip, dull yellow; iris hazel (*Hume*).

Length about 10; tail 4; wing 5·4; tarsus 1·3; bill from gape 1·1.

Distribution. A winter visitor to the Himalayas from Kashmir to Assam. This species has also been observed in winter at Goalpara in Assam, Sylhet, Cachar, the Khási hills, and Manipur.

In winter this Ouzel occurs on the west in Afghanistan, and on the east in China. It summers in Siberia and Central Asia.

676. Merula boulboul. *The Grey-winged Ouzel.*

Lanius boulboul, *Lath. Ind. Orn.* i, p. 80 (1790).
Turdus poecilopterus, *Vigors, P. Z. S.* 1831, p. 54; *Gould, Cent.* pl. xiv.
Merula boulboul (*Lath.*), *Blyth, Cat.* p. 162; *Horsf. & M. Cat.* i, p. 196; *Jerd. B. I.* i, p. 525; *Stoliczka, J. A. S. B.* xxxvii, pt. ii, p. 35; *Hume, N. & E.* p. 234; *id. Cat.* no. 361; *Scully, S. F.* viii, p. 285; *Seebohm, Cat. B. M.* v, p. 248; *Hume, S. F.* xi, p. 128; *Oates in Hume's N. & E.* 2nd ed. ii, p. 93.

The Grey-winged Blackbird, Jerd.; *Kasturi,* Hind.; *Patariya masaicha,* Beng.; *Phoyiong pho,* Lepch.; *Chomam,* Bhut.

Coloration. Male. The whole upper plumage, wings and tail,

the whole head, neck, and breast deep glossy black, except the tips
of the median coverts, the outer webs of the greater coverts and
tertiaries, and the margins of the outer webs of the later secondaries, which are silvery ashy grey with a tinge of vinaceous;
lower plumage from the breast downwards, the axillaries, and
under wing-coverts dull black, each feather narrowly margined
with whitish.

Fig. 37.—Head of *M. boulboul*.

Female. Brownish ashy throughout with an olivaceous tinge, the
marks on the wings, which are similar to those of the male in
shape and disposition, being pale rufous.

In the male the legs and feet are brownish in front, yellow
behind; bill coral-red, tip black; iris brown; edges of eyelids
orange-yellow (*Hume Coll.*): in the female the iris is hazel-red;
bill orange, horny at tip: legs burnt sienna (*Cockburn*).

Length about 11·5; tail 4·5; wing 5·7; tarsus 1·3; bill from
gape 1·25.

Distribution. A resident on the Himalayas from their bases up
to 8000 feet, the range varying according to season. This Ouzel
occurs from Murree to Sikhim; it has also been obtained in the
Bhutan Doars, the Khási hills, Cachar, and Manipur.

Habits, &c. This Ouzel builds its nest sometimes on the ground
in the hollow of a massive root or fallen trunk, and sometimes, more frequently perhaps, on a ledge of rock or on the
extremity of a thick branch, where it has been cut or broken off.
The nest is constructed of moss and leaves, and little or no mud is
used in the structure. The breeding-season lasts from April to
August. The eggs, four in number, are dingy green thickly marked
with reddish brown, and measure about 1·2 by ·87.

677. Merula atrigularis. *The Black-throated Ouzel.*

Turdus atrogularis, *Temm. Man. d'Orn.* ed.2, i, p.169, pl.(1820); *Blyth,
Cat.* p. 161; *Horsf. & M. Cat.* i, p. 195; *Hume, Cat.* no. 365.
Planesticus atrogularis (*Temm.*), *Jerd. B. I.* i, p. 529; *Stoliczka,
J. A. S. B.* xxxvii, pt. ii, p. 35; *Hume & Henders. Lah. to Yark.*
p. 192; *Scully, S. F.* iv, p. 140, viii, p. 286.
Cichloides atrogularis (*Temm.*), *Hume, S. F.* i, p. 179.

K 2

Merula atrigularis (*Temm.*), *Barnes, Birds Bom.* p. 173; *Seebohm, Cat. B. M.* v, p. 267.

The Black-throated Thrush, Jerd.; *Mach-reycha*, Beng.

Coloration. Male. After the autumn moult the lores, cheeks, chin, throat, breast, and sides of the neck are black, each feather with a broad white margin; rest of the underparts white, the sides of the body with ashy streaks; under wing-coverts dull orange-brown; axillaries rufous-grey; under tail-coverts dark brown tipped with white; ear-coverts, the whole upper plumage, and the visible portions of the closed wings and tail greyish brown, the feathers of the crown centred with dark brown. Soon after the moult the white margins of the head, neck, and breast become reduced in width, and are altogether lost by summer, causing the parts to appear uniformly black.

Female. Sides of the head and neck greyish brown like the upper plumage: chin and throat whitish streaked with dark brown; breast ashy brown spotted with black; otherwise as the male.

Legs and feet greyish brown; bill blackish brown, dusky yellow at base of lower mandible; iris blackish brown (*Butler*).

Length about 10; tail 3·8; wing 5·2; tarsus 1·3; bill from gape 1.

Distribution. A winter visitor to the Himalayas and the plains of Upper India. This species extends throughout the Himalayas from Hazára to Assam. In the plains it is found as far south as Karáchi, Cutch, Delhi, and Dacca. From Assam it ranges south through the hill-tracts to Manipur.

Jerdon speaks of this Ouzel as inhabiting the higher ranges of the Himalayas in summer. This statement has received no confirmation since he made it; but it is not improbably correct, as I have seen a specimen killed at Simla on the 14th August and one killed in Kashmir in May. The bulk of these Ouzels, however, if not all, retire north to Siberia to breed. In winter they are found in Central Asia and Afghanistan, but not to the east of Assam.

678. Merula unicolor. *Tickell's Ouzel.*

Turdus unicolor, *Tick. J. A. S. B.* ii, p. 577 (1833).
Petrocincla homochroa, *Hodgs. in Gray's Zool. Misc.* p. 83 (1844).
Turdus dissimilis, *Blyth, J. A. S. B.* xvi, p. 144, part. (1847).
Geocichla dissimilis, *Blyth, Cat.* p. 163; *Horsf. & M. Cat.* i, p. 191.
Geocichla unicolor (*Tick.*), *Blyth, Cat.* p. 163; *Jerd. B. I.* i, p. 519; *Hume & Henders. Lah. to Yark.* p. 192; *Hume, N. & E.* p. 230; *Ball, S. F.* ii, p. 408, vii, p. 213; *Hume, Cat. no.* 356; *Scully, S. F.* viii, p. 283; *Barnes, Birds Bom.* p. 171.
Merula unicolor (*Tick.*), *Seebohm, Cat. B. M.* v, p. 271; *Oates in Hume's N. & E.* 2nd ed. ii, p. 96.

The Dusky Ground-Thrush, Jerd.; *Desi pawai*, Hind.; *Machasah*, Beng.; *Poda palisa*, Tel.

Coloration. Male. The upper plumage, sides of head and neck, and the visible portions of the closed wings and tail ashy grey;

lower plumage slaty grey, paler on the chin and becoming white on the abdomen, vent, and under tail-coverts ; axillaries ashy grey, generally tinged with buff ; under wing-coverts chestnut-brown.

Female. Upper plumage and sides of neck olive-brown ; wings and tail dark brown, the outer webs suffused with olive-brown ; lores blackish with a pale band above ; sides of the head mixed brown and fulvous ; chin and throat white, the sides streaked with black ; breast olivaceous, the upper part spotted with black ; sides of the body pale ochraceous ; abdomen, vent, and under tail-coverts white; axillaries and under wing-coverts chestnut-brown.

In the male the iris is reddish, legs and feet light brownish (*Hume Coll.*); in the female the bill is yellow with a few dusky cloudings ; iris brown ; eyelid greenish yellow ; feet vivid orange-yellow ; claws yellowish horny (*Scully*).

Length about 9 ; tail 3·4 ; wing 4·7 ; tarsus 1·2 ; bill from gape 1·1.

Distribution. Found in summer throughout the Himalayas from Murree to Sikhim up to about 7000 feet. In the winter this species occurs throughout the plains of Northern India from Sind to Bengal. So far as is known it extends at this season to Khandála, Raipur, and Orissa, and Jerdon records it even from the Eastern Ghâts, a specimen from this latter locality being now in the British Museum. An Ouzel obtained at Belgaum in March, now in the Hume Collection, and referred to *M. unicolor*, is undoubtedly a specimen of *M. obscura.*

Habits, &c. Breeds in the Himalayas in May and June, constructing a nest of moss and fibres in trees. The eggs, three or four in number, are greenish white, spotted and freckled with rufous and measure 1·06 by ·78.

679. Merula protomomelæna. *The Black-busted Ouzel.*

Turdus dissimilis, *Blyth, J. A. S. B.* xvi, p. 144, ♂ (1847) ; *Godw.-Aust. J. A. S. B.* xli, pt. ii, p. 142 ; *Seebohm, S. F.* viii, p. 437.
Geocichla dissimilis (*Blyth*), *Blyth, Cat.* p. 163, ♂ ; *Jerd. Ibis,* 1872, p. 136, pl. vii ; *Hume, Cat.* no. 358 ; *id. S. F.* ix, p. 103, xi, p. 126.
Turdulus cardis (*Temm.*), *apud Jerd. B. I.* i, p. 521.
Turdus protomomelas, *Cabanis, Journ. f. Orn.* 1867, p. 280.
Geocichla tricolor, *Hume, Ibis,* 1871, p. 411 : *id. S. F.* iii, p. 400.
Merula protomomelæna (*Cab.*), *Seebohm, Cat. B. M.* v, p. 265.

The Variable Pied Blackbird, Jerd.

Coloration. Male. The whole head, neck, and upper breast black ; upper plumage, wings, and tail dark slate-colour ; sides of the lower breast, sides of the body, axillaries, and under wing-coverts bright orange-ferruginous ; middle of lower breast, abdomen, vent, and under tail-coverts white.

Younger males have the upper wing-coverts tipped with rufous and some black spots on the red of the lower parts of the plumage.

Female. The whole upper plumage olive-brown tinged with slaty ; wings and tail brown, suffused with olive on the outer webs ; sides of the head ashy brown, the shafts of the ear-coverts whitish; chin and upper throat white streaked with brown, the streaks increasing in number at the sides; upper breast olivaceous, spotted with black; middle of lower breast, abdomen, vent, and under tail-coverts white; sides of breast, sides of body, axillaries, and under wing-coverts bright orange-ferruginous.

In the male the bill and orbital skin are yellow; iris deep brown; legs and feet dusky orange-yellow (*Cripps*). In the female the legs, feet, bill, and eyelids are wax-yellow (*Hume*); iris deep brown (*Scully*).

Length about 9; tail 3·3; wing 4·7; tarsus 1·3; bill from gape 1·1.

The synonymy of *M. protomomelæna* has been determined entirely by a careful perusal of the various original descriptions of the bird, which fortunately are sufficiently in detail to render the identification certain. Judging from Hume's remarks (S. F. ix, p. 103), any appeal to Blyth's types in the Indian Museum on this point must prove useless if not misleading. Blyth applied the name *Turdus dissimilis* to specimens of both *M. unicolor* and *M. protomomelæna*, confounding the two together, and consequently it is advisable to discard this name.

Distribution. I have examined specimens of this Ouzel from Dibrugarh in Assam, the Tipperah hills and Manipur. Blyth appears to have procured it from the neighbourhood of Calcutta, and I know of no other locality for this species, which is probably a constant resident in the above-mentioned places.

Scully (S. F. viii, p. 284) records a specimen of this Ouzel from Nepal, but judging from his description, in which a supercilium is mentioned, and the sides of the breast and flanks are referred to as ferruginous, there can be little doubt that the bird was *M. obscura*, which Hodgson procured in Nepal, one of his specimens being now in the British Museum.

680. **Merula obscura.**   *The Dark Ouzel.*

Turdus obscurus, *Gmel. Syst. Nat.* i, p. 816; *Hume & Dav. S. F.* vi, p. 251; *Hume, Cat.* no. 369 bis; id. *S. F.* xi, p. 130.
Turdus pallens, *Pall. Zoogr. Rosso-Asiat.* i, p. 457 (1811).
Turdus rufulus, *Drap. Dict. Class. d'Hist. Nat.* x, p. 443 (1826): *Horsf. & M. Cat.* i, p. 401.
Turdus modestus, *Eyton, P. Z. S.* 1833, p. 103.
Turdus javanicus?, *Horsf., Blyth, Cat.* p. 161.
Geocichla dissimilis (*Blyth*), *Scully, S. F.* viii, p. 284.
Merula obscura (*Gm.*), *Seebohm, Cat. B. M.* v, p. 273; *Oates, B. B.* i, p. 1.
Geocichla unicolor (*Tick.*), *apud Butler, S. F.* ix, p. 399.

Coloration. Male. Upper plumage olive-brown; the forehead, crown, and nape in old birds tinged with ashy; lores black; a broad white supercilium from the lores to the nape; chin, a patch at base of bill, and under the eye white; ear-coverts and the whole throat dark slaty brown; breast and sides of the body chestnut-brown;

abdomen, vent, and under tail-coverts white, the last basally margined with brown ; wings and tail brown, suffused with olive on
the outer webs ; axillaries and under wing-coverts slaty grey.

Female. Resembles the male in general coloration, but has the
crown always of the same colour as the upper plumage ; the lores
and ear-coverts pale, the latter with whitish shafts ; the middle of
the chin and throat white with a few minute brown streaks.

Iris olive-brown ; eyelids greenish ; upper mandible dark brown ;
lower mandible and gape yellow ; inside of mouth yellow ; legs
yellowish brown ; claws horn-colour.

Length nearly 9 ; tail 3·5 ; wing 4·8 ; tarsus 1·2 ; bill from
gape 1·1.

Distribution. A winter visitor, more or less abundant, to the
whole of Burma, the Andamans, Manipur, Shillong, Sikhim, and
Nepal. An occasional straggler visits the plains of India, and in
the Hume Collection there is a specimen procured at Belgaum in
March. In the winter this species extends to China and to the
Malay peninsula and islands, and it summers in Siberia.

681. Merula subobscura. *Salvadori's Ouzel.*

Merula subobscura, *Salvadori, Ann. Mus. Cir. Gen.* (2)   i, p. 413
(1889).

Coloration. Similar to *Merula obscura* but larger, with the white
superciliary band less conspicuous, the sides of the body paler
ochraceous, and the proportion of the primaries different.

The type of this species, the only specimen known, was procured
by Mr. Fea at Tahò in the Karen hills, north-east of Toungngoo,
in March. It is an adult male.

The measurements of this specimen are : length 10 ; tail 3·8 ;
wing 5·25 ; tarsus 1·2 ; bill from gape 1.

The third and fourth primaries are subequal and longest ; the
second shorter than the fifth and longer than the sixth. In
M. obscura the third primary is the longest, the fourth is rather
shorter than the third, and the second is between the fourth and
fifth.

I have examined the type of this species and I have failed to
find any example of this Thrush from Burma in the British Museum
series.

682. Merula feæ. *Fea's Ouzel.*

Turdus chrysolaus, *Temm., apud Godw.-Aust. J. A. S. B.* xxxix,
pt. ii, p. 102, xli, pt. ii, p. 143.
Turdulus pallens (*Pall.*), *apud Godw.-Aust. J. A. S. B.* xliii, pt. ii,
p. 178.
Turdus pallidus, *Gmel., apud Godw.-Aust. J. A. S. B.* xlv, pt. ii,
p. 195; *Hume & Dav. S. F.* xi, p. 253 ; *Hume, Cat. no.* 368 ter ;
id. S. F. xi, p. 130.
Merula pallida, *Gmel., apud Oates, B. B.* i, p. 2.
Merula feæ, *Salvadori, Ann. Mus. Cir. Gen.* (2) v, p. 514 (1887),
p. 610 (1888).
Turdus subpallidus, *Hume, S. F.* xi, p. 132 (1888).

Coloration. Resembles *M. obscura*, but differs in the following respects :—the breast and sides of the body in both sexes are slaty grey, not chestnut-brown ; the upper plumage in both sexes is russet-brown, not olive-brown ; the crown in the adult male is never darker than the other upper parts ; the sides of the head and the sides of the chin and throat are russet-brown, not slaty brown ; and in the adult male the throat itself is slaty grey, not slaty brown.

Iris deep chocolate ; bill black ; legs pale cloudy brown (*Wardlaw Ramsay*) ; legs and feet brownish yellow ; bill blackish brown, yellow at gape and on base of lower mandible ; iris brown (*Hume*).

Of the same dimensions as *M. obscura*.

This Ouzel resembles *M. pallida*, Gmelin, but may be instantly distinguished from that species by the presence of a supercilium, which is altogether absent in *M. pallida*. The latter inhabits Eastern Asia and may occasionally visit Burma.

Distribution. Shillong and Cherra Poonjee ; Japvo peak in the Nága hills at 10,000 feet ; Manipur ; Karennee at 5000 feet ; Muleyit mountain in Tenasserim.

All the specimens of this species that I have examined from the above localities were procured in the winter months, but this Ouzel is not unlikely to prove a resident species in those parts.

Genus **GEOCICHLA**, Kuhl (*teste* Gould), 1836.

In the Thrushes of this genus the sexes are different and the under wing-coverts and axillaries are each of two colours, the position of the two colours on the under wing-coverts being transposed on the axillaries.

From *Merula* this genus differs in having a somewhat blunter wing and shorter tail. The underside of the wing presents a pattern formed by the white bases of many of the quills.

Key to the Species.

a. No chestnut on lower plumage.
   a'. Upper tail-coverts margined with white .. *G. wardi*, p. 137.
   b'. No white on upper tail-coverts .... *G. sibirica*, p. 138.
b. Lower plumage almost entirely chestnut.
   c'. Median wing-coverts broadly tipped with white.
      a''. Chin and throat white .... *G. cyanonotus*, p. 139.
      b''. Chin and throat chestnut like the breast. *G. citrina*, p. 140.
   d'. Median wing-coverts without white tips.
      c''. Chin and throat chestnut .... *G. innotata*, p. 141.
      d''. Chin and throat white .... *G. albigularis*, p. 142.
      e''. Chin white, throat chestnut .... *G. andamanensis*,
                                          [p. 142.

683. **Geocichla wardi.** *The Pied Ground-Thrush.*

Turdus wardii, *Jerd. J. A. S. B.* xi, p. 882 (1842); *id. Ill. Ind. Orn.* pl. viii; *Legge, Birds Ceyl.* p. 453.
Merula wardii *(Jerd.), Blyth, Cat.* p. 163; *Horsf. & M. Cat.* i, p. 402.
Turdulus wardii *(Jerd.), Jerd. B. I.* i, p. 520; *Hume, Cat.* no. 357; *Barnes, Birds Bom.* p. 172.
Cichloselys wardii *(Jerd.), Hume, N. & E.* p. 231.
Oreocincla pectoralis, *Legge, S. F.* iv, p. 244.
Geocichla wardi *(Jerd.), Seebohm, Cat. B. M.* v, p. 178; *Oates in Hume's N. & E.* 2nd ed. ii, p. 97.

Ward's Pied Blackbird, Jerd.

Coloration. Male. The whole head, neck, breast, upper plumage, wings, and tail black; the lesser and median wing-coverts very broadly tipped with white; the greater wing-coverts and quills tipped with white, except the earlier primaries, which, with the primary-coverts, are partially margined with white; the rump and upper tail-coverts with crescentic white tips; tail with a considerable amount of white, increasing in extent from the middle feathers to the outer; a white supercilium to the nape; abdomen, vent, and under tail-coverts white; sides of the body and the axillaries white, each feather with a subterminal black bar; under wing-coverts black tipped white.

Female. Upper plumage and wings olive-brown, all the wing-coverts and tertiaries with buff tips, the outer webs of the quills suffused with russet, the longer feathers of the rump and upper tail-coverts tipped with dull white; tail olive-brown, the portion next the shafts darker, the four outer pairs of feathers tipped white; a broad buff supercilium to the nape; sides of the head and of the throat mixed buff and black; chin nearly plain white; middle of throat and the upper breast pale buffish white, each feather margined with dark brown; lower breast, upper abdomen, and sides of the body barred with olivaceous and suffused with ochraceous; middle of abdomen, vent, and under tail-coverts white.

Iris brown; bill ochre-yellow, the tip of upper mandible black; legs and feet fleshy ochre *(Hume).*

Length about 8·5; tail 3·3; wing 4·5; tarsus 1; bill from gape 1·15.

Distribution. Summers in the Himalayas from the Sutlej valley to Sikhim and the Bhutan Doars up to 6000 or 7000 feet; winters in Southern India and Ceylon. The chief winter-quarters of this species appear to be the Nilgiris and other hill-ranges down to Cape Comorin and Ceylon. It must necessarily occur over a great part of India when migrating, but it has seldom been observed at that period. Major Lloyd records it from the Konkan, and Jerdon from Nellore in the Carnatic.

Habits, &c. Brooks remarks that this species has a strange song of two notes and quite unmusical. It breeds in the Himalayas from May to July, constructing a nest of moss and fibres, with or without mud, in the branch of a tree, and lays four eggs, which are described as being pale green marked with purple and brownish red, and measuring about 1 by ·72.*

684. Geocichla sibirica. *The Siberian Ground-Thrush.*

Turdus sibiricus, *Pall. Reis. Russ. Reich.* iii, p. 694 (1776); *Hume, Cat.* no. 369 quat.
Oreocincla inframarginata, *Blyth, J. A. S. B.* xxix, p. 106 (1860); *Ball, S. F.* i, p. 70; *Hume, S. F.* ii, p. 223.
Turdulus davisoni, *Hume, S. F.* v, pp. 63, 136 (1877).
Turdulus sibiricus (*Pall.*), *Hume & Dav. S. F.* vi, pp. 255, 513; *Hume, S. F.* xi, p. 132.
Geocichla sibirica (*Pall.*), *Seebohm, Cat. B. M.* v, p. 180; *Oates, B. B.* i, p. 4.

Coloration. Male. The fully adult has the whole plumage slaty black, the margins of the feathers paler; the outer three pairs of tail-feathers narrowly tipped with white; a broad white supercilium to the nape; the under tail-coverts tipped with white; axillaries white tipped with dark ashy, and the under wing-coverts ashy tipped white. Males after the second autumn moult are bluish slaty instead of slaty black, but the middle of the abdomen

---

* GEOCICHLA AVENSIS.

Turdus avensis, *Gray, Griffith's ed. Cuvier,* vi, p. 530, pl. (1829).
Geocichla avensis (*Gray*), *Hume, S. F.* viii, p. 39; *Seebohm, Cat. B. M.* v, p. 167.

Coloration. Forehead, crown, nape, and hind neck bright chestnut; upper plumage, wings, and tail dark slaty brown, the lesser and median wing-coverts almost entirely white, and the greater coverts tipped with white; lores, cheeks, and a portion of the ear-coverts white; remainder of the head, throat, and upper breast black; lower breast, abdomen, and sides of the body white spotted with black; middle of abdomen, vent, and under tail-coverts white; axillaries white tipped black; under wing-coverts black tipped white. Wing 4·2; tail 2·5. It is not known how the sexes differ. The above description probably applies to the male only.

The only record of the occurrence of this species within Indian limits is the statement of Gray that the plate of *G. avensis* in his work was taken by Mr. Crawfurd from a specimen procured at Ava. Until this habitat is confirmed, I think it preferable merely to notice this species and thus draw attention to it. This species has never been observed in Burma again since Crawfurd's time.

I have little doubt that *G. avensis* is the same bird as *G. interpres,* Kuhl. The two species are said to differ only in one slight respect. *G. avensis* has the greater wing-coverts plain, and *G. interpres* has them tipped with white; but as all we know of the former bird is derived from Gray's figure, too much reliance must not be placed on this character.

Hume received a specimen of a Thrush from the Malay peninsula (Rumbow) which he identified with *G. avensis.* On examining this specimen, which is now in the British Museum, I find that the greater wing-coverts are wanting or in part moulting, and that the new sprouting feathers of this part appear to be tipped with white. The specimen is by no means a good one for the purpose of deciding the question of the identity or difference of the two species, which must for the present remain unsettled.

is white ; and the distribution of white marks is the same as in the fully adult. Males after the first autumn moult are similar to those just described, but the centres of the feathers of the upper abdomen and sides of the body are white and the tips darker than the other parts, causing a barred appearance; they have also a rufous band across the breast, the remains of the nestling plumage. The nestling is unknown.

Female. The whole upper plumage, wings, and tail olive-brown with a slaty tinge on the rump; the wing-coverts tipped with buff; the outer webs of quills tinged with rufescent; the outer tail-feathers narrowly tipped white: an indistinct buff supercilium to the nape ; sides of the head mixed brown and buff; cheeks buff bordered below by a dusky stripe; chin and throat buff; breast pale buff, the feathers tipped and margined with brown; middle of abdomen white; sides of the body olivaceous brown obsoletely barred darker; under tail-coverts white with basal brown margins; axillaries white tipped olive-brown; under wing-coverts olive-brown tipped white.

Adult males have the bill black ; iris deep brown : front of legs, feet, and claws greenish yellow ; back of legs dirty yellow. Females have the iris dark brown ; the upper mandible very dark brown ; the lower mandible and gape to angle of gonys dirty yellow ; legs, feet, and claws orange-yellow (*Hume & Davison*).

Length about 9; tail 3·6 : wing 4·8 ; tarsus 1·1 ; bill from gape 1·1.

Distribution. A winter visitor to the eastern portions of the Empire. This species has been obtained on Muleyit and Nwalabo mountains in Tenasserim ; at Toungngoo ; in Karennee ; and in Manipur. It has also occurred in the Andamans, a female specimen from these islands having been named *O. inframarginata* by Blyth. In winter this bird is found from China to Java, and it summers in Siberia and Japan.

685. Geocichla cyanonotus. *The White-throated Ground-Thrush.*

Turdus cyanotus, *Jard. & Selby, Ill. Orn.* i, pl. xlvi (1828).
Geocichla cyanotus (*J. & S.*), *Blyth, Cat.* p. 163; *Horsf. & M. Cat.* i, p. 191 ; *Jerd. B. I.* i, p. 517 ; *Blanf. J. A. S. B.* xxxviii, pt. ii, p. 179; *Hume, J. A. S. B.* xxxix, pt. ii, p. 118; *id. N. & E.* p. 229 ; *id. Cat.* no. 354; *Davison, S. F.* x, p. 374 ; *Seebohm, Cat. B. M.* v, p. 172 ; *Barnes, Birds Bom.* p. 171 ; *Oates, in Hume's N. & E.* 2nd ed. ii, p. 98.

Coloration. Male. Forehead, crown, nape, hind neck and sides of the neck, breast, abdomen, and sides of the body golden rufous, the crown tinged with greenish ; vent and under tail-coverts white ; back, rump, upper tail-coverts, scapulars, and wing-coverts slaty blue; the median wing-coverts broadly tipped with white ; quills dark brown, margined on the outer webs with pale slaty ; tail slaty blue, the outer feathers tipped pale ; lores, cheeks, chin, and throat white; an oblique brown band from the eye downwards, succeeded by a band of white behind it running down the neck, and by another brown band running through the middle of the ear-coverts, followed again by a narrow white patch; axillaries white, tipped with ashy fulvous; under wing-coverts slaty blue, tipped with white; a large patch of white on the underside of the quills.

Female. Differs from the male in having the back, scapulars, the outer webs of the secondaries, and many of the wing-coverts suffused with olive-green.

Iris dark brown; bill black; feet fleshy; claws dusky (*Fairbank*).

Length about 8·5; tail 3; wing 4·1; tarsus 1·2; bill from gape 1·1.

Distribution. The southern half of the peninsula of India, from about north latitude 24° to Travancore. This species appears to be resident or very locally migratory within the above-defined area, and to be found up to 4000 feet.

Habits, &c. Breeds from June to September, making a nest apparently very similar to that of *G. citrina*, but using mud in its construction. The eggs are pale bluish or greenish white marked with rufous, and measure about 1 by ·75.

686. Geocichla citrina. *The Orange-headed Ground-Thrush.*

Turdus citrinus, *Lath. Ind. Orn.* i, p. 350 (1790).
Geocichla citrina (*Lath.*), *Blyth, Cat.* p. 163; *Horsf. & M. Cat.* i, p. 189; *Jerd. B. I.* i, p. 517; *Hume, N. & E.* p. 229; *Hume & Dav. S. F.* vi, p. 250; *Hume, Cat. no. 355; Legge, Birds Ceyl.* p. 457; *Scully, S. F.* viii, p. 283; *Seebohm, Cat. B. M.* v, p. 172; *Oates, B. B.* i, p. 3; *Barnes, Birds Bom.* p. 171; *Oates in Hume's N. & E.* 2nd ed. ii, p. 100.
Geocichla layardi, *Wald. A. M. N. H.* (4) v, p. 416 (1870); *Hume, S. F.* iii, p. 401.

Coloration. Male. The whole head, neck, and lower parts as far as the vent orange-chestnut, darker on the crown and paler beneath; vent, thighs, and under tail-coverts pure white; back, scapulars, rump, upper tail-coverts, and lesser wing-coverts bluish grey, the edges of the feathers paler; median wing-coverts broadly tipped white, forming a conspicuous spot; remaining coverts and the quills dark brown, edged exteriorly with bluish grey; tail ashy brown indistinctly cross-barred; axillaries white, tipped with grey; under wing-coverts ashy tipped with white; a large white patch on the underside of the quills.

Female. Of a paler chestnut throughout; the back and scapulars greenish brown with yellowish margins; upper tail-coverts and the outer webs of the feathers of the wings and tail suffused with green.

Bill very dark brown, the gape and the base of the lower mandible flesh-colour ; inside of mouth flesh-colour ; eyelids slate-colour ; iris dark hazel ; legs fleshy pink ; claws pink.

Length nearly 9 ; tail 3 ; wing 4·6 ; tarsus 1·3 ; bill from gape 1·1.

Distribution. Found in summer throughout the Himalayas from Murree to the extreme east of Assam up to 5000 or 6000 feet. At other times of the year this Thrush occurs sparingly in the plains of India, extending occasionally to Ceylon, but it has not been known to occur in the Punjab, Rajputana, Sind, or Guzerat, and it appears to be extremely rare in the west and south of the peninsula. This bird is more abundant to the east, being found throughout the whole country stretching from Assam to Tenasserim, where a considerable number remain the whole year and breed. This species extends down the Malay peninsula as far as Tongkah, but does not otherwise occur outside the limits of the Empire.

Habits, &c. Breeds on the Himalayas and also in Burma from April to July, constructing a large nest of coarse grasses, roots, and fibres, in a bush or low tree, and laying three or four eggs, which are greenish white freckled with rufous, and measure about 1 by ·77.

687. Geocichla innotata. *The Malay Ground-Thrush.*

Geocichla innotata, *Blyth, J. A. S. B.* xv, p. 370 (1846), xvi, p. 146 ; *id. Cat.* p. 163 ; *Ball, S. F.* i, p. 69 ; *Hume & Dav. S. F.* vi, p. 250 ; *Hume, S. F.* viii. p. 60 ; *id. Cat.* no. 355 ter ; *Seebohm, S. F.* ix, p. 99 ; *id. Cat. B. M.* v, p. 176.

Coloration. Resembles *G. citrina,* and differs only in entirely wanting the white tips to the median wing-coverts.

Iris intense rich brown ; bill black, whitish plumbeous at base of lower mandible ; legs dull white tinged with pink, especially on the feet (*Wardlaw Ramsay*).

Of the same size as *G. citrina.*

I look upon this species as quite distinct from *G. citrina.* In the large series of this latter bird in the British Museum, I fail to find a single specimen from any part of India or Burma north of Amherst without the white tips to the wing-coverts. From Amherst southwards to Malacca spotless birds occur—as far as Tongkah in company with *G. citrina,* but south of that place by themselves.

Distribution. *G. innotata* occurs at Amherst, Toungya, Bankasun, and Malawun in Tenasserim ; in Karennee ; and down the Malay peninsula as far at least as Malacca. There are no grounds for the belief that this species occurs in the Andamans or Nicobars.

Young birds shot in Tenasserim in September and October show that this species breeds in Burma.

688. Geocichla albigularis. *The Nicobar Ground-Thrush.*

Geocichla albogularis, *Blyth, J. A. S. B.* xvi, p. 146 (1847); *Hume, S. F.* ii, p. 221 (part.); *id. S. F.* iv, p. 289 (part.); *id. Cat.* no. 355 bis (part.); *Seebohm, S. F.* ix, p. 90; *id. Cat. B. M.* v, p. 175.

Coloration. Resembles *G. citrina,* sex for sex, but differs in having the chin and throat white, and the lores and cheeks also whitish; the median wing-coverts are not tipped with white; the under tail-coverts are much tipped and otherwise marked with greenish or slaty brown; and the chestnut of the hind neck descends on to the upper back.

The colour of the bill &c. does not appear to have been recorded.

Length about 8·5: tail 2·8; wing 4; tarsus 1·2; bill from gape 1·05.

Distribution. The Nicobar Islands.

689. Geocichla andamanensis. *The Andaman Ground-Thrush.*

Geocichla albogularis, *Blyth, apud Wald. Ibis,* 1874, p. 138; *Hume, S. F.* ii, p. 221 (part.), iv, p. 289 (part.); *id. Cat.* no. 355 bis (part.).

Geocichla andamanensis, *Wald. A. M. N. H.* (4) xiv, p. 156 (1874); *Hume, S. F.* ii, p. 495; *Seebohm, S. F.* ix, p. 100; *id. Cat. B. M.* v, p. 175.

Coloration. Resembles *G. citrina.* Differs in having the forehead, crown, and nape suffused with brown, in having no white tips to the median wing-coverts, and in having the chin white.

From *G. albigularis* it differs in having the forehead, crown, and nape suffused with brown, and the throat chestnut.

Iris umber-brown; bill horny brown, whitish at base of lower mandible; legs fleshy white (*Wardlaw Ramsay*).

Length rather more than 8; tail 2·9; wing 4·1; tarsus 1·2: bill from gape 1.

Distribution. The Andaman Islands.

Genus PETROPHILA, Swains., 1837.

The genus *Petrophila* contains those Rock-Thrushes which have a short wing and a comparatively long tail. The males have the under wing-coverts and axillaries entirely of one colour and the lower plumage blue, chestnut, or black, or a combination of these colours. The females have the lower plumage squamated or irregularly barred and the under wing-coverts and axillaries also barred.

The Rock-Thrushes frequent open rocky ground and are generally solitary in their habits. They make their nests in holes of walls and rocks.

Key to the Species.*

a. Lower plumage of two colours, black
   or blue with chestnut.
   a′. Chin and throat black .... *P. erythrogastra* ♂, p. 143.
   b′. Chin and throat blue.
     a″. Large white patch on wing.... *P. cinclorhyncha* ♂, p. 144.
     b″. No white patch on wing ...... *P. solitaria* ♂, p. 145.
b. Lower plumage almost uniformly of
   one colour, barred or squamated with
   black or brown.
   c′. Upper plumage blue or suffused
     with blue.
     e″. Under wing-coverts and ax-
       illaries blue, narrowly tipped
       white..................... *P. cyanus* ♂, p. 146.
     d″. Under wing-coverts and axil-
       laries barred with black or   { *P. solitaria* ♀, p. 145.
       brown ....   { *P. cyanus* ♀, p. 146.
   d′. Upper plumage olive-brown.
     e″. Back and rump barred; wing 5. *P. erythrogastra* ♀, p. 143.
     f″. Back plain, rump barred; wing 4. *P. cinclorhyncha* ♀, p. 144.

690. **Petrophila erythrogastra.** *The Chestnut-bellied Rock-*
Thrush.

Turdus erythrogaster, *Vigors, P. Z. S.* 1831, p. 171; *Gould, Cent.*
  pl. xiii.
Petrocincla erythrogastra (*Vig.*), *Blyth, Cat.* p. 164; *Horsf. & M.*
  Cat. i, p. 185.
Orocetes erythrogastra (*Vig.*), *Jerd. B. I.* i, p. 514; *Wardlaw*
  Ramsay, P. Z. S. 1876, p. 677; *id. Ibis,* 1877, p. 463.
Petrophila erythrogaster (*Vig.*), *Hume, N. & E.* p. 227; *id. Cat.*
  no. 352; *Scully, S. F.* viii, p. 282; *Oates in Hume's N. & E.* 2nd
  ed. ii, p. 102.
Monticola erythrogaster (*Vig.*), *Seebohm, Cat. B. M.* v, p. 325; *Oates,*
  B. B. i, p. 10.
The Chestnut-bellied Thrush, Jerd.; *Niugri-pho,* Lepch.

Coloration. Male. After the autumn moult, the lores, sides of
the head and neck, and the mantle black, each feather margined
with whitish; remaining upper plumage brilliant cobalt-blue;
lesser and median coverts brown edged with cobalt-blue; greater
coverts and quills brown edged with duller blue; tail bluish brown;
chin and throat black overlaid with blue; remainder of lower
plumage maroon-chestnut. The white edges to the black portions
of the plumage soon wear off, and in spring and summer these
parts are usually black.

Female. Dull olive-brown, the feathers of the back, scapulars,
rump, and upper tail-coverts with wavy black bars and paler fringes;

* I cannot identify *Petrocincla castaneocollis*, Lesson, Rev. Zool. 1840, p. 166,
described as occurring in the Himalayas. Seebohm does not refer to this species
in his Catalogue of the Thrushes. Stoliczka, J. A. S. B. xxxvii, p. 34 note,
suggests that the type was a young *Monticola saxatilis.*

lores, centre of chin and throat, a patch on the side of the neck, and a broad but ill-defined cheek-stripe buff, each feather more or less fringed with black ; ear-coverts black with mesial buff streaks ; greater coverts and quills more or less margined white ; tail plain brown ; lower plumage, axillaries, and under wing-coverts barred with black and buff.

In the male the bill is black ; gape yellow ; iris dark brown ; feet vinous brown or black ; claws blackish : in the female the bill is dusky ; mouth and gape yellow ; iris brown ; tarsus dark brown ; toes blackish (*Scully*).

Length about 9·5 ; tail 4·2 ; wing 4·9 ; tarsus 1·1 ; bill from gape 1·2.

Distribution. A permanent resident in the Himalayas from Chamba to Bhutan ; the Khási hills ; Cachar ; Manipur ; the mountains east of Toungngoo. This species extends into Western China.

Habits, &c. Breeds from April to July, constructing a nest on the ground under a rock or stump or in a hole in a bank, and laying three eggs, which measure about 1 by ·75, and are described by Hodgson as being somewhat buff-coloured.

691. Petrophila cinclorhyncha.   *The Blue-headed Rock-Thrush.*

Petrocincla cinclorhyncha, *Vigors*, *P. Z. S.* 1831, p. 172.
Phœnicura cinclorhyncha (*Vig.*), *Gould, Cent.* pl. xix.
Monticola cinclorhyncha (*Vig.*), *Blyth, Cat.* p. 164 ; *Seebohm, Cat. B. M.* v, p. 320 ; *Oates, B. B.* i, p. 9 ; *Barnes, Birds Bom.* p. 170.
Orocetes cinclorhynchus (*Vig.*), *Horsf. & M. Cat.* i, p. 188 ; *Jerd. B. I.* i, p. 515 ; *Blanf. J. A. S. B.* xxxviii, pt. ii, p. 179.
Petrophila cinclorhynchus (*Vig.*), *Hume, N. & E.* p. 227 ; *id. Cat.* no. 353 ; *Scully, S. F.* viii, p. 282 ; *Oates in Hume's N. & E.* 2nd ed. ii, p. 103.

The Blue-headed Chat-Thrush, Jerd. ; *Krishen patti*, Nepal.

Coloration. Male. Head, from the nostrils to the nape, and the lesser wing-coverts, the chin, throat, and cheeks cobalt-blue ; lores, under the eye, ear-coverts, sides of neck, back, and scapulars black ; primaries black, all but the first two edged exteriorly with blue : secondaries black, each with a white patch on the outer web ; tertiaries wholly black ; greater coverts black, edged with faint blue ; rump, upper tail-coverts, and lower plumage, with the axillaries and under wing-coverts, chestnut ; tail blackish, edged faintly on the outer webs with bluish. In autumn most of the feathers of the black and blue portions of the plumage are fringed with pale buff and these fringes are dropped in spring and summer plumage.

Female. The whole upper plumage is olive-brown tinged with ochraceous, especially on the rump and upper tail-coverts, which are also barred with black ; wings brown, the quills ochraceous on the outer web, and the tertiaries and later secondaries margined with white ; chin and throat nearly white ; sides of the head mottled with white and brown ; remainder of lower plumage white, tinged with ochraceous on the breast, and the whole, with the exception of the abdomen, barred with dark brown; under tail-coverts white with blackish streaks.

Bill brownish black, the gape bright yellow; tarsi dusky slaty; the toes brownish black; claws blackish horny (*Scully*).

Length about 7·5; tail 2·8; wing 4; tarsus 1; bill from gape 1·1.

Distribution. Found in summer throughout the Himalayas from Afghanistan and Kashmir to Bhutan; in winter throughout the plains of India as far south as Coorg, the Nilgiris and probably to Cape Comorin. This species in the winter months is more frequent on the hill-ranges of Western India than elsewhere, but it is known to occur in almost all parts of the peninsula from Sind to Bengal. Blyth records it from Arrakan.

Habits, &c. Breeds in the Himalayas from 4000 to 8000 feet from April to June, constructing a cup-shaped nest of moss and dead leaves at the root of a tree, in a hole in a bank or in an old wall. The eggs, four in number, are pinkish white, densely freckled with brown and rufous, and measure about ·92 by ·72.

692. **Petrophila solitaria.** *The Eastern Blue Rock-Thrush.*

Turdus solitarius, *P. L. S. Müller, Syst. Nat., Anhang*, p. 142 (1776).
Turdus manillensis, *Gmel. Syst. Nat.* i, p. 833 (1788).
Petrocincla manillensis (*Gm.*), *Blyth, Cat.* p. 164; *Horsf. & M. Cat.* i, p. 188.
Cyanocincla solitaria (*Müll.*), *Hume & Dav. S. F.* vi, p. 218; *Hume, Cat.* no. 351 bis; *id. S. F.* xi, p. 125.
Monticola solitaria (*Müll.*), *Seebohm, Cat. B. M.* v, p. 319.

Coloration. Male. The whole head, neck, breast, upper plumage and lesser wing-coverts bright blue, most of the feathers with small white tips and subterminal black spots; median, greater, and primary coverts blackish, edged with blue and tipped with white; quills and tail black, edged with bluish and each feather very narrowly tipped white; abdomen, vent, under tail-coverts, axillaries, and under wing-coverts chestnut, with narrow white fringes and black subterminal bars; thighs and flank-feathers adjacent to them blue.

At the end of winter the white fringes and subterminal black bars on the blue parts of the plumage are entirely lost, and the marks on the chestnut parts are also removed by abrasion in great measure, but never entirely.

Female. After the autumn moult the whole upper plumage and lesser wing-coverts are a very dull blue, most of the feathers being fringed with white and with a subterminal black bar, and the feathers of the back with black shafts; quills and remaining wing-coverts dark brown, edged with dull blue and tipped white; the whole lower plumage and the sides of the head and neck pale buffy white, each feather subterminally margined with black; the under wing-coverts, axillaries, and under tail-coverts suffused with rufous and irregularly barred with black. In summer all the margins of the feathers become abraded, causing the plumage to become more uniform.

VOL. II.　　　　　　　　　　　　　　　　　　L

The nestling resembles the adult female, but has the margins of the feathers more extended, causing a squamated appearance. The young male assumes the chestnut of the adult very rapidly and acquires the greater part of it before the autumn moult.

The females and young of this and the next species cannot be discriminated with certainty; but the females of *P. solitaria* are generally suffused with rufous on the under wing- and tail-coverts.

Length about 9·5; tail 3·4; wing 4·9; tarsus 1·2; bill from gape 1·2.

Birds of this species in typical plumage are only found in Japan and the islands of the China seas. Further west the males always exhibit some admixture of blue with the chestnut of the lower parts. The only bird killed within Indian limits that I have been able to examine at all approaching a typical Japan bird is from the Andamans. On examining all the available specimens of Blue Rock-Thrushes killed in the Indian Empire, I find that out of 102 birds from the west of the longitude of Calcutta only 8 exhibit a trace of red; of 30 specimens from Assam down to Rangoon, only 7, and out of 72 Tenasserim birds only 27 show any red. This red is generally present on the under tail-coverts, and only in a few cases extends to the abdomen in varying quantities. The cause of this variation is unknown, but may be attributed either to climatic causes or to the interbreeding of *P. cyanus* with *P. solitaria*.

Distribution. Birds exhibiting red in the lower plumage are found in Nepal, Sikhim, Dacca, Cachar, the whole of Burma and the Andamans. This species visits the Empire in the winter only, and at this season is found also in Southern China, extending down to the Malayan islands. It breeds in Japan and Northern China.

693. **Petrophila cyanus.** *The Western Blue Rock-Thrush.*

Turdus cyanus, *Linn. Syst. Nat.* i, p. 296 (1766).
Petrocincla pandoo, *Sykes, P. Z. S.* 1832, p. 87; *Horsf. & M. Cat.* i, p. 186.
Petrocincla cyaneus (*Linn.*). *Blyth, Cat.* p. 164.
Petrocincla affinis, *Blyth, J. A. S. B.* xii, p. 177* (1843); *id. Cat.* p. 164; *Horsf. & M. Cat.* i, p. 187.
Petrocossyphus cyaneus (*Linn.*), *Jerd. B. I.* i, p. 511; *Hume & Henders. Lah. to Yark.* p. 190.
Cyanocincla cyanus (*Linn.*), *Hume, N. & E.* p. 226; *Hume & Dav. S. F.* vi, p. 247; *Hume, Cat.* no. 351.
Monticola cyanus (*Linn.*), *Anders. Yunnan Exped., Aves,* p. 611; *Legge, Birds Ceyl.* p. 460; *Seebohm, Cat. B. M.* v, p. 316; *Oates, B. B.* i, p. 11; *Barnes, Birds Bom.* p. 169.
Petrophila cyana (*Linn.*), *Oates in Hume's N. & E.* 2nd ed. ii, p. 105.

The *Blue Rock-Thrush*, Jerd.; *Shama*, Hind. in the South; *Pandu*, Mahr.; *Pala kachi pitta*, Tel.; *Nungri-pho*, Lepch.

Coloration. Male. After the autumn moult the whole plumage is bright blue, most of the feathers with white fringes and subterminal dark bars; a supercilium, the cheeks, throat, and ear-coverts brighter than the other parts; lores blackish; wings and

tail dark brown, the quills all tipped with white and edged with
bluish. In summer most, if not all, of the whitish fringes and
subterminal bars are cast, and the bird is nearly uniform blue.

Female. After the autumn moult the upper plumage, together
with the wings and tail, resemble the same parts in the male, but
are of a very dull blue; the lower plumage is pale buffy white,
each feather subterminally margined with black, the under wing-
coverts, axillaries, and under tail-coverts barred with black. In
summer most of the whitish fringes and black bars are lost.

The nestling closely resembles the adult female, but has the
white fringes to the feathers broader.

Iris hazel; eyelids plumbeous; bill blackish horn; mouth yellow;
feet black; claws dark horn.

Length about 9·5; tail 3·4; wing 4·9; tarsus 1·2; bill from
gape 1·2.

Distribution. This species, without any admixture of red in the
lower plumage, is found in the winter throughout the whole Empire. It extends to Southern Europe and Northern Africa. The
birds which are found in India and Burma appear to breed in
Afghanistan, Kashmir, and probably other parts of the Himalayas,
Turkestan, Tibet, and Western China.

Habits, &c. This Rock-Thrush frequents open, and by preference
rocky, country, and it is not unfrequently found near buildings.
Colonel C. H. T. Marshall found a nest of this bird at Murree in
a low stone wall in June. The eggs are described as pale blue
speckled with brownish red, and measured about 1·1 by ·75.

Genus **MONTICOLA**, Boie, 1822.

The genus *Monticola* differs from *Petrophila* in the proportion
between the length of the wing and tail, the former in *Monticola*
being twice the length of the latter. The Thrushes of the two
genera are quite alike in habits and in general type of coloration.

694. **Monticola saxatilis.** *The Rock-Thrush.*

Turdus saxatilis, *Linn. Syst. Nat.* i, p. 294 (1766).
Monticola saxatilis (*Linn.*), *Blyth, Cat.* p. 165; *Blanf. Ibis,* 1870,
p. 466; *Hume & Henders. Lah. to Yark.* p. 190; *Scully, S. F.* iv,
p. 139; *Hume, S. F.* xii, p. 379; *id. Cat. no.* 354 ter; *Biddulph,
Ibis,* 1881, p. 53; *Scully, Ibis,* 1881, p. 430; *Seebohm, Cat. B. M.*
v, p. 313.
Orocetes saxatilis (*Linn.*), *Horsf. & M. Cat.* i, p. 189.

Coloration. Male. The entire head and neck blue; back, scapulars, lesser wing-coverts, and rump blackish blue; the centre of
the back occupied by a large white patch; lower plumage, upper
tail-coverts, and tail chestnut, the middle pair of tail-feathers with
their terminal half brown; median and greater coverts and quills
dark brown, the coverts and secondaries narrowly tipped with

L 2

whitish. After the autumn moult the lower parts are fringed with white, and some of the feathers of the head with black.

Female. Upper plumage brown, each feather with a blackish shaft-streak and a subterminal dark bar with a pale tip; upper tail-coverts chestnut, similarly barred and tipped; tail chestnut, the middle pair of feathers brown on their terminal half; lower plumage dull white, suffused with rufous everywhere except on the throat, each feather with a wavy interrupted cross-bar near the tip; under tail- and wing-coverts and the axillaries chestnut with indistinct white tips.

Bill dusky, lower mandible yellow at base; iris brown; legs, feet, and claws black (*Scully*).

Length about 7·5; tail 2·5; wing 4·7; tarsus 1·1; bill from gape 1·1.

Distribution. Occurs in Gilgit at the autumn migration, the birds met with in this locality being chiefly young. Stoliczka obtained a specimen, probably of this species, near Dras. Blanford records this species from the banks of the Irrawaddy, near Ava, in Upper Burma.

This Rock-Thrush has an extensive range from Northern Africa and Southern Europe through Asia to China. Its migration appears to be of very limited extent.

Genus **TURDUS**, Linn., 1766.

The genus *Turdus* contains those Thrushes in which the sexes are alike and the under wing-coverts and axillaries of one colour. The three Indian species of this genus are found in Europe (and in England) and are among the best known birds of the tribe.

In *Turdus* both the wing and tail are long, and the latter is slightly graduated; the bill is small, and there is no pattern on the underside of the wing.

The Thrushes of this genus are good songsters; they are found in well-wooded country; they make cup-shaped nests in trees, using mud in the construction, and they feed largely on berries and fruit.

Key to the Species.

a. Under wing-coverts and axillaries white.
   a'. Crown and mantle brown .... *T. viscivorus*, p. 148.
   b'. Crown blue, mantle rufous .... *T. pilaris*, p. 150.
b. Under wing-coverts and axillaries rufous ...... *T. iliacus*, p. 150.

695. **Turdus viscivorus.** *The Missel-Thrush.*

Turdus viscivorus, *Linn. Syst. Nat.* i, p. 291 (1766); *Blyth, Cat.* p. 160; *Horsf. & M. Cat.* i, p. 194; *Hume, Cat.* no. 368; *Seebohm, Cat. B. M.* v, p. 194; *Biddulph, Ibis*, 1881, p. 53; *Scully, Ibis*, 1881, p. 439; *Oates in Hume's N. & E.* 2nd ed. ii, p. 106.

Turdus hodgsoni, *Homeyer, Rhea*, ii, p. 159 (1849); *Jerd. B. I.* i, p. 531; *Stoliczka, J. A. S. B.* xxxvii, pt. ii, p. 36; *Hume, N. & E.* p. 236; *Brooks, S. F.* iii, p. 247, viii, p. 471.

The Himalayan Missel-Thrush, Jerd.

Fig. 38.—Head of *T. viscivorus*

Coloration. Upper plumage greyish brown, the edges of the feathers paler, a tinge of ochraceous running through the rump and upper tail-coverts; tail ashy brown, the exterior webs narrowly edged with white, and all the feathers tipped whitish, the middle pair narrowly, the others more and more; wings brown, all the quills and coverts edged and tipped with fulvous white; lores pale fulvous; a whitish ring round the eye; ear-coverts brown streaked with fulvous; lower plumage pale buff, the chin and middle of the throat nearly spotless, the sides of the throat and the whole breast with triangular black spots, the abdomen and sides of the body with roundish spots; the under tail-coverts broadly margined at the base with brown; axillaries and under wing-coverts pure white. Birds in the summer with worn plumage are paler and greyer.

Bill dark horny brown, paler on lower mandible, which is yellowish along the margins; iris deep brown; legs and feet pale yellowish brown; claws dark horny brown (*Hume*).

Length nearly 12; tail 4·8; wing 6·4 to 6·8; tarsus 1·4; bill from gape 1·2.

Birds from Europe have the wing generally under 6 inches and the bill slightly smaller, but do not otherwise differ from Himalayan examples.

Distribution. Occurs in the Himalayas from Kashmir to Nepal. All the dated specimens that I have seen from India were killed in the summer months. Scully states that this species is met with in the Gilgit district in summer at elevations of over 9000 feet, where it breeds; and Biddulph writes that it was tolerably common in Gilgit during the severe winter of 1877-78, but seldom comes so low down, keeping generally to the higher valleys, where he found it in July at 10,000 feet. The Missel-Thrush occurs in Europe, North Africa, and a considerable part of Asia.

Habits, &c. Breeds in the Himalayas from April to June above 6000 feet. The nest is a large deep cup made of grass and dry leaves, with clay and mud, placed in trees. The eggs vary from pink to greenish grey; they are marked with brownish red and purplish pink, and measure about 1·2 by ·9.

696. **Turdus pilaris.** *The Fieldfare.*

Turdus pilaris, *Linn. Syst. Nat.* i, p. 291 (1766); *Blyth, Cat.* p. 161; *Horsf. & M. Cat.* i, p. 194; *Hume, Cat.* no. 367; *Seebohm, Cat. B. M.* v, p. 205.
Planesticus pilaris (*Linn.*), *Jerd. B. I.* i, p. 530.

Coloration. Forehead, crown, nape, and hind neck slaty grey, the feathers with narrow brown tips and darker shafts; back and scapulars chestnut-brown, with pale edges; rump and upper tail-coverts slaty grey; tail dark brown, the outer feathers very narrowly tipped white; wing-coverts dull rufous-brown with greyish margins; winglet, primary-coverts, and primaries dark brown with narrow grey margins; secondaries with the outer webs rufous, the inner brown; lores and under the eye dark brown; ear-coverts slaty grey; traces of a pale supercilium extending as far as the ear-coverts; chin, throat, and breast bright buff streaked with black; abdomen white; sides of the body white, with large roundish rufous-brown spots; axillaries and under wing-coverts pure white.

The plumage of this bird in summer differs little from the plumage in winter, the loss of the margins of the feathers causing but little change.

Bill yellow; feet and legs black; iris very dark brown (*Seebohm*). Length about 11; tail 4; wing 5·5; tarsus 1·3; bill from gape 1·1.

Distribution. The Fieldfare, according to Jerdon, has occurred once at Simla, and Adams records it from Kashmir. The only specimen I have ever seen from India is one obtained by Dr. Jameson at Saháranpur, and presented by him to the Indian Museum, from which it passed to the British Museum. It can only be considered a very rare winter visitor to the north-west of India.

The Fieldfare has a wide range, being found from the Atlantic to the Yenesay river in Siberia, and coming south in winter as far as Turkestan on the east and North Africa on the west.

697. **Turdus iliacus.** *The Redwing.*

Turdus iliacus, *Linn. Syst. Nat.* i, p. 292 (1766); *Blyth, Cat.* p. 161; *Jerd. B. I.* i, p. 532; *Hume, Cat.* no. 369; *Seebohm, Cat. B. M.* v, p. 189.

The Redwing Thrush, Jerd.

Coloration. The whole upper plumage and tail olive-brown; the wings dark brown, all the feathers edged with olive-brown, the

greater coverts most conspicuously so; a broad pale buff supercilium from the bill to the nape; lores black; ear-coverts brown
streaked with buff; chin, throat, and breast pale buff streaked
with blackish; middle of the abdomen white; sides of the abdomen
white streaked with brown; flanks, axillaries, and under wing-
coverts chestnut; under tail-coverts white, basally margined with
brown.

The summer plumage, resulting from the wear of the feathers
at their margins, does not differ very much from the winter
plumage.

Bill dark brown; legs pale; iris brown (*Scebohm*).

Length about 8·5; tail 3·3; wing 4·6; tarsus 1·2; bill from
gape 1.

Distribution. I have not been able to examine any specimen of
Redwing procured in India, and I admit the species on the authority
of Jerdon, who states that at the time he wrote it had been lately
found in the N.W. Himalayas, but very rarely. " But at Kohát,"
he adds, " as I am assured by Mr. Blyth, according to a very good
observer, the late Lieut. Trotter, it is a regular winter visitant in
large flocks."

The Redwing has even a larger range than the Fieldfare, being
found in the Northern parts of Europe and Asia in summer from
the Atlantic to the Pacific, and wandering south in the winter as
far as Turkestan and Persia on the east, and Southern Europe on
the west.

Genus OREOCINCLA, Gould, 1837.

In the genus *Oreocincla* the sexes are alike, the under wing-
coverts and axillaries are each of two colours, those on the axillaries
being transposed or reversed in order on the under wing-coverts;
the lower plumage is distinctly barred or spotted, never squamated,
and the rictal bristles are few and confined to the gape. The tail
is typically short, and the upper tail-coverts very ample. There
is a distinct pattern on the underside of the wing.

The Thrushes of this genus are permanent residents in the tracts
they inhabit, or very locally migratory. They are found in thickly
wooded parts.

The bill of the Thrushes of this genus varies much in shape and
size. In *O. dauma*, *O. mollissima*, and *O. dixoni* it is as small as
in *Turdus*; in *O. spiloptera* it is larger and very deep; and in
O. imbricata and *O. nilgiriensis* it is extremely large and coarse,
resembling the bill of *Zoothera*.

Key to the Species.

a. Feathers of upper plumage boldly tipped with
   crescentic black bars.
  a'. Ground-colour of lower plumage white
    a''. Third and fourth quills equal and longest;

second and fifth nearly equal, and about
a quarter inch shorter than longest;
wing 5·5 .... *O. dauma*, p. 152.
　　b'. Third, fourth, and fifth quills equal and
longest; second rather shorter than
sixth, and about half inch shorter than
longest; wing 5 .... *O. nilgiriensis*, p. 153.
　　b'. Ground-colour of lower plumage ochraceous
buff .... *O. imbricata*, p. 154.
t. Feathers of upper plumage plain without
darker margins or tips.
　　c'. Lower plumage with black crescentic tips:
wing over 5.
　　　e''. Wing-coverts not tipped; tail not exceeding 4·3 .... *O. mollissima*, p. 154.
　　　d''. Wing-coverts tipped; tail about 4·7 .. *O. dixoni*, p. 155.
　　d'. Lower plumage with black triangular
spots; wing about 4 .... *O. spiloptera*, p. 155.

698. Oreocincla dauma. *The Small-billed Mountain-Thrush.*

Turdus dauma, *Lath. Ind. Orn.* i, p. 362 (1790).
Oreocincla dauma (*Lath.*), *Blyth, Cat.* p. 160; *Horsf. & M. Cat.* i,
p. 193; *Jerd. B. I.* i, p. 533; *Hume, N. & E.* p. 236; *Ball, S. F.*
ii, p. 408; *Hume, S. F.* iii, p. 115; *id. & Dav. S. F.* vi, p. 256;
Ball, S. F. vii, p. 213; *Hume, Cat.* no. 371; *Oates in Hume's N.
& E.* 2nd ed. ii, p. 107.
Geocichla dauma (*Lath.*), *Seebohm, Cat. B. M.* v, p. 154; *Oates, B.
B.* i, p. 6.

Coloration. After the autumn moult the whole upper plumage
is ochraceous brown, each feather with a crescentic black bar at
the tip, preceded by a fulvous patch; wing-coverts with large
bright fulvous tips, the median series blackish above the tips;
primary-coverts black, with a broad band of fulvous on the outer
webs; quills dark brown, margined on the outer web with fulvous;
the four middle tail-feathers olive-brown, the next three pairs
blackish with white tips, the outermost feathers blackish, with the
terminal third fulvous; sides of the head pale fulvous variegated
with black; chin, middle of throat, and middle of abdomen white;
remainder of lower plumage white, tinged with fulvous, each
feather with a terminal band of black, and with a subterminal
lighter patch; under tail-coverts white, some of the feathers tipped
with black; axillaries with basal half white and terminal half
black; under wing-coverts black and terminally white.

In summer the plumage becomes very dull, the fulvous parts
fading to olive-brown.

Upper mandible and middle of lower dark brown, remainder of
bill pale brown, the gape tinged with orange; inside of mouth
yellowish; eyelid and ocular region plumbeous; iris dark hazel-
brown; legs and claws fleshy white.

Length about 10·5; tail 3·8; wing 5·6; tarsus 1·3; bill from
gape 1·2.

Distribution. The Himalayas from Hazára and Kashmir to Assam, and thence down to the central parts of Tenasserim. This Thrush is also found in the plains of India, where it has been recorded from the North-West Provinces, Behar, Bengal, Chutia Nagpur, Orissa, and Central India, extending, according to Jerdon, as far south as the Wynaad.

It is doubtful to what extent this Thrush is migratory. It breeds throughout the Himalayas, and also occurs in those mountains in winter, and it is found throughout the year in the Dhoon. From the plains of India and from Assam to Tenasserim I have seen no specimens that were killed in the summer months; but this is not improbably due to the inactivity of collectors during the latter part of the hot season and during the rains. On the whole I am inclined to think that this bird is resident on all the hill-ranges within its area of distribution, and merely descends to the adjoining plains in the winter.

Habits, &c. Breeds in the Himalayas in May and June up to 7000 feet at least. The nest is a cup, constructed of moss and lined with fern-leaves, placed in a tree. The eggs, probably three in number as a rule, are greenish white, marked with brownish and reddish purple, and measure about 1·23 by ·91.*

699. **Oreocincla nilgiriensis.** *The Nilgiri Thrush.*

Oreocincla nilgiriensis, *Blyth, J. A. S. B.* xvi, p. 141 (1847); *id. Cat.* p. 160; *Jerd. B. I.* i, p. 534; *Hume, S. F.* iv, p. 380; *id. Cat.* no. 372; *Davison, S. F.* x, p. 374; *Terry, S. F.* x, p. 474; *Oates in Hume's N. & E.* 2nd ed. ii, p. 107.
Geocichla nilgiriensis (*Blyth*), *Seebohm, Cat. B. M.* v, p. 157.

Coloration. Resembles *O. dauma* in general appearance, but has the wing shorter and more rounded, the third, fourth, and fifth quills being about equal and longest, the second rather shorter than the sixth; has a much larger bill, resembling that of *Zoothera*; has the upper plumage more rufous, with the subterminal pale patches hardly indicated, and the lower plumage less tinged with fulvous and whiter throughout.

Legs, feet, and claws dark fleshy; iris dark brown; upper mandible blackish; lower mandible brown, palest at base; gape yellow (*Davison*).

---

* 698 a. OREOCINCLA VARIA. *White's Thrush.*

Turdus varius, *Pall. Zoogr. Rosso-As.* i. p. 449 (1811).
Geocichla varia, *Seebohm, Cat. B. M.* v. p. 154.

So similar to *O. dauma* as to require no separate description, but much larger and with 14 tail-feathers.
Iris brown; upper mandible brown, lower pale; legs whitish brown (*Wardlaw Ramsay*).
Length about 12; tail 4·5; wing 6·4; tarsus 1·35; bill from gape 1·5.
Distribution. South-eastern Siberia and North China in summer; South Japan, South China, and the Philippine Islands in winter. A male specimen was procured by Wardlaw Ramsay at Toungngoo on January 11th, 1876.

Length about 10·5; tail 3·6; wing 5·2; tarsus 1·2; bill from gape 1·5.

Distribution. The hill-ranges of Southern India, from the Nilgiris to Travancore, above 2000 feet, where this Thrush appears to be a permanent resident.

Habits, &c. Represented to be a very fine songster. Breeds from March to June, constructing a nest of green moss in trees, and laying three eggs, which are greenish blue speckled with rusty brown, and measure 1·21 by ·82.

700. Oreocincla imbricata. *The Ceylon Thrush.*

Zoothera imbricata, *Layard, A. M. N. H.* (2), xiii, p. 212 (1854).
Oreocincla gregoriana, *Nevill, Hume, S. F.* i. p. 457 (1873).
Oreocincla imbricata (*Layard*), *Hume, Cat. no.* 372 quat.; *Legge, Birds Ceyl.* p. 455, pl. xix.
Geocichla imbricata (*Layard*), *Seebohm, Cat. B. M.* v, p. 159.

Coloration. Resembles *O. dauma* in general appearance, but has the upper plumage darker olive-brown, the tips of the feathers blacker, and the subterminal pale bars absent; tips of wing-coverts inconspicuous; the whole lower plumage a rich ochraceous buff, with the bars very black; the bill is much larger, resembling that of a *Zoothera.*

Iris brown; bill blackish brown, paling at the base of the lower mandible; legs and feet fleshy brown, some with a bluish tinge; claws brownish at the tips (*Legge*).

Length about 9·5; tail 3·2; wing 4·9; tarsus 1·1; bill from gape 1·4.

Distribution. Confined to Ceylon, above 3000 feet.

701. Oreocincla mollissima. *The Plain-backed Mountain-Thrush.*

Turdus mollissimus, *Blyth, J. A. S. B.* xi, p. 188 (1842).
Oreocincla mollissima (*Blyth*), *Blyth, Cat.* p. 160; *Horsf. & M. Cat.* i, p. 193; *Jerd. B. I.* i, p. 533; *Stoliczka, J. A. S. B.* xxxvii, pt. ii, p. 36; *Hume, Cat. no.* 370; *id. S. F.* xi, p. 132; *Oates in Hume's N. & E.* 2nd ed. ii, p. 108.
Geocichla mollissima (*Blyth*), *Seebohm, Cat. B. M.* v. p. 159.

Phanaiok-kiok-pho, Lepch.; *Teliakanria,* Bhut.

Coloration. The whole upper plumage rich olive-brown; wings dark brown, the feathers all broadly edged with olive-brown, the median and greater coverts very narrowly and indistinctly tipped with fulvous; the four middle tail-feathers olive-brown, the next three pairs blackish faintly tipped with white, the outermost pair black on the basal two thirds of the inner web, olive-brown elsewhere and tipped with white; a ring of fulvous feathers round the eye; sides of the head mingled black and fulvous; chin, throat, breast, and sides of the body ochraceous, each feather with a terminal black band; abdomen white, with similar bands; under tail-coverts greenish brown streaked with white; axillaries white

broadly tipped with black ; under wing-coverts black, broadly tipped
with white.

The young after the first autumn moult appear to be tinged with
rufous, and to have the spots on the wing-coverts larger than in
the adult.

Legs and feet yellowish fleshy ; claws horny brown ; bill dull
black ; base of lower mandible pale fleshy brown ; iris deep brown
(*Hume*).

Length about 11 ; tail 3·8 to 4·2 ; wing 5·5 to 6 ; tarsus 1·4 :
bill from gape 1·2.

Distribution. I have examined specimens of this Thrush from
numerous localities in the Himalayas from Chamba to Darjiling,
and from no other part. Godwin-Austen records it from the
Khási hills, and Hume is under the impression that he recognized
this species in Manipur, but he was unable to secure a specimen.

This Thrush ascends the Himalayas up to 6000 or 8000 feet in
summer, and probably descends to the lower valleys in winter.

Habits, &c. The eggs of this species are described as being
white marked with two shades of red, and measuring about 1·35
by ·88.

702. Oreocincla dixoni. *The Long-tailed Mountain-Thrush.*

Oreocincla mollissima (*Blyth*), *apud Walden in Blyth's Birds Burm.*
p. 100 ; *Hume & Dav. S. F.* vi, p. 256 ; *Wardlaw Ramsay, Ibis,*
1877, p. 463.
Geocichla dixoni, *Seebohm, Cat. B. M.* v, p. 161 ; *Oates, B. B.* i, p. 7.

Coloration. Resembles *O. mollissima*, but differs in having a
much longer tail and large and distinct fulvous tips to the median
and greater wing-coverts at all ages.

Iris brown ; bill brown ; gape yellowish ; legs dull brownish
yellow (*Wardlaw Ramsay*).

Of the same size as *O. mollissima*, with the exception of the tail,
which measures about 4·7.

Distribution. I have examined specimens of this species procured
in the hills north of Mussooree, in Nepal, at Darjiling, and in
Karennee.

703. Oreocincla spiloptera. *The Spotted-wing Thrush.*

Oreocincla spiloptera, *Blyth, J. A. S. B.* xvi, p. 142 (1847) ; *id. Cat.*
p. 160 ; *Legge, S. F.* iii, p. 367 ; *Hume, S. F.* vii, p. 382 ; *id. Cat.*
no. 372 ter ; *Oates in Hume's N. & E.* 2nd ed. ii, p. 109.
Turdus spiloptera (*Blyth*), *Legge, Birds Ceyl.* p. 451, pl. xix.
Geocichla spiloptera (*Blyth*), *Seebohm, Cat. B. M.* v, p. 167.

Coloration. Upper plumage and lesser wing-coverts rich olive-
brown tinged with russet ; tail rusty olive-brown ; wings dark
brown, broadly edged with olive-brown, the median and greater
coverts more or less black tipped with white : lores whitish ; ear-
coverts mixed white and black ; lower plumage white, tinged with

grey on the sides of the breast and body; the cheeks, sides of the throat and neck, the whole breast, and the upper half of the abdomen with fan-shaped or triangular black spots; axillaries

Fig. 39.—Head of *G. spiloptera.*

white, tipped with dark brown; under wing-coverts dark brown, tipped with white.

Iris brown; eyelid leaden grey; bill blackish, pale at gape; legs and feet dusky bluish grey or greyish fleshy; claws dusky horn (*Legge*).

Length about 8·5; tail 3·2; wing 4; tarsus 1·4; bill from gape 1·15.

Distribution. Confined to Ceylon up to 4000 feet.

Habits, &c. A nest of this bird was found by Legge in January containing two eggs bluish green freckled with rufous, and measuring 1·19 by ·79.

Genus **ZOOTHERA**, Vigors, 1831.

In the genus *Zoothera* the sexes are alike, the under wing-coverts and axillaries of two colours, the colours in the one part transposed or reversed in the other, the lower plumage squamated, not distinctly barred nor spotted, and the rictal bristles very numerous and long. The anterior or supplementary bristles extend over the nostrils as in the Flycatchers, and *Zoothera* is the only genus of Thrushes in which this feature is present. The bill is very long and strongly curved near the tip, and the edges of the mandibles are frequently serrated by wear and tear, but never originally so.

The Thrushes of this genus are non-migratory, and they are found in thickly-wooded tracts.

Key to the Species.

a. Upper plumage dark slaty: feathers of breast tipped with brown; wing 5·5 .... *Z. monticola*, p. 157.
b. Upper plumage olive-brown: feathers of breast margined with brown; wing 5 .... *Z. marginata*, p. 157.

704. **Zoothera monticola.** *The Large Brown Thrush.*

Zoothera monticola, *Vigors, P. Z. S.* 1831, p. 172; *Gould, Cent.*
pl. xxii; *Blyth, Cat.* p. 160: *Horsf. & M. Cat.* i, p. 192; *Jerd. B.*
I. i, p. 509; *Stoliczka, J. A. S. B.* xxxvii, pt. ii, p. 33; *Blanf. J. A.*
S. B. xli, pt. ii, p. 49; *Godw.-Aust. J. A. S. B.* xlv, pt. ii, p. 72;
Hume, Cat. no. 350.
Geocichla monticola (*Vig.*), *Seebohm, Cat. B. M.* v, p. 161.

Coloration. The whole upper plumage dark slaty, each feather
narrowly but distinctly margined darker, giving a scaly appearance
to the feathers; forehead and crown tinged with rufous; wings
dark brown, the quills margined with fulvous-brown and the
median and greater coverts tipped with ochraceous; sides of the
head rufescent brown, with ochraceous shafts to the feathers;
chin and middle of the throat white, each feather with a triangular
brown tip; breast and sides of the throat dark olive-brown, with
black subterminal spots, the feathers of the middle of the breast
with a good deal of fulvous at their centres; abdomen white, irregularly spotted and barred with brown; under tail-coverts dark
olive-brown, broadly tipped white; sides of the body dark olivaceous; axillaries basally white, then black, and narrowly tipped
white; under wing-coverts black, tipped white; tail blackish, obsoletely cross-barred, and the outer feathers paler.

Fig. 10.—Head of *Z. monticola*.

Legs and feet light brown; iris dark brown; bill dark brown
(*Hume*).
Length between 11 and 12; tail 3·5; wing 5·5; tarsus 1·3;
bill from gape 1·8.
Distribution. The Himalayas from the Sutlej valley to the
Daphla hills in Assam, where this bird is a permanent resident
up to 10,000 feet.

705. **Zoothera marginata.** *The Lesser Brown Thrush.*

Zoothera marginata, *Blyth, J. A. S. B.* xvi, p. 141 (1847); *id. Cat.*
p. 160; *Horsf. & M. Cat.* i, p. 192; *Godw.-Aust. J. A. S. B.* xxxix,
pt. ii, p. 268; *Hume, N. & E.* p. 226; *Blyth & Wald. Birds Burm.*
p. 100; *Hume & Dav. S. F.* xi, p. 246; *Hume, Cat.* no. 350 bis;

Bingham, S. F. viii, p. 195, ix, p. 177 ; *Oates, B. B.* i, p. 8; *Hume, S. F.* xi, p. 124 ; *Oates in Hume's N. & E.* 2nd ed. ii, p. 109.
Zoothera monticola, Bl., apud Godw.-Aust. J. A. S. B. xli, pt. ii, p. 142 (*fide Wald. in Bl. Birds Burm.* p. 100 *u*).
Geocichla marginata (Blyth), Seebohm, Cat. B. M. v, p. 162.

Coloration. The whole upper plumage olive-brown tinged with rufous, especially on the outer webs of the quills, all the feathers margined darker, causing a scaly appearance ; tips of the median and greater coverts ochraceous : tail brown, indistinctly cross-rayed, and the outer feathers paler ; sides of the head mixed fulvous and dark brown ; chin, throat, middle of breast, and abdomen white, with brown tips and margins ; sides of the throat, breast, and body dark olive-brown, the feathers of the sides of the throat and upper breast with darker centres : under tail-coverts pale buff, the feathers broadly margined with olivaceous at the sides ; axillaries basally white, then black ; under wing-coverts basally black, and then white.

Bill very dark horny brown, lower mandible reddish from angle of gonys to base ; iris dark brown ; legs and feet dark brown (*Hume Coll.*).

Length about 10 ; tail 3 ; wing 4·9 ; tarsus 1·1 ; bill from gape 1·5.

Distribution. Sikhim ; Bhutan ; the whole of the Eastern portion of the Empire from the Brahmaputra river to the extreme south of Tenasserim. This species is probably a permanent resident throughout its range.

Habits, &c. Mr. Gammie found the nest of this Thrush in Sikhim at the end of May—a cup composed of moss and fibres and placed on a low branch of a tree. The eggs were three in number, and one of these is described by Hume as being very pale greenish white much marked with ferruginous-brown and pinky purple. It measured 1·05 by ·79.

Genus COCHOA, Hodgson, 1836.

The genus *Cochoa* contains two species, the position of which remained doubtful for many years. An examination of the young of these birds, however, clearly proves, as shown by Hume, that they belong to the *Turdinæ.*

In *Cochoa* the sexes differ and the plumage of both sexes is very brilliant. The bill is short and very broad at the base ; the nostrils are large exposed ovals ; the rictal bristles are obsolete ; the wing is long and pointed, and the first primary minute ; the tail of moderate length, and the tarsus short.

The Thrushes of this genus inhabit forests, go in pairs or small flocks, feed both on the ground and on trees, and have a harsh note. They make cup-shaped nests in trees, and lay spotted eggs. They are non-migratory.

Key to the Species.

a. Crown of head lavender-blue .... *C. purpurea*, p. 159.
b. Crown of head cobalt-blue .... *C. viridis*, p. 160.

706. **Cochoa purpurea.** *The Purple Thrush.*

Cochoa purpurea, *Hodgs. J. A. S. B.* v, p. 359 (1836); *Blyth, Cat.* p. 195; *Horsf. & M. Cat.* i, p. 309; *Jerd. B. I.* ii, p. 243; *Hume, N. & E.* p. 388; *Hume & Dav. S. F.* vi, p. 357; *Hume, Cat.* no. 607; *Sharpe, Cat. B. M.* iv, p. 3; *Oates, B. B.* i, p. 136; *id. in Hume's N. & E.* 2nd ed. ii, p. 110.

The Purple Thrush-Tit, Jerd.; *Cocho,* Nep.; *Lit-nyam-pho,* Lepch.

Coloration. Male. Forehead, crown, and nape lavender-blue; a narrow frontal band, lores, and sides of the head black, this colour extending narrowly round the hind neck; the upper plumage ashy purple; wing-coverts, winglet, tertiaries, and the basal half, or more, of the outer webs of most of the primaries and of all the secondaries dull lavender; primary-coverts and remainder of wing black; tail lavender-blue, the inner webs of all but the median pair of feathers mostly black, and all tipped black; the whole lower plumage purplish brown, inclining to black on the chin and throat.

Female. Forehead, crown, and nape lavender-blue; a narrow frontal band, lores, and sides of the head black, this colour extending narrowly round the hind neck; upper plumage reddish brown; wing-coverts, tertiaries, and the basal half, or more, of the outer webs of the secondaries reddish brown; the basal portion of the outer webs of the primaries blue; remainder of the wing dark brown; winglet and the larger coverts near the edge of the wing and the bases of the primary-coverts suffused with blue; tail as in the male; lower surface reddish brown, paler than the upper plumage.

The young male has the wings and tail like the adult male; the forehead, crown, and nape black, barred with white; the upper plumage black, with fulvous streaks on the scapulars and lesser wing-coverts; the sides of the head black, with a white patch on the ear-coverts; the whole lower plumage bright reddish brown cross-barred with black. A young male in August has nearly lost the fulvous streaks on the scapulars and lesser wing-coverts; the back is turning to ashy purple; the head is still barred with white, and the lower plumage is still barred as in the young bird above described. This plumage is probably retained throughout the winter.

The young female has the wings and tail like the adult female; the forehead, crown, and nape black, each feather with a broad subterminal white band; the upper plumage reddish brown, with pale fulvous stripes, which become elongated drops on the scapulars; sides of the head black; ear-coverts centrally white; lower plumage reddish brown cross-barred with black.

Legs and feet dark plumbeous, shaded with black; claws dark horny-brown; bill black; gape dark plumbeous; iris red-brown; eyelids dark plumbeous (*Hume & Davison*).

Length about 11; tail 4·5; wing 5·7; tarsus 1·1; bill from gape 1·3.

Distribution. The Himalayas from Kumaun to Sikhim; Muleyit mountain and the Thoungyeen valley in Tenasserim. This species is no doubt a permanent resident up to at least 8000 feet.

Habits, &c. A nest said to belong to this species, and found by Mr. Horne at Binsur, in Kumaun, was a cup made of moss placed in a small tree. The egg was greenish thickly blotched with brown.

707. **Cochoa viridis.** *The Green Thrush.*

Cochoa viridis, *Hodgs. J. A. S. B.* v, p. 359 (1836); *Blyth, Cat.* p. 194; *Jerd. B. I.* ii, p. 243; *Hume, N. & E.* p. 389; *id. Cat.* no. 608; *Sharpe, Cat. B. M.* iv, p. 2; *Hume, S. F.* xi, p. 239; *Oates in Hume's N. & E.* 2nd ed. ii, p. 111.

The Green Thrush-Tit, Jerd.

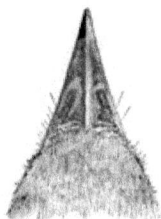

Fig. 41.—Bill of *C. viridis*.

Coloration. Male. Forehead, crown, nape, and hind neck brilliant cobalt-blue; lores and a short supercilium black; ear-coverts indigo-blue; cheeks greenish blue; the whole upper plumage and scapulars varying in different individuals, according to age, from deep green to golden-yellow, the feathers of the back and scapulars fringed with black; median pair of tail-feathers purplish blue tipped black, the next four pairs black on the inner web, blue on the outer and tipped black, the outermost pair entirely black; lesser wing-coverts green narrowly tipped black, the median with broader tips; winglet black; primary-coverts with the inner webs black, the basal two thirds of the outer blue and the terminal third black; greater coverts with the inner webs greenish brown tipped black, the outer with the basal two thirds green, broadly edged with blue, the remainder black; first two primaries black; the remaining quills black, with a long patch of blue at the base of each outer web; the whole lower plumage green, suffused with blue on the abdomen and with oil-yellow on the breast and flanks.

Female. Differs from the male in the coloration of the wing only. The secondaries and tertiaries have the basal patch on the outer webs yellowish brown instead of blue, and this colour occupies the major portion of the outer webs of the greater coverts, the blue edgings being very narrow.

The young bird (perhaps the female nestling only, the male resembling the adult male probably in the coloration of the wing) has the crown, forehead, and nape bluish brown, each feather tipped black and with a subterminal band of white; the upper plumage greenish brown, with large buff spots; supercilium black; ear-coverts white, with blackish tips; the whole lower plumage buff, irregularly and narrowly cross-barred with black; the wings and tail as in the adult female.

A nearly adult bird killed in Sikhim in March, and one in Manipur in May, have the cheeks, ear-coverts, and a demi-collar at the side of the neck pure white; the chin and throat also are whitish, and the lower plumage bright chestnut-brown. According to Hume this is the second stage of plumage. There does not appear to be anything analogous to it, however, in *C. purpurea*.

Iris brown, brownish orange, dull brownish maroon; legs brown, with a varying amount of pinkish tinge; bill black; gape and orbital skin pink (*Hume Coll.*).

Length about 11; tail 4·8; wing 5·6; tarsus 1·1; bill from gape 1·2.

Distribution. The Himalayas from Kumaun to Sikhim up to 11,000 feet; Cachar; Manipur. This species extends into China. Hume gives the Bhágirati valley as the western limit of this Thrush, but there is no specimen from this locality now in his collection, and I have seen none from further west than Jeoli below Naini Tal.

Habits, &c. The nest of this species is described as being a cup composed of fine twigs and roots coated externally with moss, and built on branches of large trees ten to twenty feet from the ground. An egg was greyish green marked with red, and measured 1·03 by ·75.

Subfamily CINCLINÆ.

The *Cinclinæ* or Dippers appear to be allied to the Thrushes, but to have undergone some modification of structure to adapt them to a different mode of life.

In the Dippers the bill is about as long as the head, narrow and straight, the tip slightly bent down and notched; the nostrils are covered by a large membrane and the rictal bristles are entirely absent; the wing is very short and rounded; the tail exceedingly short; the tarsus long and smooth.

The sexes are alike and the young are spotted. These do not assume the adult plumage till the first spring of their life, and the change is effected by the casting-off of the margins of the feathers.

162   TURDIDÆ.

The Dippers are aquatic in their habits, and they are admirably fitted for obtaining their food in the water. The plumage is everywhere very dense and even the eyelids are clothed with feathers; the head is narrowed in front, and the feathers of the forehead are very short and lie flat.

The Dippers frequent mountain-streams, and the Indian species do not migrate. They build large domed nests of moss amongst rocks or between the roots of trees near the water, and they lay numerous white eggs.

Genus CINCLUS, Bechst., 1802.

The characters of the genus are the same as those of the subfamily.

Key to the Species.

a. Throat and breast white . . . .   C. kashmiriensis, p. 162.
b. Throat and breast brown, uniform with remainder of lower plumage.
  a'. Plumage light-coloured . . . .   C. asiaticus, p. 163.
  b'. Plumage dark-coloured . . . .   C. pallasi, p. 164.
c. Throat and breast brown, but conspicuously paler than remainder of lower plumage ..   C. sordidus, p. 165.

708. Cinclus kashmiriensis.   The White-breasted Asiatic Dipper.

Cinclus cashmeriensis, Gould, P. Z. S. 1859, p. 494; Salvin, Ibis, 1867, p. 117; Blanf. J. A. S. B. xli, pt. ii, p. 48; Hume, Cat. no. 348; Sharpe, Cat. B. M. vi, p. 312; Scully, Ibis, 1881, p. 438.
Hydrobata cashmeriensis (Gould), Jerd. B. I. i, p. 507; Blyth, Ibis, 1866, p. 374; Stoliczka, J. A. S. B. xxxvii, pt. ii, p. 35; Hume & Henders. Lah. to Yark. p. 189.
The White-breasted Cashmere Dipper, Jerd.

Coloration. Forehead, crown, nape, lores, sides of the head and neck, and the whole mantle chocolate-brown; remainder of the upper plumage slate-colour, each feather distinctly margined with black; the mantle and back blending gradually together, the brown of the former suffusing the upper part of the latter; wings dark brown, the outer webs edged with slate-colour, and the secondaries and tertiaries tipped with white; wing-coverts dark brown, broadly edged with slaty; tail slaty, the shafts dark; cheeks, chin, throat, and breast pure white; remainder of the lower plumage chocolate-brown, gradually turning to dark brown or blackish towards the tail.

The young have the whole upper plumage slate-colour with black margins; the wing-coverts tipped white; the quills more broadly tipped with white than in the adult; the whole lower plumage white with numerous irregular cross-lines of brown. After the autumn moult the young resemble the adult, but the abdomen is dark brown without any tinge of chocolate immediately next the white of the breast, as is the case in the adult, and each feather of the vent, abdomen, and under tail-coverts narrowly margined with white; tips to quills broader than in the fully adult. Early in the first spring the white margins and tips are cast and the full plumage donned.

Legs and feet dark brown; bill black (*G. Henderson*).

Length about 8; tail 2·1; wing 3·8; tarsus 1·15; bill from gape ·95.

This species extends westward to Asia Minor, and is closely allied to the three races of Dipper which are found in Europe. It may be distinguished from them by the absence of rufous on the abdomen immediately next the white breast, and further by the brown of the mantle extending some distance down the back and blending with the colour of the latter. To the north, the present form tends to run into *C. leucogaster*, in typical examples of which the whole lower parts are white. In some specimens there is a tendency towards *C. sordidus*, the white of the throat and breast being infuscated and occasionally these parts are quite brown.

Distribution. The Himalayas from Gilgit to Sikhim from 9000 to 14,000 feet or higher, according to season. This Dipper extends on the west to Asia Minor and on the east to China, and it has a very extended range through Central Asia.

709. Cinclus asiaticus. *The Brown Dipper.*

Cinclus asiaticus, *Swains. Faun. Bor.-Amer., Birds*, p. 174 (1831); *Salvin, Ibis*, 1867, p. 120; *Blanf. J. A. S. B.* xli, pt. ii, p. 48; *Hume, Cat.* no. 347; *Sharpe, Cat. B. M.* vi, p. 314; *Scully, S. F.* viii, p. 281; *id. Ibis*, 1881, p. 457; *Oates in Hume's N. & E.* 2nd ed. ii, p. 112.

Hydrobata asiatica (*Swains.*), *Blyth, Cat.* p. 158; *Horsf. & M. Cat.* i, p. 185; *Jerd. B. I.* i, p. 506; *Hume, N. & E.* p. 225; *Hume & Henders. Lah. to Yark.* p. 188; *Godw.-Aust. J. A. S. B.* xlv, pt. ii, p. 203; *Biddulph, Ibis*, 1881, p. 52.

The Brown Water-Ouzel, Jerd.; *Namboay karriak*, Lepch.; *Chubia rakba*, Bhut.

Fig. 42.—Head of *C. asiaticus*.

Coloration. The whole plumage chocolate-brown, the edges of the feathers somewhat paler in places; the eyelids covered with white feathers; wings and tail dark brown, edged with the same colour as that of the plumage in general, the later quills tipped white.

The young in March, just fledged, have the upper plumage grey, the feathers tipped with black and the subterminal portion more or less whitish; wings black, the quills and coverts all tipped and margined with white; tail very dark brown, tipped white; sides of the head grey with white shaft-streaks; the lower parts grey and marked in the same manner as the upper plumage, but with more conspicuous black tips and subterminal white patches. In August the nearly adult plumage is assumed by a moult, but the new feathers of each side of the head, of the throat, breast, and middle of the abdomen, and sometimes those of the crown and back, are fringed with white, and the wing-coverts and quills are conspicuously margined with white. The fully adult plumage is assumed in the first spring by the casting of the white fringes.

Bill black; legs pale brown; soles of the feet yellow; iris dark brown (*Jerdon*).

Length about 8; tail 2·2; wing 3·9; tarsus 1·15; bill from gape 1·05.

Distribution. The Himalayas from Afghanistan and Gilgit to Bhutan at elevations from 1000 to 14,000 feet, according to season. This species occurs in the North Khási hills, specimens from that locality being in the National Collection. It extends into Turkestan.

Habits, &c. Breeds throughout the Himalayas at all levels from December to May. The nest is a rounded ball of moss lined with ferns and roots, with an opening at the side, wedged into a cleft of a rock near water and not far above its surface. The eggs, generally five in number, are white and measure about 1 by ·72.

710. Cinclus pallasi. *Pallas's Dipper.*

Cinclus pallasii, *Temm. Man. d'Orn.* ed. 2, i, p. 177 (1820); *Salvin, Ibis,* 1867, p. 118; *Hume, S. F.* vii, p. 378; *id. Cat.* no. 349 bis; *Sharpe, Cat. B. M.* vi, p. 316; *Hume, S. F.* xi, p. 124.

Coloration. The whole plumage with the lesser wing-coverts a very rich dark chocolate-brown; the eyelids clothed with white feathers; the abdomen blackish; greater wing-coverts dark brown, edged with chocolate-brown; wings and tail blackish, suffused with chocolate-brown on the outer webs.

The young differ markedly from those of *C. asiaticus*. The whole upper plumage and the sides of the head and neck are blackish brown with subterminal rufous margins; the wings and coverts with white, or on some of the feathers slightly rufous, edges; tail black, narrowly tipped with white; the whole lower plumage blackish brown, with ashy fringes to all the feathers.

Another bird, which has just completed its first autumn moult, resembles the adult, but the throat, breast, and middle of the abdomen are mottled with white and the wings retain their white edges.

Iris hazel; bill horny; legs plumbeous in front, dusky behind (*Cockburn*).

Length about 8; tail 2·1; wing 3·9; tarsus 1·15; bill from gape 1·15.

Distribution. In the British Museum there is an adult procured by Cockburn at Shillong and two young birds obtained by A. W. Chennell in the North Khási hills. The latter were shot in March and are those described above. In the collection of Godwin-Austen are three specimens of this species, two adults, procured at Shillong in April and May respectively, and a young bird procured in North Cachar without date. I have seen no other specimens from Indian limits. This species is found in China and the whole of North-eastern Asia.

711. Cinclus sordidus. *The Sombre Dipper.*

Cinclus sordidus, *Gould, P. Z. S.* 1859, p. 494; *Salvin, Ibis,* 1867, p. 118; *Blanf. J. A. S. B.* xli, pt. ii, p. 48; *Hume, Cat.* no. 349; *Sharpe, Cat. B. M.* vi, p. 317.
Hydrobata sordida (*Gould*), *Jerd. B. I.* i, p. 507.

The Black-bellied Cashmere Dipper, Jerd.

Coloration. Forehead, crown, nape, mantle, sides of the head and neck chocolate-brown; chin, throat, and upper part of breast the same but strikingly paler; upper plumage and wing-coverts blackish brown; wings and tail blackish, the outer webs somewhat paler; lower plumage very dark chocolate-brown; the eyelids are probably clothed with white feathers as in the other members of the genus, but the type specimen, the only one I have been able to examine, exhibits no trace of them.

Length about 7; tail 2; wing 3·5; tarsus 1·1; bill from gape ·9.

Distribution. This species was founded on a specimen procured in Kashmir. Blanford appears to have met with this Dipper in Sikhim at 15,000 feet, and Hume records its occurrence in the tract of country traversed by Dr. Henderson (Lah. to Yark. p. 189). I can find no other notice of it, and this species appears to be very rare. It has been met with in Kansu and Northern Tibet.

Subfamily ACCENTORINÆ.

The *Accentorinæ* or Accentors comprise a number of birds the position of which has been much disputed. Looking to their habits and to the colour of the nestling, their position appears to me to be among the Thrushes.

The Accentors have a bill about half the length of the head, wide at base, compressed somewhat abruptly in the middle, the culmen nearly straight, the upper mandible terminating in rather a fine point and slightly notched; the nostrils large, diagonal, and

covered by a large membrane ; the rictal bristles few and weak ;
the feathers of the forehead slightly disintegrated ; the tail nearly
square or sometimes slightly forked ; the tarsus strongly scutellated.

The sexes are alike, and some species have a seasonal change of
plumage caused by the wearing away, in winter, of the margins of
the feathers. The young moult into adult plumage at the first
autumn. The nestlings of the various species resemble each other
closely, and may be described as pale rufous below, densely streaked
with dark brown, especially on the breast and sides of the body,
the chin being frequently barred. The upper plumage is dark
brown, each feather edged with rufous.

The majority of the Accentors inhabit mountains at considerable
elevations ; others, like the common Accentor or Hedge-Sparrow
of England, inhabit gardens and cultivated spots. They feed on
insects, and also, it is said, on small seeds. They build their nests
in bushes or in holes of rocks, and lay blue eggs.

Key to the Genera.

a. Wing large and pointed, longer than tail by
   more than length of tarsus.............. ACCENTOR, p. 166.
b. Wing small and blunt, longer than tail by
   much less than length of tarsus.......... THARRHALEUS, p. 168.

Genus ACCENTOR, Bechst., 1802.

The genus *Accentor*, in addition to the characters of the subfamily,
has a long and pointed wing. The wing is longer than the tail by
more than the length of the tarsus, and the secondaries fall short
of the tip of the wing also by about the length of the tarsus.

The Accentors of this genus are more or less migratory.

Key to the Species.

a. Breast uniformly greyish brown.......... *A. nepalensis*, p. 166.
b. Breast rufous, feathers edged with white .. *A. himalayanus*, p. 168.

712. Accentor nepalensis. *The Eastern Alpine Accentor.*

Accentor nipalensis, *Hodgs., Blyth, J. A. S. B.* xii, p. 958 (1843);
Blyth. Cat. p. 130; *Horsf. & M. Cat.* i. p. 359; *Jerd. B. I.* ii, p. 286;
Blanf. J. A. S. B. xli, pt. ii, p. 63; *Hume & Hend. Lah. to Yark.*
p. 234; *Hume, Cat.* no. 652; *Biddulph, Ibis,* 1881, p. 74; 1882,
p. 284; *Scully, Ibis,* 1881, p. 568; *Sharpe, Cat. B. M.* vii, p. 664.
Accentor cachariensis, *Hodgs. P. Z. S.* 1845, p. 34.

The Large Himalayan Accentor, Jerd.

Coloration. Forehead, crown, nape, and hind neck greyish

brown, with indistinct darker streaks ; back dark brown, edged
with rufous-brown ; rump and upper tail-coverts pale rufous, with
blackish shaft-streaks ; lesser wing-coverts greyish brown ; all the
other coverts blackish, tipped with white ; scapulars and tertiaries
black, edged with ferruginous ; the other quills dark brown,

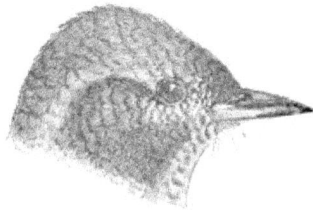

Fig. 13. — Head of *A. nepalensis.*

narrowly edged and tipped with rufous ; tail dark brown, each
inner web tipped with a spot, which is white on the outer feathers
and gradually turns to rufous on the inner ; chin and throat white
barred with black ; sides of head, sides of neck, and the breast
greyish brown ; the region of the eye speckled with white ; middle
of the abdomen rufous-grey, barred with white and brown ; under
tail-coverts chestnut-brown, broadly edged with white ; sides of
body and flanks dark ferruginous, some of the feathers near the
thighs narrowly margined with white.

Base of upper mandible from nostril to gape, the gape, and base
of lower mandible bright yellow ; rest of bill black ; iris very dark
brown ; legs and feet very pale reddish brown, almost fleshy
(*Hume*).

Length about 7 ; tail 2·8 ; wing 4 ; tarsus ·95 ; bill from gape
·65.

This species is allied to the European *A. collaris*, from which it
differs in being very richly coloured, and in having the second
primary equal to or shorter than the sixth, whereas in *A. collaris*
the second primary is much longer than the sixth. *A. nepalensis*
has, moreover, few white margins on the flanks, and is not so much
barred beneath. Gilgit examples of *A. nepalensis* are paler than
typical birds, but do not otherwise differ. Afghan specimens are
still paler, and in Asia Minor an intermediate race is found. *A.
rufilatus*, Severtzow, from Turkestan, appears to me to be identical
with *A. nepalensis*. *A. erythropygius*, Swinhoe, from China, differs
from the present species in having very rufous upper tail-coverts,
and is doubtfully distinct.

Distribution. The Himalayas from Afghanistan and Gilgit to
Sikhim, at very high elevations, Blanford recording this species
from 14,000 feet. I have seen specimens from Sikhim killed in
every month of the year, but in Gilgit this Accentor is represented
to be merely a winter visitor.

713. **Accentor himalayanus.** *The Altai Accentor.*

Accentor himalayanus, *Blyth*, J. A. S. B. xi. p. 187 (1842).
Accentor altaicus, *Brandt*, *Bull. Acad. St. Petersb.* i, p. 365 (1843);
　Jerd. B. I. ii, p. 287; *Stoliczka*, J. A. S. B. xxxvii, pt. ii, p. 52;
　Hume, Cat. no. 653; *Biddulph, Ibis*, 1881, p. 74; *Sharpe, Cat. B.
　M.* vii, p. 600.
Accentor variegatus, *Blyth*, J. A. S. B. xii, p. 958 (1843); *id. Cat.*
　p. 131; *Horsf. & M. Cat.* i, p. 359.

The Himalayan Accentor, Jerd.

Coloration. Forehead, crown, nape, and hind neck greyish brown, with darker shaft-streaks; a pale but distinct greyish supercilium; ear-coverts rufous, with pale shafts; all the feathers under the eye speckled with white; back, scapulars, and tertiaries black edged with rufous; rump greyish brown, with obsolete darker shaft-streaks; upper tail-coverts and tail dark brown, edged with rufous, the inner web of each feather tipped with white or rufous; wing-coverts blackish, more or less edged with rufous and tipped with white; quills dark brown, edged with rufous; middle of chin and throat pure white; sides of these parts banded with black; feathers of lower throat tipped with black, forming a small collar; sides of neck greyish brown; remainder of lower plumage ferruginous, each feather edged with white; the middle of the abdomen almost pure white; the feathers of the flanks and under tail-coverts with broader white edges.

Base of bill at gape and the gape fleshy; rest of bill dull black; legs and feet brownish fleshy; claws dull black; iris carmine-red or cinnabar-red (*Hume*).

Length about 6; tail 2·4; wing 3·7; tarsus ·85; bill from gape ·6.

Distribution. The Himalayas from Chamba and Gilgit to Sikhim. This species occurs in Sikhim throughout the year, but probably at various altitudes according to season. At Simla it appears to be found only in winter, and it visits Gilgit at the same season. It occurs throughout a considerable portion of Central Asia.

Genus **THARRHALEUS**, Kaup, 1829.

The genus *Tharrhaleus* contains those Accentors which have a blunt and feeble wing. The wing is longer than the tail, but by a distance much less than the length of the tarsus, and the secondaries fall short of the tip of the wing by a distance equal to about half the length of the tarsus.

The Accentors of this genus do not migrate to such an extent as those of the genus *Accentor*, and some are resident.

Key to the Species.

a. Upper plumage un-treaked .... *T. immaculatus*, p. 169.
b. Upper plumage streaked.
  a'. No supercilium .... *T. rubeculoides*, p. 169.
  b'. A supercilium.
    a''. Chin and throat black.............. *T. atrigularis*, p. 170.
    b''. Chin and throat white or fulvous.
      a'''. Breast of same colour as remainder
        of lower plumage............... *T. fulvescens*, p. 171.
      b'''. Breast ferruginous, quite different
        to other parts of lower plumage.
        a⁴. Supercilium and breast deep ferru-
          ginous ; crown and rump dis-
          tinctly streaked .... *T. strophiatus*, p. 171.
        b⁴. Supercilium and breast pale ferru-
          ginous ; crown and rump obso-
          letely streaked .... *T. jerdoni*, p. 172.

714. **Tharrhaleus immaculatus.** *The Maroon-backed Accentor.*

Accentor immaculatus, *Hodgs. P. Z. S.* 1845, p. 34 ; *Horsf. & M.
Cat.* i, p. 361 ; *Jerd. B. I.* ii, p. 286 ; *Hume, Cat.* no. 651 ; *Sharpe,
Cat. B. M.* vii, p. 656.
Accentor mollis, *Blyth, J. A. S. B.* xiv, p. 581 (1845) ; *id. Cat.* p. 131.

Coloration. Forehead, crown, and nape dark ashy, the feathers
of the forehead margined with white ; upper back olive-brown
tinged with rufous, gradually changing to maroon on the lower
back, rump, scapulars, and outer webs of tertiaries and later
secondaries ; upper tail-coverts olive-brown tinged with rufous ;
tail greyish brown ; wing-coverts dark grey ; primary-coverts and
winglet black ; quills dark brown, the earlier primaries edged with
grey on the outer webs above the emarginations ; lores black ;
ear-coverts slaty brown ; chin, throat, breast, sides of neck, upper
abdomen, and sides of body ashy brown ; lower abdomen, vent,
and under tail-coverts dark chestnut ; thighs slaty.

Iris whitish yellow ; bill black ; legs horny (*Hume Coll.*).

Length about 6 ; tail 2·5 ; wing 3·2 ; tarsus ·9 ; bill from
gape ·6.

Distribution. Nepal and Sikhim. The specimens of this species
that I have examined from the latter country were killed from
October to April. It extends to Western China and Moupin.

715. **Tharrhaleus rubeculoides.** *The Robin Accentor.*

Accentor rubeculoides, *Hodgs., Moore, P. Z. S.* 1854, p. 118 ; *Horsf.
& M. Cat.* i, p. 361 ; *Jerd. B. I.* ii, p. 288 ; *Stoliczka, J. A. S. B.*
xxxvii, pt. ii, p. 53 ; *Blanf. J. A. S. B.* xli, pt. ii, p. 64 ; *Hume &
Henders. Lah. to Yark.* p. 234 ; *Hume, Cat.* no. 656 ; *Sharpe,
Cat. B. M.* vii, p. 657.

Coloration. Forehead, crown, nape, and sides of head brown ;
back, scapulars, and rump reddish brown with broad black streaks ;
upper tail-coverts nearly plain brown ; tail brown, edged paler ;

lesser and median wing-coverts ashy, centred darker and tipped whitish ; greater coverts dark brown, broadly edged with rufous and tipped whitish : quills dark brown, edged with rufous ; chin, throat, and sides of neck ashy brown with the bases of the feathers darker, causing a mottled appearance ; breast deep ferruginous : abdomen whitish or pale fulvous : sides of the body and under tail-coverts rufous streaked with brown.

Iris clear pale brown ; bill black ; legs reddish brown (*Blanford*).

Length about 6·5 ; tail 2·7 ; wing 3·2 ; tarsus ·9 ; bill from gape ·6.

Distribution. The Himalayas from Eastern Kashmir and Ladák to Sikhim. Blanford found this species above 14,000 feet, and I have seen specimens from Sikhim killed in June and October and in the adjoining parts of Tibet in February, March, and September : it breeds in Tibet, as Mandelli procured quite a young bird in September. Dr. Henderson met with this species at Sukti in July and at Tankse in October.

716. Tharrhaleus atrigularis. *The Black-throated Accentor.*

Accentor atrigularis, *Brandt, Bull. Acad. St. Petersb.* ii, p. 140 (1844) ; *Blyth, Cat.* p. 131 ; *Hume, Cat.* no. 655 ; *Biddulph, Ibis,* 1881, p. 75 ; *Scully, Ibis,* 1881, p. 569 ; *Sharpe, Cat. B. M.* vii, p. 656.
Accentor huttoni, *Moore, P. Z. S.* 1854, p. 119 ; *Horsf. & M. Cat.* i, p. 360 ; *Jerd. B. I.* ii, p. 288 ; *Stoliczka, J. A. S. B.* xxxvii, pt. ii, p. 53 ; *Scully, S. F.* iv, p. 155.

Coloration. After the autumn moult the forehead, crown, nape, back, and scapulars dark brown, each feather with lateral fulvous margins ; rump and upper tail-coverts ashy brown ; a broad black band on each side of the crown, below which runs a broad buff supercilium ; lores, cheeks, ear-coverts, and under the eye dark brown or blackish ; a narrow moustachial streak buff, joining the breast and enclosing the chin and throat, which are black with narrow white margins ; the whole breast and sides of the throat ochraceous buff, with partially concealed black bases to all the feathers ; middle of the abdomen creamy white ; sides of the body and under tail-coverts buff streaked with rich brown ; tail brown, edged paler ; quills and coverts brown, margined with fulvous or rufous and the coverts more or less tipped paler. In summer the buff tips of the moustachial streaks wear away, as also the whitish margins of the chin and throat, which parts become quite black and are joined to the black of the sides of the head. The front part of the supercilium in a similar manner becomes black and the hinder part bleaches to white.

Iris hazel-brown ; base of upper and lower mandibles at gape, and the gape, legs, and feet brownish fleshy : rest of bill dull black ; claws horny brown (*Hume*).

Length about 6 ; tail 2·6 ; wing 2·9 ; tarsus ·8 ; bill from gape ·55.

Distribution. The Himalayas from Afghanistan and Gilgit to Garhwál. Jerdon records this species from the Punjab Salt-Range. This Accentor is a winter visitor to the Himalayas, summering in Turkestan and other parts of Central Asia.

717. Tharrhaleus fulvescens. *The Brown Accentor.*

Accentor fulvescens, *Severtz. Turkest. Jevotu.* pp. 66, 132 (1873); *id.
S. F.* iii, p. 428; *Biddulph, Ibis,* 1881, p. 75; *Scully, Ibis,* 1881,
p. 569; *Biddulph, Ibis,* 1882, p. 281, pl. viii; *Sharpe, Cat. B. M.*
vii, p. 655.
Accentor montanellus (*Pall.*), *apud Scully, S. F.* iv, p. 155; *Hume,
Cat.* no. 655 bis.

Coloration. Forehead, crown, and nape uniform dark brown, or sometimes dark brown with the feathers edged paler; a broad supercilium white or buffy white; lores, ear-coverts, and under the eye dark brown; back and scapulars ashy brown, sometimes tinged with fulvous, each feather with a broad dark brown streak; rump and upper tail-coverts plain ashy brown; wings and tail brown, edged with fulvous; the wing-coverts and tertiaries tipped with buffy white; lower plumage rich ochraceous buff, the flanks with a few darker streaks.

Bill black, brownish at base below; iris very dark brown; legs and feet fleshy; claws dusky, yellowish at tips (*Scully*).

Length about 6; tail 2·6; wing 3; tarsus ·8; bill from gape ·6.

T. montanellus, Pallas, from Lake Baikal and Amurland, differs from the present species in having the back and scapulars rich chestnut-brown, the feathers edged with pale ashy, and the breast-feathers with partially concealed black bases. In both species the colour of the supercilium is liable to variation from buff to white.

Distribution. Gilgit and Sikhim in winter only. Mandelli procured this species in the country north of Sikhim. It summers in Turkestan, Mongolia, and Southern Siberia.

718. Tharrhaleus strophiatus. *The Rufous-breasted Accentor.*

Accentor strophiatus, *Hodgs., Blyth, J. A. S. B.* xii, p. 959 (1843);
Blyth, Cat. p. 131; *Horsf. & M. Cat.* i, p. 369; *Jerd. B. I.* ii,
p. 287; *Hume, N. & E.* p. 401; *id. Cat.* no. 654; *Sharpe, Cat.
B. M.* vii, p. 658.
Accentor multistriatus, *David, A. M. N. H.* (4), vii, p. 256 (1871).
Tharrhaleus strophiatus (*Hodgs.*), *Oates in Hume's N. & E.* 2nd ed.
ii, p. 113.

Phoocking-pho, Lepch.

Coloration. The whole upper plumage rufous-brown streaked with black; wings dark brown edged with rufous, the coverts tipped with fulvous on the outer web; tail brown; lores, cheeks, and ear-coverts black; a broad supercilium, white in front of the eye, deep ferruginous behind, bordered above by a black band; chin and throat white, spotted with black chiefly at the sides; sides

of neck ashy streaked with black ; breast deep ferruginous ; middle
of abdomen whitish ; remainder of lower plumage reddish brown
streaked with black.

Many birds have the breast streaked with black ; they are probably birds of the first winter.

Bill black ; legs reddish brown ; iris dark brown (*Jerdon*).

Length about 5·5 ; tail 2·3 ; wing 2·7 ; tarsus ·8 ; bill from
gape ·55.

Distribution. Sikhim and Nepal, extending west along the
Himalayas to Kotgarh. This species appears to be a constant
resident in the higher portions of the Himalayas and to breed
both in Nepal and Sikhim. It is found throughout Tibet and
Western China.

Habits, &c. According to Hodgson this species makes a cup-
shaped nest of grass-roots and moss, lined with wool and hair, in
tufts of grass, and lays three or four eggs, which measure about
·74 by ·54.

719. Tharrhaleus jerdoni. *Jerdon's Accentor.*

Accentor jerdoni, *Brooks, J. A. S. B.* xli, pt. ii, p. 327 (1872) ; *Hume,*
  N. & E. p. 408 ; *id. S. F.* iv, p. 491 ; *id. Cat.* no. 654 bis ; *Bid-*
  dulph, Ibis, 1881, p. 75 ; 1882, p. 281 ; *Scully, Ibis,* 1881, p. 569 ;
  Sharpe, Cat. B. M. vii, p. 660.
Accentor strophiatus, *Hodgs., Stoliczka, J. A. S. B.* xxxvii, pt. ii,
  p. 53 ; *Hume & Henders. Lah. to Yark.* p. 234.
Tharrhaleus jerdoni (*Brooks*), *Oates in Hume's N. & E.* 2nd ed. ii,
  p. 114.

Coloration. Resembles *T. strophiatus.* Differs in having the
whole upper plumage greyish brown, with only the back streaked,
the few traces of streaks visible on the crown being obscure or
obsolete ; the lateral black bands on the crown are broader and
more massive ; the hinder part of the supercilium and the breast
are pale rufous, not deep ferruginous.

Iris dark brown ; base of bill, legs, and feet fleshy white ; upper
mandible and base of lower dull black ; rest of lower mandible and
claws pale brown (*Hume*).

Length 5·5 ; tail 2·2 ; wing 2·7 ; tarsus ·8 ; bill from gape ·55.

Distribution. The Himalayas from Gilgit to Mussooree. This
species appears to be a summer visitor to Gilgit, where it breeds
at 10,000 feet and upwards, and a winter visitor to the lower
portions of the mountains.

Habits, &c. Captain Cock found the nest of this bird at Sona-
marg in Kashmir in June, a cup of pine-needles and grass placed
in a low bough of a pine-tree. The eggs measure about ·75 by
·55.

Fig. 11. *Ploceus baya.*

Family PLOCEIDÆ.

The intrinsic muscles of the syrinx fixed to the ends of the bronchial semi-rings; the edges of both mandibles perfectly smooth; the hinder part of the tarsus longitudinally bilaminated, the laminæ entire and smooth; wing with ten primaries, the first notably small; nostrils pierced within the line of the forehead or closely outside it, the space between the nostril and the edge of the mandible greater than the space between the nostril and the culmen; bill conical and entire, the notch being absent or obsolete.

174   PLOCEIDÆ.

The *Ploceidæ* are divisible into two subfamilies, the *Ploceinæ* or Weaver-birds and the *Viduinæ* or Munias.

First primary about as long as the tarsus; a
   partial spring moult .... *Ploceinæ*, p. 174.
First primary very minute, much shorter than
   tarsus; no spring moult .... *Viduinæ*, p. 181.

Subfamily PLOCEINÆ.

The *Ploceinæ* or Weaver-birds comprise a large number of birds which are found in Africa and South-eastern Asia. They are Finch-like in structure and appearance, but they differ from the Finches in having ten primaries and in undergoing a partial spring-moult.

The Weaver-birds are gregarious, breeding in company, and associating at other seasons in large flocks. They construct elaborate nests of grass which are suspended from the branches of trees or attached to the stalks of tall reeds. The eggs are either two or three in number, in the genus *Ploceus* pure white, in *Ploceëlla* of various colours.

The males of these birds have a distinct summer and winter plumage, and the former is acquired by a moult of the feathers of those parts which undergo a change of colour. The moult in the spring is thus apparently partial.

All the Weaver-birds are sedentary in their habits, fearless of man in the breeding-season, but more wary at other times. They feed largely on grain and seeds. They have no song, but they keep up a ceaseless chirping in the breeding-season, especially when the building of the nest is in progress.

The Asiatic Weaver-birds form two well-defined genera, differing in structure and their mode of nidification, as well as in the colour of their eggs.

Key to the Genera.

a. Bill considerably longer than it is high: no
   nuchal hairs: difference between length of
   wing and length of tail more than length of
   tarsus .... PLOCEUS, p. 174.
b. Bill as long as it is high; nuchal hairs present;
   difference between length of wing and length
   of tail much less than length of tarsus ...... PLOCEËLLA, p. 179.

Genus PLOCEUS, Cuvier, 1817.

The genus *Ploceus* contains the true Weaver-birds, which construct flask-shaped nests with a tubular entrance, varying in length from two feet to a few inches. The eggs are in all cases white. In this genus the males acquire a yellow crown in the spring, and

in the autumn resume their plain appearance, which resembles
that of the female.

In *Ploceus* the bill is thick with the culmen curved, and the
length of the bill is considerably more than its height; the wings
are of moderate length, and the first primary is as long as the
tarsus, and slightly curved inwards; the tail is short and mode-
rately rounded, of twelve feathers; the tarsus is strong and
scutellated, and the claws are of considerable length.

Key to the Species.

a. Crown of head yellow (breeding-males).
    a'. Breast yellow .... P. baya, p. 175.
    b'. Breast fulvous .... P. megarhynchus, p. 176.
    c'. Breast black, or black with fulvous
       fringes .... P. bengalensis, p. 177.
    d'. Breast fulvous, boldly streaked with
       black .... P. manyar, p. 179.
b. Crown of head brown (females at all seasons
   and males in winter).
    e'. Lower plumage plain fulvous .... { P. baya, p. 175.
                              { P. megarhynchus, p. 176.
    f'. Breast black, or black fringed with
       fulvous .... P. bengalensis, p. 177.
    g'. Breast boldly streaked with black .... P. manyar, p. 179.

720. Ploceus baya *. *The Baya.*

? Loxia philippina, *Linn. Syst. Nat.* i, p. 305 (1766).
Ploceus baya, *Blyth, J. A. S. B.* xiii, p. 945 (1844); *Horsf. & M.
Cat.* ii, p. 515 (part.); *Jerd. B. I.* ii, p. 343 (part.); *Blanf. J. A.
S. B.* xli, pt. ii, p. 167; *Hume, N. & E.* p. 436 (part.); *id. & Dav.
N. F.* vi, p. 389; *Sharpe, Cat. B. M.* xiii, p. 488; *Oates in Hume's
N. & E.* 2nd ed. ii, p. 114.
Ploceus philippinus (*Linn.*), *Blyth, Cat.* p. 115 (part.); *Legge, Birds
Ceyl.* p. 641; *Hume, Cat.* no. 694; *Barnes, Birds Bom.* p. 259.

The Common Weaver-bird (Jerdon); *Baya*, Hind.; *Chinbora*, Hind.
in Bengal; *Baui, Tulbabi*, Beng.; *Pursupu-pitta*, Tel.; *Manja-kuruvi*,
Tam.; *Thookaam kuruvi*, Tam. in Ceyl.; *Tatta kurula, Wadu kurulla*,
Ceyl.

Coloration. Male. After the autumn moult the whole upper
plumage is fulvous streaked with blackish brown, the streaks becoming obsolete on the lower rump and upper tail-coverts; wing-
coverts, quills, and tail dark brown, each feather edged with
fulvous, the edges of the primaries and tail-feathers also being
tinged with greenish; a clear fulvous supercilium; sides of the
head pale fulvous-brown; the whole lower plumage fulvous,

---

* Linnæus's name, even if it applied to the Continental race of Weaver-bird,
which is very doubtful, is inappropriate, as no bird of this genus is known to
occur in the Philippine Islands. I prefer, therefore, to follow Sharpe in
adopting Blyth's well-known name for this species.

darker on the breast and flanks, the feathers of which parts are frequently streaked with narrow shaft-lines of brown. After the spring moult the appearance of the bird is much changed: the forehead, crown, and nape become bright yellow; the back and scapulars are black, each feather broadly margined with bright yellow; the sides of the head, the chin, and throat dark blackish brown, and the breast bright yellow; the other parts of the plumage remain unchanged.

Fig. 45.—Head of *P. baya*.

Female. At all times resembles the male in winter plumage so closely as to require no separate description.

The intensity of the fulvous tinge on these birds varies much according to age, and in some degree according to the time which has elapsed since the moult.

In the male in summer the bill is dark horny brown, yellowish at gape and base of lower mandible; legs and feet flesh-colour; iris brown. The female in summer and both sexes in winter have the bill yellowish horn-colour.

Length about 6; tail 2; wing 2·9; tarsus ·8; bill from gape ·65.

Distribution. Ceylon and the whole of India proper from the extreme south to the base of the Himalayas as far east as the 85th degree of longitude, about which boundary this species meets the next.

Habits, &c. Breeds from April to September, constructing a hanging flask-shaped nest of grass, strongly woven, suspended from a branch of a tree generally growing over water. The nest terminates in a long funnel of grass, sometimes nearly two feet in length, through which the bird enters the nest proper. The eggs, either two or three in number, measure about ·82 by ·59.

721. Ploceus megarhynchus. *The Eastern Baya.*

Ploceus atrigula, *Hodgs. in Gray's Zool. Misc.* p. 84 (1844, desc. null.); *Sharpe, Cat. B. M.* xiii, p. 491 (1890).
Ploceus philippinus (*Linn.*), *Blyth, Cat.* p. 115 (pt.).
Ploceus baya, *Blyth, Horsf. & M. Cat.* ii, p. 515 (pt.); *Jerd. B. I.* ii, p. 343 (pt.); *Hume, N. & E.* p. 436 (pt.); *id. & Dav. S. F.* vi, p. 398; *Anders. Yunnan Exped., Aves,* p. 597; *Hume, Cat. no. 694 bis; Oates, B. B.* i, p. 358.

Ploceus megarhynchus, *Hume*, S. F. iii, p. 406 (1875) ; *id. Cat.*
no. 694 ter; *Oates in Hume's N. & E.* 2nd ed. ii, p. 119.
Ploceus passerinus, *Hodgs. MS., Reichenow, Zool. Jahrb. Jena*, i, p. 156
(1886).

Took-ra, Assam.

Coloration. The male in winter and the female at all seasons
resemble the same sexes of *P. baya* at the same seasons. In
summer the male has the forehead, crown, and nape bright yellow;
the whole upper plumage fulvous streaked with blackish brown,
the streaks becoming obsolete on the lower rump and upper tail-
coverts ; wing-coverts, quills, and tail dark brown edged with
fulvous ; sides of the head, chin, and throat dark blackish brown ;
the remaining lower plumage fulvous or tawny, becoming albescent
on the abdomen ; the breast and flanks occasionally with narrow
shaft-streaks.

The present species varies as much as does the last in the intensity of the fulvous tinge, and towards the southern portion of
the bird's range the fulvous changes to a rich tawny.

The male in summer has the bill black, the inside of the mouth
flesh-colour : eyelid grey ; iris dark brown ; legs flesh-colour ;
claws pinkish horn-colour. The male in winter and the female at
all seasons have the bill yellowish horn-colour.

Length about 6 ; tail 2 ; wing 2·9 ; tarsus ·8 ; bill from gape
·75.

This species increases in size from south to north, and attains
its greatest size in the Himalayan Terai. Here the tarsus frequently reaches to a length of ·95, and the bill, from gape to tip,
·85. To this larger bird Hume gave the name of *megarhynchus.*
It is, however, impossible to separate this larger race from the
form inhabiting Burma and the Malay peninsula, and consequently
Hume's name will stand for the species, no previously-imposed
name being available.

Distribution. Bengal and the base of the Himalayas from the
85th degree of longitude to Assam, and southwards through the
whole of Burma to the southern end of the Malay peninsula, extending to Sumatra and Java. This species, in a somewhat larger
form, is found along the base of the Himalayas as far west as the
Dhoon, in company with *P. baya.*

Habits, &c. Breeds from April to September, constructing a
nest very similar to that of *P. baya.* The nest is not unfrequently
attached to the thatched eaves of native houses, or even to the
ends of the loose leaves inside the verandahs. The eggs measure
about ·82 by ·6.

722. Ploceus bengalensis. *The Black-throated Weaver-bird.*

Loxia benghalensis, *Linn. Syst. Nat.* i, p. 305 (1766).
Ploceus bengalensis (*Linn.*), *Blyth, Cat.* p. 115 ; *Horsf. & M. Cat.*
ii, p. 515; *Jerd. B. I.* ii, p. 349; *Hume, N. & E.* p. 441; *id Cat.*
no. 696 ; *Oates, B. B.* i, p. 361 ; *Sharpe, Cat. B. M.* xiii, p. 493;

Barnes, Birds Bom. p. 261; *Hume, S. F.* xi, p. 270; *Oates in Hume's N. & E.* 2nd ed. ii, p. 120.

Sarbo baya, Hind. ; *Shor baya, Kantawala baya,* Beng.

Coloration.   Male. After the autumn moult the forehead, crown, and nape are dark brown, each feather narrowly margined with fulvous-brown ; back, scapulars, and all the feathers of the wings dark blackish brown, each feather broadly margined with fulvous ; rump and upper tail-coverts dull fulvous, streaked with pale brown ; tail brown, narrowly margined paler ; a distinct supercilium, widening and becoming paler at the nape, yellow ; a large yellow patch on the side of the neck ; a pale yellow patch under the eye ; remainder of side of the head brown ; cheeks, chin, and throat pale yellow ; a narrow moustachial streak black ; fore neck and upper breast black, each feather very broadly margined with fulvous ; remainder of lower plumage fulvous, paler on the abdomen, and the sides of the body narrowly streaked with brown.   Soon after the moult the broad fulvous margins on the fore neck and upper breast commence to wear away, and by February and March these parts are almost uniform black.

After the partial spring moult the forehead and crown become bright golden yellow surrounded by a black band ; the chin and throat become whitish or whity brown, and the entire side of the head and neck become uniform brown ; the other parts of the plumage remain unchanged, except that the pale margins to the feathers of the back and wings get worn away, causing those parts to become very dark and more uniform brown.

Female. Very similar to the male in winter plumage, but with all the yellow marks on the head paler, and with much less black on the fore neck and upper breast.   The pale margins on these parts wear away as in the male, but these parts apparently never become quite so black in the breeding-season as in the male.

Bill pearly white or pale plumbeous at all seasons and in both sexes ; iris light brown ; legs flesh-colour.

Length about 5·5 ; tail 1·8 ; wing 2·8 ; tarsus ·8 ; bill from gape ·65.

Distribution. Occurs sparingly throughout Northern India from Sind to Bengal, and up to the foot of the Himalayas.   The Southern limit appears to be a line drawn from Bombay to Bastar, south of which line this species does not appear to have been recorded. Eastwards it occurs throughout Assam, and it is found commonly as far south as Manipur.   Blanford records this Weaver-bird from Ava and Thayetmyo, but its occurrence in Burma has not been confirmed.

Habits, &c. Breeds in the rains.   The nest resembles that of *P. baya* in general shape, but the funnel is very short.   The nest is invariably attached to the leaves of elephant-grass, with which it is well incorporated and by which it is more or less supported. The eggs measure about ·83 by ·58.

723. Ploceus manyar. *The Striated Weaver-bird.*

Fringilla manyar, *Horsf. Trans. Linn. Soc.* xiii, p. 160 (1820).
Ploceus manyar (*Horsf.*), *Blyth, Cat.* p. 115; *Horsf. & M. Cat.* ii,
 p. 514; *Jerd. B. I.* ii, p. 348; *Hume, N. & E.* p. 440; *Anders.*
 Yunnan Exped., Aves, p. 598; *Legge, Birds Ceyl.* p. 646; *Hume,*
 Cat. no. 685; *Oates, B. B.* i, p. 360; *Barnes, Birds Bom.*
 p. 260; *Sharpe, Cat. B. M.* xiii, p. 496; *Oates in Hume's N. & E.*
 2nd ed. ii, p. 121.

Bamani baya, Hind. in the Deccan; *Telia baya,* Beng.; *Bawoyi* in
Rungpore.

Coloration. Male. After the autumn moult the forehead and
crown are yellow; throat, cheeks, ear-coverts and sides of neck
black or brownish black; lower plumage fulvous, each feather
striated with black down the centre, except on the abdomen and
under tail-coverts; upper plumage dark brown edged with pale
fulvous; wings and tail brown edged with yellowish; the throat is
sometimes coloured a pale brown, and the intensity of the fulvous
on the lower parts varies much.

After the partial spring moult the forehead, crown, and nape
become deep yellow; the supercilium and the spot behind the ear-
coverts disappear, and the whole of the sides of the head and neck
together with the cheeks, chin, and throat become blackish brown

Female. Resembles the male in winter plumage.

The male in summer has the bill bluish black, paler at gape;
the female in summer and both sexes in winter have the bill yellowish horn-colour; the iris is at all times brown, legs flesh-colour, and
claws pinkish horn.

Length nearly 6; tail 1·8; wing 2·7; tarsus ·8; bill from
gape ·7·

Distribution. The whole Empire from the foot of the Himalayas
southward to Ceylon on the one hand, and to about the latitude
of Moulmein in Tenasserim on the other. This species also occurs
in Java.

Habits, &c. Breeds throughout the rains, attaching its nest, which
resembles that of *P. baya,* but has a shorter funnel, to the extremities
of several leaves of elephant-grass. The eggs measure about ·8
by ·58.

Genus PLOCEËLLA, Oates, 1873.

The genus *Ploceëlla* differs from *Ploceus* in many important particulars. The bill is much shorter, being no longer than it is high;
the nape is furnished with a few short hairs, and the tail is much
longer and more rounded. The plumage of the male in summer
is largely yellow. The nest differs much from that of a *Ploceus,*
being supported at the sides by reeds and not suspended from their
tips; it has no tubular entrance, and the exterior surface of the
nest is of quite a different appearance, being rough and coarse

N 2

instead of smooth. The eggs are of all colours and but seldom pure white.

Ploceëlla approximates more to the African Weaver-birds than does *Ploceus*. Only one species of this genus is known.

724. Ploceëlla javanensis. *The Golden Weaver-bird.*

Loxia javanensis, *Less. Tr. d'Orn.* p. 446 (1831).
Ploceus hypoxanthus (*Daud.*), *Blyth, Cat.* p. 114; *Horsf. & M. Cat.*
   ii, p. 513; *Hume, N. & E.* p. 442; *id. S. F.* iii, p. 154.
Ploceëlla javanensis (*Less.*), *Oates, S. F.* v, p. 160; *Hume, Cat.*
   no. 696 bis; *Oates, B. B.* i, p. 362; *Sharpe, Cat. B. M.* xiii, p. 474;
   Oates in Hume's N. & E. 2nd ed. ii, p. 124.
Ploceus chryseus, *Hume, S. F.* vi, p. 399 note (1878).

Coloration. Male. After the autumn moult the whole upper plumage is rufous-brown streaked with dark brown; wings and tail brown, margined with rufous-brown; a supercilium, the sides of the head and neck, and the whole lower plumage tawny buff, becoming paler on the abdomen. After the spring moult, the cheeks, ear-coverts, chin, and throat become deep black; the remainder of the lower plumage, the sides of the neck, forehead, crown, nape, rump, and upper tail-coverts rich golden yellow; back and scapulars black, each feather edged with bright golden yellow; wings and tail black, edged with pale yellowish, the latter also tipped with the same colour.

Female. Resembles the male in winter plumage.

The male in summer has the bill black with the underside of the lower mandible dark horn; inside of the mouth flesh-colour; iris brown; eyelids grey; legs pinkish flesh-colour; claws horn-colour. The female in summer and both sexes in winter have the bill fleshy brown, dark on the upper and pale on the lower mandible, and the other parts as in the male in summer.

Length nearly 6; tail 2·3; wing 2·7; tarsus ·8; bill from gape ·55; the female has the tail about 1·9 and the wing about 2·5.

Distribution. Upper Burma and Pegu between the Irrawaddy and Sittoung rivers from Mandalay down to the Gulf of Martaban; Northern Tenasserim; Siam, Cochin China, and Java.

Habits, &c. Commences to breed in May and June, making a cylindrical nest of woven grass attached to several stalks of elephant-grass or sometimes placed in a thorny bush or tree. The eggs, either two or three in number, vary much in colour, being white, greenish white, grey, or purplish, either unmarked or marked with grey, greenish brown, and neutral tint. They measure about ·73 by ·54.

Subfamily VIDUINÆ.

The *Viduinæ* or Munias differ from the Weaver-birds in having a very minute first primary and in having no spring moult.

The Munias associate in large flocks during the winter months, but they separate and are no longer sociable in the breeding-season, although several nests may be found near each other. They have a rapid flight. The sexes of these birds are usually quite alike in plumage, and when they differ, the differences between the sexes are not very great.

The Munias construct large round nests of grass with an opening at the side. The nest is placed in a bush or in a clump of grass, and some species approach man and build their nests in houses, under the eaves or in the trellis-work of a verandah. The eggs are numerous, being frequently six or more, and they are invariably pure white.

None of the Munias are known to migrate. They feed on the ground or else cling to the heads of flowering grass or corn, and they consume large quantities of grain.

Key to the Genera.

a. Middle pair of tail-feathers narrow and pointed.
  a'. Tail rounded and very slightly graduated; difference between wing and tail quite equal to tarsus; crown black, different from back .... MUNIA, p. 181.
  b'. Tail wedge-shaped and much graduated; difference between wing and tail much less than tarsus; crown of much the same colour as back .... UROLONCHA, p. 183.
  c'. Tail wedge-shaped and much graduated; tail in male longer than wing; colours green and crimson................. ERYTHRURA, p. 190.
b. Middle tail-feathers broad and rounded.
  d'. Difference between outer and middle tail-feathers less than tarsus; plumage green .... STICTOSPIZA, p. 190.
  e'. Difference between outer and middle tail-feathers equal to tarsus; plumage red. SPORÆGINTHUS, p. 192.

Genus MUNIA, Hodgs. 1836.

The genus *Munia* contains two Indian species, which are characterized by a short and rounded tail, having the middle pair of feathers very narrow and pointed. The tail is shorter than the wing by a distance quite equal to the length of the tarsus. The plumage is chiefly black and chestnut, and the sexes are absolutely alike.

Key to the Species

a. Lower breast and sides of body white .... *M. malacca,* p. 182.
b. Lower breast and sides of body chestnut .. *M. atricapilla,* p. 183.

Fig. 46.—Tail of *M. malacca.*

Fig. 47.—Head of *M. malacca.*

725. Munia malacca. *The Black-headed Munia.* ·

Loxia malacca, *Linn. Syst. Nat.* i, p. 302 (1766).
Munia malacca (*Linn.*), *Blyth, Cat.* p. 116; *Horsf. & M. Cat.* ii,
    p. 507; *Jerd. B. I.* ii, p. 352; *Hume, N. & E.* p. 443; *Legge, Birds
    Ceyl.* p. 652; *Sharpe, Cat. B. M.* xiii, p. 350; *Oates, in Hume's
    N. & E.* 2nd ed. ii, p. 126.
Amadina malacca (*Linn.*), *Hume, Cat.* no. 697; *Barnes, Birds Bom.*
    p. 362.

Nakal-nor, Hind.; *Nalla jinawayi,* Tel.; *Wé-kurulla,* Ceyl.; *Tinna
kururi,* Tam.

Coloration. The whole head, neck, and upper breast, the middle
of the abdomen, vent, thighs, and under tail-coverts deep black;
remainder of lower plumage white; back, upper rump, scapulars,
and the whole of the wings chestnut; lower rump and upper tail-
coverts glistening maroon; tail dull chestnut, edged with glistening
maroon.

The young bird has the whole upper plumage, wings, and tail
rufous-brown: the whole lower plumage pale buff.

Bill pale lavender; legs and feet leaden blue; iris dark brown
(*Butler*).

Length about 5; tail 1·6; wing 2·2; tarsus ·6; bill from gape
·45.

Distribution. Ceylon and the southern half of India up to about
the latitude of Bhandára and Raipur in the Central Provinces.

---

* MUNIA ORYZIVORA (Linn.).

The Java Sparrow has been introduced into parts of India and is now to be
met with in the wild state in Madras and, according to Blyth, in Tenasserim.
I do not, however, propose to include it in my list. The following is a description of the bird. The sexes are alike :—

Cheeks and ear-coverts white; chin, throat, a line bordering the ear-coverts,
the forehead, and the whole top of the head black; neck, breast, upper abdomen,
back, scapulars, and wings bluish grey; rump, upper tail-coverts, and tail black;
abdomen, sides, thighs, and vent vinous, paler down the middle; under tail-
coverts white.

Jerdon states that this species occasionally extends to Bengal, but I have not seen any specimen from that province.

Habits, &c. Breeds from April to October. The eggs measure about ·64 by ·47.

726. **Munia atricapilla.** *The Chestnut-bellied Munia.*

Loxia atricapilla, *Vieill. Ois. Chant.* p. 84, pl. 53 (1805).
Munia rubronigra, *Hodgs. As. Res.* xix, p. 153 (1836); *Blyth, Cat.* p. 116; *Horsf. & M. Cat.* ii, p. 507; *Jerd. B. I.* ii. p. 354; *Legge, Birds Ceyl.* p. 652.
Munia atricapilla (*Vieill.*). *Hume, N. & E.* p. 444; *Anders. Yunnan Exped., Aves,* p. 598; *Sharpe, Cat. B. M.* xiii. p. 334; *Oates in Hume's N. & E.* 2nd ed. ii. p. 129.
Amadina rubronigra (*Hodgs.*), *Hume, Cat.* no. 698; *Barnes, Birds Bom.* p. 262; *Hume, S. F.* xi, p. 272.
Amadina atricapilla (*Vieill.*), *Oates B. B.* i, p. 366.

Coloration. The whole head, neck, and upper breast deep black; middle of the abdomen, vent, thighs, and under tail-coverts dull black; remainder of the lower plumage, the back, scapulars, upper rump, and the whole of the wings chestnut; lower rump and upper tail-coverts rich maroon, the latter tipped with glistening golden fulvous; tail brown, margined with glistening golden fulvous.

Towards the southern part of its range this species has the back washed with silvery grey.

The young bird is plain rufous-brown above and pale buff below.

Bill leaden blue; iris dark brown; legs dark plumbeous.

Length about 4·5; tail 1·5; wing 2·1; tarsus ·55; bill from gape ·45.

Distribution. The country skirting the base of the Himalayas from the Dehra Dún to the extreme east of Assam; Lower Bengal; Chutia Nagpur and Sambalpur in the Central Provinces; the whole eastern portion of the Empire from Assam down to Tenasserim. This species has been recorded from Ceylon, but it is probable that the specimens observed there had escaped from captivity. *M. atricapilla* has a wide range, being found in China, Siam, and the Malay Peninsula, Sumatra, and Borneo.

Habits, &c. Breeds from June to September in swampy localities. The eggs measure about ·63 by ·43.

Genus **UROLONCHA**, Cabanis, 1851.

The genus *Uroloncha* contains numerous species of Munias in which the tail is much longer than in typical *Munia*, the difference between the tail and the wing being much less than the length of the tarsus; the graduation of the tail-feathers is also much greater.

In this genus the sexes are absolutely alike, and the colour of the crown of the head is very closely, or quite, the same as that of the back.

Key to the Species.

a. Rump white.
    a'. Abdomen white mottled with brown....   *U. acuticauda*, p. 184.
    b'. Abdomen plain, unmottled white.
        a". Feathers of upper plumage with white
            shafts.
            a'''. Fore neck and breast deep black ..   *U. striata*, p. 185.
            b'''. Fore neck and breast brown with
                rufous margins.................   *U. semistriata*, p. 186.
        b". Feathers of upper plumage without
           white shafts ....   *U. fumigata*, p. 186.
b. No white on rump.
    c'. Shafts of feathers of upper plumage pale;
        upper tail-coverts black or tipped with
        glistening fulvous.
        c". Abdomen white, sides of body black..   *U. leucogastra*, p. 186.
        d". Abdomen and sides of the body plain
           pinkish brown ....   *U. pectoralis*, p. 187.
        e". Abdomen and sides of the body cross-
           barred with brown.
            c'''. Chin and throat black............   *U. kelaarti*, p. 187.
            d'''. Chin and throat chestnut ....   *U. punctulata*, p. 189.
    d'. Shafts of feathers of upper plumage of
        same colour as feathers; upper tail-
        coverts white ....   *U. malabarica*, p. 188.

Fig. 48.—Tail of *U. acuticauda*.

727. Uroloncha acuticauda. *Hodgson's Munia.*

Munia acuticauda, *Hodgs. As. Res.* xix, p. 153 (1836); *Horsf. & M.
Cat.* ii, p. 510; *Jerd. B. I.* ii, p. 356; *Hume, N. & E.* p. 450.
Munia mohecca (*Linn.*), *Blyth, Cat.* p. 117.
Amadina acuticauda (*Hodgs.*), *Hume, Cat.* no. 702; *Oates, B. B.* i,
   p. 364; *Hume, S. F.* xi, p. 273.
Uroloncha acuticauda (*Hodgs.*), *Sharpe, Cat. B. M.* xiii, p. 356;
   Oates in Hume's N. & E. 2nd ed. ii, p. 131.

The Himalayan Munia, Jerd.; *Sumprek-pho*, Lepch.; *Namprek*, Bhut.

Coloration. The whole upper plumage chocolate-brown, with a
band of white across the rump, the upper tail-coverts washed with
rufous, all the shafts of the feathers of the upper plumage white;
tail black; wing-coverts brown with white shafts; wings blackish

brown ; feathers round the bill, the chin, and upper throat black ; ear-coverts and sides of neck rufous, with paler margins and white shafts ; lower throat, fore neck, and breast chocolate-brown, with pale rufous or whitish margins and white shafts ; abdomen and sides of the body white mottled with brown ; vent, thighs, and under tail-coverts chocolate-brown with white shaft-streaks.

Young birds have the upper plumage somewhat similar to that of the adult, and the lower plumage almost uniform dull greyish white mottled with brown.

Upper mandible blackish, the lower plumbeous : iris dark brown : legs plumbeous ; claws horny.

Length about 4·5 ; tail 1·75 ; wing 2 ; tarsus ·55 ; bill from gape ·4.

This bird varies slightly in plumage throughout its great range.

Distribution. The lower ranges of the Himalayas from Garhwál to Assam, up to 5000 feet ; the whole eastern portion of the Empire from Assam to Tenasserim : sparingly distributed. This species extends to Southern China, Siam, the Malay peninsula, and Sumatra.

Habits, &c. Breeds from about June to November or December. The eggs measure about ·61 by ·42.

728. Uroloncha striata. *The White-backed Munia.*

Loxia striata, *Linn. Syst. Nat.* i, p. 306 (1766).
Munia striata (*Linn.*), *Blyth, Cat.* p. 117 ; *Horsf. & M. Cat.* ii, p. 511 ; *Jerd. B. I.* ii, p. 356 ; *Hume, N. & E.* p. 448 ; *Legge, Birds Ceyl.* p. 660.
Amadina striata (*Linn.*), *Hume, Cat.* no. 701 ; *Oates, B. B.* i, p. 365 ; *Barnes, Birds Bom.* p. 263.
Uroloncha striata (*Linn.*), *Sharpe, Cat. B. M.* xiii, p. 359 ; *Oates in Hume's N. & E.* 2nd ed. ii, p. 133.

Shakari munia, Beng.; *Wé-kurulla,* Ceyl.; *Tinna kururi,* Tam.

Coloration. Forehead and crown blackish, with the shafts indistinctly white ; upper plumage chocolate-brown, becoming blacker on the upper tail-coverts. all the feathers with a white shaft ; a broad white band across the rump ; tail black : wing-coverts black with whitish shafts : quills black ; lores, round the eye, cheeks, chin, throat, and breast deep black ; ear-coverts and sides of the neck chocolate-brown with pale shaft-streaks ; abdomen and sides of the body white ; vent, thighs, and under tail-coverts dark brown or blackish ; under wing-coverts and axillaries white.

Young birds resemble the adult, but have the chin, throat, and breast brown with paler margins and shafts.

Upper mandible blackish, the lower bluish ; iris reddish brown ; legs greenish horny.

Length about 4·5 ; tail 1·8 ; wing 2 ; tarsus ·5 ; bill from gape ·4.

This bird is subject to slight variations in plumage according to age and locality.

Distribution. Ceylon and the southern half of India up to a line

drawn roughly from Bombay to Sambalpur and Manbhoom in
S.W. Bengal. Blyth recorded this species from Arakan, but probably by an oversight.

Habits, &c. Breeds from April to December and probably in
the remaining months of the year. The eggs measure about ·61
by ·44.

729. Uroloncha semistriata. *The Nicobar White-backed Munia.*

Munia semistriata, *Hume, S. F.* ii, p. 257 (1874).
Amadina semistriata (*Hume*), *Hume, Cat.* no. 701 quat.
Uroloncha semistriata (*Hume*), *Sharpe, Cat. B. M.* xiii. p. 361.

Coloration. Resembles *U. striata* but is smaller ; the striations or
white shafts on the upper plumage less distinct ; and the feathers
of the fore neck and breast dark brown, not deep black, with
conspicuous rufous-brown margins.

Length about 4·2; tail 1·6; wing 1·8; tarsus ·5; bill from
gape ·4.

Distribution. The Nicobar Islands.

730. Uroloncha fumigata. *The Andaman White-backed Munia.*

Munia fumigata, *Walden, A. M. N. H.* (4) xii, p. 488 (1873); *id.
Ibis,* 1874, pp. 144, 145.
Munia nonstriata, *Hume, S. F.* ii, p. 257 (1874).
Munia striata (*Linn.*), *Hume, S. F.* ii, p. 497.
Amadina fumigata (*Wald.*), *Hume, Cat.* no. 701 ter.
Uroloncha fumigata (*Wald.*), *Sharpe, Cat. B. M.* xiii, p. 361 ; *Oates
in Hume's N. & E.* 2nd ed. ii, p. 135.

Coloration. Resembles *U. striata.* Differs in entirely wanting
any white shafts on the upper plumage, in having the upper tail-
feathers broadly margined with rufous-brown, and in the feathers
of the sides of the breast being also margined with rufous, these
margins sometimes extending across the lower breast.

Iris reddish brown ; upper mandible black, the lower leaden blue ;
legs and feet plumbeous green or greenish horny (*Hume*).

Length about 4·5; tail 1·8; wing 2; tarsus ·5; bill from gape ·4.
Distribution. The Andaman Islands.

Habits, &c. Davison remarks that when he arrived at the Andamans in December, the young birds of this species had left the
nest. This bird probably breeds, therefore, in the latter part of
the rains.

731. Uroloncha leucogastra. *The White-bellied Munia.*

Amadina leucogastra, *Blyth, J. A. S. B.* xv, p. 286 (1846); *Hume,
Cat.* no. 701 bis; *Oates, B. B.* i, p. 367.
Munia melanictera (*Gm.*), *Blyth, Cat.* p. 117.
Munia leucogastra (*Blyth*), *Davison, S. F.* v, p. 400 ; *Hume & Dav.
S. F.* vi, p. 402.
Uroloncha leucogastra (*Bl.*), *Sharpe, Cat. B. M.* xiii, p. 362; *Oates
in Hume's N. & E.* 2nd ed. ii, p. 135.

Coloration. Upper plumage, scapulars, and wing-coverts chocolate-brown, the shaft of each feather white; upper tail-coverts black; tail blackish, with broad shiny fulvous margins; wings blackish; lores, cheeks, chin, throat, breast, sides of the body, thighs, and under tail-coverts black; abdomen and under wing-coverts white, the white of the abdomen forming an angle in the breast.

Legs and feet dusky plumbeous or dull smalt-blue; lower mandible dull smalt- or pale blue; upper mandible brownish black or black; iris dark brown (*Hume & Dav.*).

Length about 4·5; tail 1·8; wing 1·9; tarsus ·55; bill from gape ·45.

Distribution. The extreme south of Tenasserim, ranging down the Malay peninsula and to Borneo.

Habits, &c. Davison took a nest of this species in Tenasserim in April. The eggs appear to measure about ·65 by ·44.

732. Uroloncha pectoralis. *The Rufous-bellied Munia.*

Munia pectoralis, *Jerd. B. I.* ii, p. 355 (1863); *Hume, N. & E.* p. 418; *id. S. F.* iii, p. 263, iv, p. 403.
Amadina pectoralis (*Jerd.*), *Hume, Cat.* no. 700.
Uroloncha pectoralis (*Jerd.*), *Sharpe, Cat. B. M.* xiii, p. 365; *Oates in Hume's N. & E.* 2nd ed. ii, p. 156.

Coloration. Upper plumage chocolate-brown with pale shafts, the head a darker brown; the rump blackish, with broad triangular streaks of pale buff; upper tail-coverts glistening fulvous; tail black; wings dull black; lores, cheeks, chin, throat, and fore neck black; ear-coverts buff with pale shafts; sides of the fore neck, sides of breast, breast, sides of the body, and the whole abdomen pinkish brown; vent and under tail-coverts blackish, with pinkish-brown streaks.

The young are plain chocolate-brown above with fulvous upper tail-coverts, and have the lower plumage buff with pale shaft-streaks.

Bill, legs, and feet slaty; iris brown (*Miss Cockburn*).

Length nearly 5; tail 1·8; wing 2·2; tarsus ·5; bill from gape ·45.

Distribution. The hills on the western coast of India from the Wynaad to Travancore.

Habits, &c. Appears to breed chiefly in June and July. The eggs measure about ·62 by ·44.

733. Uroloncha kelaarti. *The Ceylon Munia.*

Munia kelaarti, *Blyth, Jerd. B. I.* ii, p. 356 (1863); *Blyth, Ibis,* 1867, p. 299; *Hume, S. F.* vii, p. 410; *Legge, Birds Ceyl.* p. 650, pl. xxvii, fig. 2.
Amadina kelaarti (*Blyth*), *Hume, Cat.* no. 700 bis.
Uroloncha kelaarti (*Blyth*), *Sharpe, Cat. B. M.* xiii, p. 366.

Wé-kurulla, Ceyl.; *Tinna kuruvi,* Tam.

Coloration. Upper plumage chocolate-brown with pale shafts,

188 PLOCEIDÆ.

the head darker ; rump black, with white cruciform marks ; upper tail-coverts glistening fulvous ; tail black ; wings dark brown or blackish ; cheeks, lores, chin, throat, and fore neck black; sides of the neck and of the breast pinkish brown with pale shafts ; breast and whole lower plumage pinkish white, irregularly but closely cross-barred with black.

The young are plain chocolate-brown above, without pale shafts or white marks on the rump, and pale buff below with white shafts and irregular cross-bands of brown.

Iris sepia-brown ; bill blackish leaden, bluish at the base of the lower mandible ; legs and feet plumbeous, in some with a greenish tinge (*Legge*).

Length about 4·5 ; tail 1·8 ; wing 2·1 ; tarsus ·5 ; bill from gape ·45.

Distribution. Confined to the island of Ceylon.

734. Uroloncha malabarica. *The White-throated Munia.*

Loxia malabarica, *Linn. Syst. Nat.* i, p. 305 (1766).
Munia malabarica (*Linn.*), *Blyth, Cat.* p. 117 ; *Horsf. & M. Cat.* ii, p. 508 ; *Jerd. B. I.* ii, p. 357 ; *Hume, N. & E.* p. 451 ; *Legge, Birds Ceyl.* p. 662.
Munia similaris, *Stoliczka, J. A. S. B.* xxxvii, pt. ii, p. 56 (1868).
Amadina malabarica (*Linn.*), *Hume, Cat.* no. 703 ; *Barnes, Birds Bom.* p. 263.
Aidemosyne malabarica (*Linn.*), *Sharpe, Cat. B. M.* xiii, p. 369.
Uroloncha malabarica (*Linn.*), *Oates in Hume's N. & E.* 2nd ed. ii, p. 136.

The Plain Brown Munia, Jerd.; *Charchara* in the N.W. Prov.; *Piddari,* Southern and Central India; *Sar-munia,* Beng.; *Jinuwayi,* Tel.

Coloration. Upper plumage, wing-coverts, secondaries, and tertiaries earthy brown ; primaries and winglet black ; upper tail-coverts white, the outer webs of the exterior feathers partially black ; tail dark brown, margined with rusty ; sides of the head and lower plumage pale buffy white, the sides of the body faintly cross-barred with rusty.

The young closely resemble the adult.

Upper mandible plumbeous horn-colour; lower mandible lavender ; legs and feet pale purplish pink ; iris dark brown (*Butler*).

Length about 4·5 ; tail 1·9 ; wing 2·1 ; tarsus ·55 ; bill from gape ·4.

Distribution. The whole continent of India from the Himalayas, which this species ascends up to 5000 feet, to Cape Comorin and Ceylon. The most easterly locality at which this bird appears to have been observed is Kooshtea on the Ganges, Bengal, where Godwin-Austen obtained it (J. A. S. B. xliii, pt. ii, p. 171). To the west it ranges into Afghanistan.

Habits, &c. Breeds throughout the greater part of the year. The eggs measure ·6 by ·47.

735. Uroloncha punctulata. *The Spotted Munia.*

Loxia punctulata, *Linn. Syst. Nat.* i, p. 302 (1766).
Loxia undulata, *P. L. S. Müll. Syst. Nat., Anhang*, p. 151 (1776).
Munia undulata (*Lath.*), *Blyth, Cat.* p. 117; *Horsf. & M. Cat.* ii, p. 506; *Jerd. B. I.* ii, p. 354.
Munia punctulata (*Linn.*), *Hume, N. & E.* p. 444; *Legge, Birds Ceyl.* p. 656; *Sharpe, Cat. B. M.* xiii, p. 346.
Munia subundulata, *Godw.-Aust. P. Z. S.* 1874, p. 48; *Hume, S. F.* iii, p. 398.
Munia superstriata, *Hume, S. F.* ii, p. 481 (1874); *Hume & Dav. S. F.* vi, p. 402.
Munia inglisi, *Hume, S. F.* v, p. 39 (1877).
Amadina punctulata (*Linn.*), *Hume, Cat. no. 699; Oates, B. B.* i, p. 368; *Barnes, Birds Bom.* p. 262.
Amadina subundulata (*Godw.-Aust.*), *Hume, Cat. no. 699 bis; id. S. F.* xi, p. 272.
Amadina superstriata (*Hume*), *Hume, Cat. no. 699 ter.*
Amadina inglisi (*Hume*), *Hume, Cat. no. 699 quat.*
Uroloncha punctulata (*Linn.*), *Oates in Hume's N. & E.* 2nd ed. ii, p. 141.

Telia munia, Hind. in the North; *Sing-baz, Shiabaz*, Hind. in the Deccan and at Mussooree; *Shubz munia*, Beng.; *Kakkara jinuwayi*, Tel.; *Wé-kurulla*, Ceyl.; *Tinna kuruvi*, Tam.

Coloration. Upper plumage dull chocolate-colour with the shafts pale; lower rump barred irregularly with brown and yellowish and streaked with white; upper tail-coverts glistening yellowish fulvous; tail fulvous-yellow: wings chocolate, the coverts with pale shafts; sides of the head, chin, and throat rich chestnut; lower plumage white, each feather, with the exception of those on the abdomen, submarginally banded with fulvous-brown; under tail-coverts fulvous white mottled with black.

The above description applies to birds from the Continent of India, which, however, vary considerably among themselves in the shade of colouring of the rump and the amount and distinctness of the bars on this part. Birds from Assam, southwards to Burma, are less distinctly barred on the rump, the general colour of which and of the upper tail-coverts and tail is more olivaceous. Many species have been established on these slight differences, but I am unable to recognize them even as races, the differences being by no means constant over the same small areas.

Bill bluish black, paler and somewhat plumbeous on the lower mandible; iris deep reddish brown; legs plumbeous; claws horny.

Length nearly 5; tail 1·7; wing 2·1; tarsus ·6; bill from gape ·45.

Young birds are rufous-brown above and pale buff below without marks of any kind.

Distribution. The whole continent of India, except Sind, the Punjab, and portions of Rajputana and the N.W. Provinces, ascending the Himalayas up to about 5000 feet; Ceylon; the eastern part of the Empire from Assam to about the latitude of

Tavoy. To the east, in China, this species is replaced by an allied race *M. topela*.

Habits, &c. Breeds almost throughout the year. The eggs measure about ·65 by ·46.

Genus ERYTHRURA, Swains., 1837.

The genus *Erythrura* contains one species of Munia, the prevailing colours of which are green and crimson. The sexes are slightly different. In the male the tail is longer than the wing; in the female it is considerably shorter. The middle pair of tail-feathers is very narrow and pointed.

736. Erythrura prasina. *The Long-tailed Munia.*

Loxia prasina, *Sparrm. Mus. Carls.* pls. 72, 73 (1788).
Erythrura prasina (*Sparrm.*), *Blyth, Cat.* p. 118.
Erythrura prasina (*Sparrm.*), *Horsf. & M. Cat.* ii, p. 503; *Hume & Dav. S. F.* vi, p. 405; *Hume, Cat.* no. 703 ter; *Oates, B. B.* i, p. 370; *Sharpe, Cat. B. M.* xiii, p. 381.

Coloration. Male. Lores and a narrow line to the nostril black; forehead, cheeks, round the eye, chin, and throat blue; upper plumage, wing-coverts, and tertiaries bright green; lower rump and upper tail-coverts crimson; middle pair of tail-feathers dull red, the others brown tipped with greenish; primaries and secondaries black, margined with green; ear-coverts and sides of the neck green; lower plumage buff, except the middle of the abdomen, which is crimson.

Female. Resembles the male in general appearance, but has the blue of the forehead, cheeks, and round the eye replaced by green with a slight blue tinge, and the blue of the chin and throat replaced by greenish buff; the crimson on the abdomen is absent, that part being buff like the remainder of the lower plumage.

The young resemble the female in general appearance, but have the upper tail-coverts and middle pair of tail-feathers yellowish, not crimson.

Legs, feet, and claws fleshy pink; bill black; iris dark brown (*Hume & Dav.*).

Length of male about 6; tail 2·8; wing 2·3; tarsus ·6; bill from gape ·55. Length of female nearly 5; tail 1·6; other parts as in the male.

Distribution. The extreme south of Tenasserim, extending down the Malay peninsula and to Sumatra, Java, and Borneo.

Genus STICTOSPIZA, Sharpe, 1890.

The genus *Stictospiza* contains a single species of Munia of a green colour. The female differs from the male chiefly in being paler. In this genus the middle tail-feathers are broad and rounded, and not narrow and pointed as in the preceding genera.

737. **Stictospiza formosa.** *The Green Munia.*

Fringilla formosa, *Lath. Ind. Orn.* i, p. 441 (1790).
Estrelda formosa (*Lath.*), *Blyth, Cat.* p. 119 ; *Jerd. B. I.* ii, p. 361 ;
    Hume, N. & E. p. 456 ; *Butler, S. F.* iii, p. 496 ; *Hume, Cat.* no.
    705 ; *Reid, S. F.* x, p. 56 ; *Barnes, Birds Bom.* p. 265.
Stictospiza formosa (*Lath.*), *Sharpe, Cat. B. M.* xiii, p. 287 ; *Oates
    in Hume's N. & E.* 2nd ed. ii, p. 145.

The Green Wax-bill, Jerd. ; *Harre lal, Harre munia,* Hind.

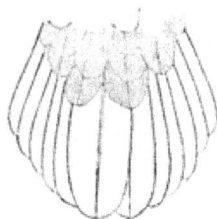

Fig. 49.—Tail of *S. formosa.*

Coloration. Male. The whole upper plumage light green, tinged
with yellow on the upper tail-coverts ; wings and their coverts
brown, each feather broadly edged with light green, the closed
wing appearing entirely of this latter colour ; tail black : sides of
the head and neck yellowish green ; lower plumage yellow, pale on
the chin, throat and fore neck, brighter on the breast, and becoming
deep on the abdomen, vent, and under tail-coverts : flanks and
sides of the body transversely barred with dark greenish brown
and white, the white bars sometimes tinged with yellow : under
wing-coverts pale yellowish.

Female. Not very dissimilar to the male, but having the green
of the upper plumage and wings duller : the chin, throat, and breast
grey barely tinged with yellow, and the yellow of the remaining
lower parts much paler.

The young bird has the upper plumage olive-brown ; the lower
plumage ochraceous, turning to pale yellow on the abdomen ; flanks
and sides of the body pale buff, uniform and unbarred : bill black.

Bill waxy red : feet plumbeous brown : iris pale brown (*Jerdon*).

Length about 4 ; tail 1·5 ; wing 1·95 ; tarsus ·5 ; bill from gape
·45.

Distribution. The Central portion of the Indian continent, the
extreme points to which this species extends being apparently
Mount Abu on the west, Palamow and Lohardugga on the east,
Jhánsi on the north, and Chánda and Ahiri on the south.

Habits, &c. Breeds apparently twice a year, once in the rains
and once in the cold season, laying five eggs, which measure about
·66 by ·47.

Genus SPORÆGINTHUS, Cabanis, 1850.

The genus *Sporæginthus* contains two Indian species of Munia, in which the males are red and the females brown, and both sexes are much spotted with white on various parts of the plumage. This genus differs from the last not only in the general coloration of the plumage but also in the shape of the tail, which in *Sporæginthus* is much more rounded.

Key to the Species *.

a. Abdomen black...................... *S. amandava* ♂, p. 192.
b. Abdomen yellowish red .... *S. flavidiventris* ♂, p. 193.

738. Sporæginthus amandava. *The Indian Red Munia.*

Fringilla amandava, *Linn. Syst. Nat.* i, p. 319 (1766).
Fringilla punicea, *Horsf. Tr. Linn. Soc.* xiii, p. 160 (1820).
Estrelda amandava (*Linn.*) *Blyth, Cat.* p. 118; *Horsf. & M. Cat.* ii,
    p. 502; *Jerd. B. I.* ii, p. 359; *Hume, N. & E.* p. 454; *Legge,
    Birds Ceyl.* p. 662; *Hume, Cat.* no. 704; *Barnes, Birds Bom.*
    p. 264.
Sporæginthus amandava (*Linn.*), *Sharpe. Cat. B. M.* xiii, p. 320;
    Oates in Hume's N. & E. 2nd ed. ii, p. 147.
The Red Wax-bill, Jerd.; *Lal munia,* Hind.; *Torra jinuvayi,* Tel.

Coloration. Male. The fully adult has the whole head, upper plumage, neck, breast, and sides of the body crimson, with the ashy or brown bases of the feathers showing through more or less; rump and upper tail-coverts, sides of the neck, breast, and body spotted with white; abdomen, vent, and under tail-coverts black, the feathers of the abdomen with crimson fringes; wings and coverts brown, each covert-feather and the tertiaries with a terminal white spot; primary-coverts and winglet plain brown; tail blackish, the outer feathers tipped white.

Female. Upper plumage and scapulars brown; upper tail-coverts dull crimson with minute white tips; tail dark brown, the lateral feathers tipped white; wings brown, the median and greater coverts with the tertiaries tipped white; lores black; chin and throat whitish; sides of the head and neck and the breast ashy brown; remainder of lower plumage dull saffron, the sides of the body more or less tinged with ashy.

The young have the whole upper plumage brown, the wing-coverts and tertiaries broadly edged with fulvous; the whole lower plumage uniform ochraceous brown; bill dark brown.

Iris orange-red; bill red, dusky at base of culmen; legs and feet brownish flesh (*Butler*).

Length about 4·5; tail 1·6; wing 1·9; tarsus ·55; bill from gape ·4.

---

* The females of the two species are not separable by any characters known to me.

Sharpe is of opinion that the male bird of this species undergoes a seasonal change of plumage. I cannot follow him in this, as all the evidence I can find in the large series of this bird in the British Museum leads me to the same conclusion I arrived at some years ago with respect to the allied Burmese race, viz., that the male is a very considerable period in acquiring his perfectly mature dress, but that having once acquired it he never changes.

The nestling male at the first autumn appears to don the female plumage, and from this point slowly advances step by step towards his complete adult plumage, which is probably not fully attained till the second autumn or a short time previously.

Distribution. The whole of India proper from Sind to Assam and from the foot of the Himalayas to Cape Comorin; Ceylon; the hill-ranges of Assam, Cachar, Sylhet, and Tipperah. This species is again found in Siam, Cochin China, Singapore, and Java.

Habits, &c. Appears to breed twice a year, once in the cold season and once in the rains, constructing its nest near the ground. The eggs measure about ·55 by ·43.

739. Sporæginthus flavidiventris. *The Burmese Red Munia.*

Estrelda flavidiventris, *Wallace, P. Z. S.* 1863, pp. 486, 495; *Wardlaw Ramsay, Ibis,* 1877, p. 461; *Anders. Yunnan Exped., Aves,* p. 600; *Hume, Cat.* no. 704 bis.
Estrelda amandava (*Linn.*), *Oates, S. F.* iii, p. 342.
Estrilda burmanica, *Hume, S. F.* iv, p. 484; *Oates, S. F.* v, p. 163.
Estrilda punicea (*Horsf.*), *Oates, B. B.* i, p. 371.
Sporæginthus flavidiventris (*Wall.*), *Sharpe, Cat. B. M.* xiii, p. 323; *Oates in Hume's N. & E.* 2nd ed. ii, p. 149.

Coloration. Very similar to *S. amandava*, the male differing from the male of that species in having the abdomen yellowish red. The females of the two species are apparently undistinguishable. The young are also alike, and the males undergo the same changes in adopting the adult plumage.

Bill deep red, the posterior half of culmen black; iris crimson; eyelids purpurescent; inside of mouth salmon-colour; legs flesh-colour; claws horny.

Length 4; tail 1·5; wing 1·8; tarsus ·55; bill from gape ·35.

Upon re-examining Horsfield's type of *Fringilla punicea* from Java, it now appears to me to be a specimen of *S. amandava* rather than of *S. flavidiventris.* Such is also Sharpe's opinion. A considerable number of specimens from Singapore are undoubtedly *S. amandava.* The distribution of the two species is thus very difficult to understand.

Distribution. Burma, from the neighbourhood of Bhámo down to the southern coast of Pegu and to Karennee and Central Tenasserim. This species occurs in the islands of Flores and Timor.

Habits, &c. Breeds in Pegu in October and November, constructing its nest in clumps of low grass. The eggs, four to six in number, measure about ·56 by ·44.

Family FRINGILLIDÆ.

The intrinsic muscles of the syrinx fixed to the ends of the bronchial semi-rings; the edges of the mandibles smooth; the hinder part of the tarsus longitudinally bilaminated, the laminæ entire and smooth; wing with nine primaries, the first and second about equal in length; secondary quills reaching about three-quarters the length of the wing; bill more or less conical; tail of twelve feathers; tarsus scutellated; nostrils pierced close to the line of the forehead and very near the culmen; rictal bristles few and short; plumage of nestling various; sexes generally dissimilar.

The *Fringillidæ* or *Finches* comprise a very large number of birds which have a considerable general resemblance to each other, and are characterized by points of structure which render their separation from other groups comparatively easy.

Although Finches have, as a rule, but one moult a year, yet their summer and winter plumages differ considerably in many of the species. In spring and summer the margins of the feathers are lost by abrasion or by being cast off, and then the colour of the parts affected becomes more uniform and frequently more brilliant.

The Finches are normally granivorous or frugivorous, but they also eat insects and the young are fed entirely on these. They are for the most part gregarious and arboreal, but they descend to the ground freely to pick up food. Many of them are good songsters, and they are all hardy and bear captivity well.

Sharpe, in the twelfth volume of the British Museum Catalogue of Birds, has treated this family in a very complete and satisfactory manner. This was the first Catalogue written by him with the combined Hume and Tweeddale Collections at his disposal. I follow him in the arrangement of this group, and I have found no reason to differ from him except in some minor matters, such as the extent of a few of the genera.

The *Fringillidæ* may be divided into three very natural sub-families by the character of the shape of the skull and bill.

Upper mandible produced backwards
beyond the front line of the bony
orbit; inferior outline of lower
mandible straight or nearly so .... *Coccothraustinæ*, p. 196.

Fig. 50.—Skull of *Coccothraustes vulgaris*.

Upper mandible not produced backwards
beyond front line of orbit; inferior
outline of lower mandible with a
slight re-entering angle; cutting-
edges of upper and lower mandibles
everywhere in contact .... *Fringillinæ*, p. 202.

Fig. 51.—Skull of *Fringilla cœlebs*.

Upper mandible not produced backwards
beyond front line of orbit; inferior
line of lower mandible greatly an-
gulate; cutting-edges of mandibles

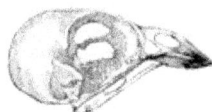

Fig. 52.—Skull of *Emberiza citrinella*.

not everywhere in contact, but
leaving a gap of greater or less
extent........................ *Emberizinæ*, p. 249.

Subfamily COCCOTHRAUSTINÆ.

The subfamily *Coccothraustinæ* contains those Finches which are characterized by a very large bill. They are birds of considerable size and rather bright coloration, and in all the Indian species the sexes differ from each other. They have only one moult a year.

The Indian Grosbeaks are chiefly inhabitants of the higher parts of the Himalayas; they live in forests, feed on stony fruits, and are mostly gregarious. They make, so far as is known, cup-shaped nests in trees and lay spotted eggs.

Key to the Genera.

a. Tips of later primaries and earlier secondaries square or sinuated; margin of upper mandible not toothed near gape .... COCCOTHRAUSTES, p. 196.
b. Tips of later primaries and earlier secondaries rounded or pointed; margin of upper mandible sinuated or toothed near gape.
　a'. Difference between wing and tail hardly equal to tarsus............ PYCNORHAMPHUS, p. 198.
　b'. Difference between wing and tail about equal to twice tarsus ...... MYCEROBAS, p. 200.

Genus COCCOTHRAUSTES, Brisson, 1760.

The genus *Coccothraustes* contains the Hawfinches, of which two species are known, one inhabiting a considerable portion of Europe and Asia, and the other the north-west portion of the Punjab and probably Afghanistan.

In *Coccothraustes* the bill is conical, with the culmen nearly straight and the cutting-edge of the upper mandible curved, but not toothed near the gape; the nostrils partially concealed by hairs; tail short and almost square; wing sharp, the primaries, commencing from the fifth, with sinuated or square tips; tarsus short.

The nestling in this genus is highly spotted and also suffused with yellow.

740. Coccothraustes humii. *Hume's Hawfinch.*

Coccothraustes vulgaris (*Pall.*), *Hume, Ibis,* 1869, p. 456; *id. S. F.* vii, p. 462; *id. Cat. no.* 728 bis; *Barnes, S. F.* ix, p. 456.
Coccothraustes humii, *Sharpe, P. Z. S.* 1886, p. 97; *id. Cat. B. M.* xii, p. 40, pl. 1.

Coloration. Male. Feathers immediately next the bill, the lores, chin, and throat black; a narrow band next these black parts dull white; forehead, crown, nape, back, scapulars, and tertiaries tawny brown; a broad ashy collar on the hind neck and sides of neck; rump, upper tail-coverts, sides of the head, and the whole lower plumage a paler but clearer tawny brown; middle of abdomen and the under tail-coverts white; tail black, the feathers with broad white tips, the middle pair frequently ashy for some distance in front of the white tip; lesser wing-coverts brown, tipped ashy; median

Fig. 53.—Head of *C. humii.*

coverts and the greater part of the outer webs of the greater coverts white; remainder of wing black, the primaries tipped with metallic blue, and each with a large white patch on the inner web; the later primaries and secondaries edged with metallic lilac or purple.

Female. Black parts of the head as in the male; remainder of head and neck ashy brown; other parts of plumage as in male, but the tawny brown everywhere very pale and dull, the wings chiefly brown with some ashy on the outer webs.

Both sexes in winter have the black feathers of the chin and throat narrowly tipped with white. These margins soon wear away.

The young of this species are unknown, but in the European ally the nestling is brown above with black tips to the feathers; the head is suffused with yellow; the lower plumage is white, each feather with a black terminal bar; the wings and tail resemble those of the adult.

Bill in winter whitish; in summer blue; legs flesh-colour.

Length about 7; tail 2·5; wing 4; tarsus ·85; bill from gape ·85.

This species differs from *C. vulgaris* of Europe in having a lighter and less richly coloured head, a paler back, and the lower plumage tawny brown, not vinaceous.

Distribution. The only specimens of this species that I have seen were collected at Attock in the Punjab in February and March. There can be little doubt that the Hawfinch procured by Barnes at Chaman in Afghanistan belonged to this species. Of it he remarks that it is a common bird and resident.

Genus **PYCNORHAMPHUS**, Hume, 1874.

In *Pycnorhamphus* the bill is very similar in shape to that of *Coccothraustes*, but the cutting-edge of the upper mandible is toothed near the gape; the tail is square and comparatively long, the difference in length between it and the wing being about equal to the tarsus; the primaries have the ordinary rounded tips.

The Grosbeaks of this genus are of rather large size and well-marked colours. They inhabit the Himalayas and but little is known of their habits.

The nestling bird appears to resemble the adult female closely in this genus, but it is doubtful whether the young male moults into adult plumage at the first autumn moult. Materials for settling this question are at present insufficient.

Key to the Species.

a. No white spot on wing.
   a'. Head black.
      a". Thighs black .... *P. icteroides* ♂, p. 198.
      b". Thighs yellow .... *P. affinis* ♂, p. 199.
   b'. Head ashy.
      c". Breast ashy grey; abdomen fawn-
         buff .... *P. icteroides* ♀, p. 198.
      d". Breast and abdomen olive-yellow .. *P. affinis* ♀, p. 199.
b. A white spot on wing .... *P. carneipes*, p. 200.

741. Pycnorhamphus icteroides. *The Black and Yellow Grosbeak.*

Coccothraustes icteroides, *Vigors, P. Z. S.* 1830, p. 8; *Gould, Cent.* pl. 45; *Blyth, Cat.* p. 125.
Hesperiphona icteroides (*Vig.*), *Horsf. & M. Cat.* ii, p. 462; *Jerd. B. I.* ii, p. 384; *Stoliczka, J. A. S. B.* xxxvii, pt. ii, p. 59; *Hume & Henders. Lah. to Yark.* p. 257; *Cock & Marsh. S. F.* i, p. 358; *Brooks, J. A. S. B.* xli, pt. ii, p. 84; *Wardlaw Ramsay, Ibis,* 1880, p. 66; *C. H. T. Marshall, Ibis,* 1884, p. 420.
Pycnorhamphus icteroides (*Vig.*), *Hume, N. & E.* p. 439; *id. Cat.* no. 725; *Sharpe, Cat. B. M.* xii, p. 41; *Oates in Hume's N. & E.* 2nd ed. ii, p. 150.

Coloration. Male. The whole head, chin and throat, wings, scapulars, sides of the back, upper tail-coverts, under wing-coverts, axillaries, and thighs dull black; remainder of plumage deep yellow, tinged with orange on the hind neck.

Female. Head, neck, chin, throat, breast, axillaries and under wing-coverts, back, scapulars, lesser and median wing-coverts, and the greater part of the outer webs of the greater coverts and secondaries ashy grey, the head darker than the other parts; rump fulvous; upper tail-coverts grey; winglet, primary-coverts, and primaries black; abdomen, sides of the body, and under tail-coverts fawn-buff.

A young male shot in August is moulting from the female to the adult male plumage.

Legs and feet fleshy pink; bill horny greenish; iris reddish brown (*Hume*). The bill becomes yellow in winter. A young bird had the bill waxy green; iris hazel; legs and feet pale fleshy (*Bingham*, August).

Length about 9; tail 3·7; wing 5·2; tarsus 1; bill from gape 1.

Distribution. The Himalayas from Murree and Central Kashmir eastwards to Garhwál, where this species is found in the hills north of Mussooree. Jerdon's statement that this bird extends into Nepal requires confirmation. This Grosbeak occurs from 5000 to 9000 feet, and according to Stoliczka not beyond the limit of the large forests.

Habits, &c. Breeds in May and June, constructing a nest of twigs and grass, lined with fern-roots, in a branch of a tree, and laying two or three eggs, which are white marked with broad longitudinal dashes of rufous-brown at the larger end, and measure from ·9 to 1·07 in length by ·77 to ·81 in breadth.

712. Pycnorhamphus affinis. *The Allied Grosbeak.*

Hesperiphona affinis, *Blyth, J. A. S. B.* xxiv, p. 179 (1855); *Jerd. B. I.* ii, p. 385; *Blyth, Ibis,* 1867, p. 43.
Pycnorhamphus affinis (*Blyth*), *Hume, Cat.* no. 726; *Sharpe, Cat. B. M.* xii, p. 46.

Coloration. Male. The whole head, chin and throat, the upper part of the fore neck, wings, scapulars, the sides of the back, and the tail deep black; the feathers of the middle line of the back black on the outer web, yellow on the inner; neck all round, rump, and entire lower plumage from the throat downwards rich yellow, tinged with orange on the rump and hind neck; upper tail-coverts black; under wing-coverts and axillaries black.

Female. The whole head, chin, and throat ashy; hind neck, sides of neck, rump, and lower plumage olive-yellow; back, scapulars, upper tail-coverts, the lesser and median wing-coverts, and the greater portion of the outer webs of the greater coverts and secondaries ashy green; remainder of wing and the tail deep black.

Males not quite adult have the head, chin, and throat dark brown with pale fringes, and the lower plumage saffron-yellow.

Bill bluish in winter, yellow in summer; feet fleshy yellow (*Jerdon*). The few dated specimens in the British Museum, however, show that the bill is blue in summer and yellow in winter, in the dried state at least.

Length about 9; tail 3·8; wing 5; tarsus 1; bill from gape ·95.

Distribution. Nepal and Sikhim, extending into Tibet and Western China. In the British Museum there is a specimen of this bird which is marked as having been procured at Dharmsála. This Grosbeak appears to be found only at high elevations.

743. **Pycnorhamphus carneipes.** *The White-winged Grosbeak.*

Coccothraustes carneipes, *Hodgs. As. Res.* xix, p. 151 (1836);
   Blyth, Cat. p. 125.
Mycerobas carneipes (*Hodgs.*), *Horsf. & M. Cat.* ii, p. 462; *Jerd.
   B. I.* ii, p. 387; *Wardlaw Ramsay, Ibis,* 1879, p. 118, 1880, p. 66;
   Biddulph, Ibis, 1881, p. 81; *Scully, Ibis,* 1881, p. 577.
Pycnorhamphus carneipes (*Hodgs.*), *Hume, Cat.* no. 728; *Sharpe, Cat.
   B. M.* xii, p. 47.

Coloration. Male. The whole head, neck, back, scapulars, wings,
chin, throat, breast, upper abdomen, upper tail-coverts, and tail
black with ashy margins; the upper tail-coverts margined with
greenish yellow: the scapulars, innermost greater coverts, and
tertiaries tipped with greenish yellow on the outer web; all but
the first primary with a white patch at base; the primaries and
secondaries narrowly margined with white on the outer web near
the tip: rump, lower abdomen, sides of body, and under tail-coverts
greenish yellow; thighs ashy brown; under wing-coverts and
axillaries pale ashy.

Female. Very similar to the male in general appearance. The
dark parts of the plumage are ashy brown, not black, and the
margins of the feathers have a greenish tinge; the cheeks and the
sides of the head are streaked with whitish; the lower abdomen,
sides of the body, and under tail-coverts are ashy yellow; the breast
is more or less streaked with white, but is occasionally quite plain.

Upper mandible brownish, the lower one whitish horn-colour;
legs pale fleshy brown; iris hair-brown (*Wardlaw Ramsay*). The
bill does not appear to undergo any seasonal change of colour.

Length 8 to 9; tail 3·5 to 4; wing 4·3 to 4·8; tarsus 1; bill
from gape ·8 to 1. The size of this species varies extremely but not
according to locality, probably according to age.

Distribution. The Himalayas from Gilgit to Sikhim, generally
above 8000 feet, but occasionally descending to 5000 feet. This
species extends to Afghanistan on the west, and to parts of Central
Asia on the north.

Genus **MYCEROBAS**, Cabanis, 1847.

In the genus *Mycerobas* the bill is of very great size, the height
at the nostrils being about equal to the length of the bill; the
cutting-edge of the upper mandible, as in *Pycnorhamphus*, is
provided with a large tooth near the gape, and the nostrils are
covered by hairs; the tail is comparatively short and decidedly
forked; and the wing-quills have ordinary rounded tips. The sexes
differ in colour.

The only member of this genus inhabits the Himalayas, and has
also occasionally been found in Manipur. Very little is known of
its habits.

714. **Mycerobas melanoxanthus.** *The Spotted-winged Grosbeak.*

Coccothraustes melanoxanthus, *Hodgs. As. Res.* xix, p. 159 (1836);
Blyth, Cat. p. 125.
Mycerobas melanoxanthus (*Hodgs.*), *Horsf. & M. Cat.* ii, p. 461;
Jerd. B. I. ii, p. 386; *Blanf. J. A. S. B.* xli, pt. ii, p. 64; *Godw.-
Aust. J. A. S. B.* xlv, pt. ii, p. 209; *Hume, Cat.* no. 727; *Sharpe,
Cat. B. M.* xii, p. 41; *Hume, S. F.* xi, p. 285.

Multam-pho. Lepch.

Fig. 54.—Head of *M. melanoxanthus.*

Coloration. Male. The whole upper plumage, sides of the head
and neck, chin and throat slaty black, each feather with an ashy
margin more or less distinct; wings black, the feathers margined
with ashy, the inner greater coverts and tertiaries with an elongated
oval pale yellow spot on the outer web near the tip; the fourth to
eighth primaries white at base: the secondaries and inner primaries
with a short white margin near the tip of the outer web; tail black:
lower plumage deep yellow; axillaries black, tipped with yellow.

Female. Upper plumage black, the feathers edged with yellowish
green, those of the head, hind neck, and back subterminally bright
yellow, causing those parts to be about equally black and yellow;
feathers of forehead and those at the side of the crown almost pure
yellow; a broad black band from the lores through the eye to the
ear-coverts, followed below by a yellow band; a black patch on the
cheeks; sides of the chin and throat, sides of neck, breast, and
sides of body deep yellow streaked with black; chin, throat,
abdomen, and under tail-coverts deep unstreaked yellow; wings
and tail much as in the male, but the feathers margined with
greenish yellow.

The nestling resembles the female in general appearance, but
has the yellow of the head and upper parts replaced by yellowish
white, and the lower plumage pale vinaceous streaked with black
and occasionally tinged with yellow. It is difficult to trace the
transition of plumage from youth to maturity in the series in the
British Museum, but Hodgson asserts that the young males retain
the plumage of the female till after the second moult.

Bill leaden blue; feet leaden grey, claws brown; iris brown
(*Hodgson*).

Length about 8·5; tail 3·1; wing 5; tarsus ·85; bill from gape 1.

Distribution. The Himalayas from the Hazára country to Sikhim at considerable elevations; Manipur.

Subfamily FRINGILLINÆ.

The *Fringillinæ* comprise the Bullfinches, the Rose-Finches, the Crossbills, the true Finches, the Sparrows, and the Mountain-Finches. They have a bill of medium size, the upper mandible not being produced behind the front line of the bony orbit, and the cutting-edges of the two mandibles are everywhere in contact.

The *Fringillinæ* have one moult a year only, but the wearing away of the margins of the feathers in parts of the plumage in the spring causes many of them to have a summer plumage, which in some cases is very different to the winter dress. The young birds resemble the adult females and probably retain this dress till the second autumn.

The *Fringillinæ* are more or less gregarious or sociable, live both on seeds and insects, and are frequently good songsters.

I have made the division of the *Fringillinæ* into genera to depend in great measure on types of colour as well as structure *.

Key to the Genera.

a. Rump white; quills and tail uniform black............................ PYRRHULA, p. 204.
b. Inner webs of tertiaries white .... PYRRHOPLECTES, p. 207.
c. Sexes (except in *Erythrospiza*) very dis-
similar; males red or pink, females
brown or greenish; no white on tail;
tail forked.
  a'. Tips of mandibles crossed............ LOXIA, p. 208.
  b'. Bill of normal shape.
    a''. Bill short and thick: culmen curved.
     a'''. Tail conspicuously short, the tip
of the wings reaching considerably
beyond middle of tail.
      a⁴. Nostrils exposed; male scarlet,
female green .... HÆMATOSPIZA, p. 209.
      b⁴. Nostrils densely plumed; sexes
nearly similar .... ERYTHROSPIZA, p. 221.
     b'''. Tail of moderate length; tip of
wing not reaching beyond middle
of tail.

---

* I cannot determine the following Finches:—
FRINGILLA PYRRHOPTERA, Less. in Bélang. Voy. p. 271 (1834).
PROPASSER MURRAYI, Blyth, J. A. S. B. xxxii, p. 458 (1863).
LINOTA PYGMÆA, Stoliczka, J. A. S. B. xxxvii. pt. ii. p. 62 (1868).
All of which were described from specimens procured in India, but too insufficiently to be recognizable.

c⁴. Males with the abdomen and
breast of quite different colours.
  a⁵. Culmen much longer than
    depth of bill at base ...... PYRRHOSPIZA, p. 211.
  b⁵. Culmen about as long as depth
    of bill at base .... PROPYRRHULA, p. 210.
d⁴. Males with the abdomen and
breast of the same colour.
  c⁵. A supercilium present in both
    sexes.................... PROPASSER, p. 212.
  d⁵. No supercilium present in
    either sex .... CARPODACUS, p. 219.
b″. Bill lengthened and slender: culmen
straight .... PROCARDUELIS, p. 223.
d. Lateral tail-feathers largely marked with
white : tail forked.
  c′. Bill long, slender and pointed ; a large
  amount of yellow on wings .... CARDUELIS, p. 225.
  d′. Bill long and thick ; wings spotted
  with white .... CALLACANTHIS, p. 226.
  e′. Bill short and pointed ; plumage
  streaked .... ACANTHIS, p. 227.
e. Both sexes with greater part of plumage
yellow or green.
  f′. Bill small and swollen; culmen curved. METOPONIA, p. 230.
  g′. Bill lengthened and sharp; culmen
  straight.
    c″. Length of culmen about equal to
    depth of bill at base ; sexes closely
    alike .... HYPACANTHIS, p. 231.
    d″. Length of culmen about twice depth
    of bill at base ; sexes dissimilar .. CHRYSOMITRIS, p. 232.
f. Sexes not strikingly dissimilar; two bars
on wing-coverts; throat and breast
generally rufous; bill lengthened and
straight .... FRINGILLA, p. 233.
g. A yellow patch on throat ; no pattern on
outer webs of earlier primaries........ GYMNORHIS, p. 235.
h. Pale margins on outer webs of earlier
primaries not of uniform width, forming
two patches, one below primary-coverts
and one above emarginations.
  h′. No yellow patch on throat and no
  white spots on tail .... PASSER, p. 236.
  i′. A yellow patch on throat and white spots
  on tail .... PETRONIA, p. 243.
i. Tail square, middle pair of feathers as long
as the others; quills and tail largely
white ; no yellow on throat .... MONTIFRINGILLA, p. 244.
k. Tail forked and not marked with white ;
sexes alike ; plumage occasionally pink
on rump and shoulder................ FRINGILLAUDA, p. 247.

Genus PYRRHULA, Briss., 1760.

The genus *Pyrrhula* comprises the Bullfinches, which are characterized by a short and very swollen bill, a white rump, and deep black wings and tail. The sexes differ considerably from each other. Three of the Indian species of Bullfinches have the tail deeply forked, but the fourth has it nearly square.

The Bullfinches inhabit well-wooded districts and are strictly arboreal. They sing well and are easily kept in captivity. They make cup-shaped nests in trees, and lay blue eggs spotted with various shades of brown. Hardly anything, however, is known of the nidification of the Indian species.

Key to the Species.

a. Middle pair of tail-feathers nearly as long
   as outermost pair .... *P. aurantiaca*, p. 204.
b. Middle pair of tail-feathers fully half an
   inch shorter than outermost pair.
   a'. A well-defined deep black band round
       base of bill.
       a''. Crown varying from greenish yellow
            to vermilion.................... *P. erythrocephala*, p. 205.
       b''. Crown ashy grey .... *P. erithacus*, p. 206.
   b'. An ill-defined brown band round base
       of bill .... *P. nepalensis*, p. 206.

745. Pyrrhula aurantiaca. *The Orange Bullfinch.*

Pyrrhula aurantiaca, *Gould, P. Z. S.* 1857, p. 222 ; *Jerd. B. I.* ii,
p. 390; *Hume & Henders. Lah. to Yark.* p. 258; *Hume, N. & E.*
p. 470; *id. Cat.* no. 732; *Biddulph, Ibis,* 1881, p. 82 ; *Scully, Ibis,*
1881, p. 577; *Sharpe, Cat. B. M.* xii, p. 455; *Oates in Hume's*
N. & E. 2nd ed. ii, p. 151.

Coloration. Male. A broad black band round the base of the bill ; rump, under tail-coverts, axillaries, and under wing-coverts white ; upper tail-coverts, tail, and quills of wing black ; lesser and median wing-coverts dusky edged with orange-rufous, the latter subterminally ashy ; greater series black, broadly tipped with orange-rufous ; the other parts of the body deep orange.

Female. A broad black band round the base of the bill exactly as in the male : crown, nape, hind neck, and sides of the head ashy brown ; back and scapulars yellowish brown ; rump white : upper tail-coverts and tail black ; lesser and median coverts like the back but with subterminal ashy bars ; greater coverts black, with broad tips of the colour of the back but paler ; throat and breast pale rufous ; abdomen dull yellowish ; under wing-coverts and axillaries white.

Young males have the sides of the head and the whole lower plumage bright yellow and the upper plumage more or less yellow. Bill black : feet fleshy ; iris dark brown (*Jerdon*).

Length about 5·5 ; tail 2·4 ; wing 3·2 ; tarsus ·65 ; bill from gape ·45.

Distribution. The Hazára country and Kashmir, extending into the adjoining native territory.

Habits, &c. Appears to breed in July, but the nest and eggs have not been taken.

746. Pyrrhula erythrocephala. *The Red-headed Bullfinch.*

Pyrrhula erythrocephala, *Vigors, P. Z. S.* 1831, p. 174; *Gould, Cent.* pl. 32; *Blyth, Cat.* p. 123; *Horsf. & M. Cat.* ii, p. 454; *Jerd. B. I.* ii, p. 389; *Stoliczka, J. A. S. B.* xxxvii, pt. ii, p. 59; *Blanf. J. A. S. B.* xli, pt. ii, p. 64; *Hume, Cat. no.* 729; *Sharpe, Cat. B. M.* xii, p. 457.

Fig. 55.—Head of *P. erythrocephala.*

Coloration. Male. A broad black band round the base of the bill, followed by a pale zone which is succeeded on the crown, nape, and hind neck by vermilion, and on the sides of the head, sides of the neck, throat, breast, and upper abdomen by paler red; back, scapulars, lesser and median wing-coverts ashy grey; rump white, with a narrow black bar between it and the lower back: greater wing-coverts black, broadly tipped with ashy grey; quills, upper tail-coverts, and tail black; lower abdomen pale ashy, passing to pure white on the under tail-coverts; under wing-coverts and axillaries white.

Female. A broad black band round the base of the bill, followed by a pale zone which is succeeded by yellowish green on the crown, nape, and hind neck, and by drab-brown on the entire lower plumage except the abdomen, under tail-coverts, axillaries, and under wing-coverts, which are white: back, scapulars, lesser and median wing-coverts drab-brown: greater coverts black, broadly tipped with drab-brown; quills, upper tail-coverts, and tail black; rump white.

The young males are at first like the female, and gradually assume the plumage of the adult male by a slow change of colour in the feathers.

Bill black; legs pale fleshy brown; iris light brown (*Jerdon*).

Length about 5·5; tail 2·6; wing 3·1; tarsus ·65: bill from gape ·45.

Distribution. The Himalayas from Chamba and Southern Kashmir to Bhutan. Stoliczka states that this species breeds near Kotgarh between 6000 and 8000 feet, and Blanford met with it in Sikhim at 11,000 feet.

747. Pyrrhula erithacus. *Beavan's Bullfinch.*

Pyrrhula erythraca, *Blyth, Ibis,* 1862, p. 389; *Jerd. B. I.* ii, p. 389; *Hume, S. F.* ii, p. 455; *id. Cat. no. 730.*
Pyrrhula erithacus, *Blyth, Ibis,* 1863, p. 441, pl. 10; *Sharpe, Cat. B. M.* xii, p. 455.

Coloration. Male. A broad black band round the base of the bill, edged posteriorly with greyish white shading off into the dark ashy grey of the crown, nape, back, and scapulars, and the paler ashy grey of the sides of the head and throat; a broad black band across the rump followed by a broader white band; upper tail-coverts and tail deep glossy black; lesser and median coverts ashy grey with dark centres; greater coverts with basal half black and terminal half ashy grey; quills black, the outer webs becoming more glossy inwardly and the tertiaries being entirely glossy black; breast, upper half of abdomen and of the sides of the body orange; remainder of lower plumage greyish white; under wing-coverts and axillaries greyish white.

Female. Differs from the male by having all the underparts, with the exception of the white abdomen, of a light chocolate-colour, and the back darker.

Iris dark brown; culmen black; tarsus body-colour (*Prjevalsky*).

Length about 6; tail 2·9; wing 3·3; tarsus ·65; bill from gape ·45.

Distribution. Sikhim, extending to Kansu and Western China.

748. Pyrrhula nepalensis. *The Brown Bullfinch.*

Pyrrhula nipalensis, *Hodgs. As. Res.* xix, p. 155 (1836); *Blyth, Cat.* p. 122; *Horsf. & M. Cat.* ii, p. 455; *Jerd. B. I.* ii, p. 390; *Blanf. J. A. S. B.* xli, pt. ii, p. 65; *Hume, Cat. no.* 731; *Scully, S. F.* viii, p. 335; *Sharpe, Cat. B. M.* xii, p. 453.

Coloration. Male. Forehead and feathers round the bill dark brown; crown and nape ashy brown, the feathers with dusky centres; hind neck, sides of the neck and head, throat, breast, and the whole lower plumage (except the under tail-coverts and middle of the abdomen which are white) plain ashy brown; back and scapulars darker ashy brown; a white patch under and behind the eye; upper part of rump purplish black, lower white; upper tail-coverts and tail black with a bronze gloss, and all the feathers tipped with velvety black; lesser and median wing-coverts dark ashy brown; greater wing-coverts pale ashy brown, the outer ones broadly margined with purplish black; quills black, margined on the outer web increasingly with purplish black, the tertiaries becoming entirely of this colour and the innermost margined exteriorly with crimson; under wing-coverts and axillaries white.

Female. Resembles the male, and differs merely in the edging to the innermost tertiary being yellow instead of crimson.

The young appear to resemble the adult female.

Bill greenish horn-colour with a black tip ; legs fleshy brown ; iris brown (*Hodgson*).

Length about 6·5 ; tail 3 ; wing 3·4 ; tarsus ·65 ; bill from gape ·5.

Distribution. The Himalayas from Garhwál to Sikhim. Blanford observed this species in Sikhim at 10,000 feet.

Genus **PYRRHOPLECTES**, Hodgs., 1844.

The genus *Pyrrhoplectes* contains one species, which resembles the Bullfinches in general appearance and structure, but it has the bill less tumid and the rump is of the same colour as the lower back. It may be recognized from all the other species of Indian *Fringillidæ* by the colour of the inner webs of the tertiaries, which are pure white. Both sexes possess this character.

749. **Pyrrhoplectes epauletta.** *The Gold-headed Black Finch.*

Pyrrhula ? epauletta, *Hodgs. As. Res.* xix, p. 156 (1836).
Pyrrhoplectes epauletta (*Hodgs.*), *Horsf. & M. Cat.* ii, p. 455 ; *Jerd. B. I.* ii, p. 392 ; *Blanf. J. A. S. B.* xli, pt. ii, p. 65 ; *Hume, Cat.* no. 733 ; *Sharpe, Cat. B. M.* xii, p. 386.
Pyrrhuloides epauletta (*Hodgs.*), *Blyth, Cat.* p. 337.

The Gold-headed Black Bullfinch, Jerd. ; *Lho samprek-pho,* Lepch.

Coloration. Male. The whole plumage black except the hinder part of the crown and nape, which are golden orange, the axillaries and a small portion of the middle abdomen, which are orange-buff, the under wing-coverts and the inner webs of the tertiaries, which are white.

Female. Forehead, anterior part of crown, lores, round the eyes, and base of cheeks ashy grey more or less suffused with olive-yellow, and turning to this colour entirely on the remainder of the crown, the ear-coverts, and the sides of the head ; hind neck, sides of neck, and upper back ashy grey, turning to chestnut-brown, which is the colour of the lower back, rump, upper tail-coverts, scapulars, and visible portions of the wing-coverts ; wings and tail dark brown, the tertiaries rufous on the outer, white on the inner, webs ; entire lower plumage chestnut-brown except the axillaries, which are orange-buff, and the under wing-coverts, which are whitish.

The nestling resembles the adult female and moults into female adult plumage at the first autumn moult ; the young males acquire the adult plumage during the first winter and by a very gradual process extending over several months.

Bill dusky horny ; legs brown ; iris brown (*Jerdon*).

Length nearly 6 ; tail 2·4 ; wing 3·4 ; tarsus ·75 ; bill from gape ·5.

Distribution. The Himalayas from the eastern side of the valley of the Satlej to Sikhim. Blanford met with this species in Sikhim at 11,000 feet.

Habits, &c. Little is recorded of the habits of this bird. Hodgson says it is shy, adhering to the forests.

Genus LOXIA, Linn., 1766.

The genus *Loxia* contains one Indian species of Finch which may be recognized at a glance by the peculiar structure of the bill, in which the tips of the mandibles cross each other. The plumage of the male is red and that of the female greenish. In *Loxia* the wing is very long, reaching to a considerable distance beyond the middle of the tail.

Fig. 56.— Head of *L. himalayana*.

The Crossbills feed chiefly on seeds from the cones of various pine-trees, for the extraction of which their bill is specially adapted.

750. Loxia himalayana. *The Himalayan Crossbill.*

Loxia himalayana, *Hodgs., Gray, Zool. Misc.* p. 85 (1844); *Horsf. & M. Cat.* ii, p. 453; *Jerd. B. I.* ii, p. 393; *Stoliczka, J. A. S. B.* xxxvii, pt. ii, p. 60; *Hume, Cat. no.* 734.
Loxia himalayensis, *Hodgs. J. A. S. B.* xiii, p. 952 (1844); *Blyth, Cat.* p. 123.
Loxia curvirostra, *Linn., Sharpe, Cat. B. M.* xii, p. 455 (part.).

Coloration. Male. Forehead, crown, nape, and hind neck red, the rump brighter red; back and scapulars brown, the feathers broadly fringed with red; wing-coverts brown, margined with rufous-brown; primary-coverts, winglet, and quills blackish with very narrow rufous margins; upper tail-coverts and tail dark brown margined with rufous; sides of the head dark brown, more or less mixed with red; lower plumage red; under tail-coverts brown, broadly edged with whitish; under wing-coverts and axillaries ashy brown washed with rufous.

Female. Upper plumage brown, each feather edged with olive-yellow; the rump purer yellow; wings and tail dark brown, margined narrowly with olive-yellow; chin, throat, and sides of the head and neck ashy, more or less mottled and washed with dull yellow; abdomen ashy; remainder of lower plumage dull yellow.

Young birds are ashy brown tinged with yellow and densely streaked all over with dark brown.

Bill and feet brown; iris dark hazel.

Length about 5·5; tail 2·2; wing 3·4; tarsus ·65; bill from gape ·75.

The Crossbills of the Himalayas form a very small race which I think it is advisable to keep distinct. There is a very marked difference in size between the Himalayan birds and *L. curvirostra*, from Northern Europe, on the one hand, and *L. japonica*, from Japan, on the other; and the only Crossbills which approach the Indian birds in size are some from America. Sharpe's view that all these Crossbills form but one species is no doubt correct; at the same time the Himalayan Crossbills are in my opinion quite distinguishable from all others in size, and it is consequently more convenient to retain them as distinct.

Distribution. The Himalayas from Chini and Lahul to Sikhim, extending into Tibet and Western China.

Habits &c. Inhabits the pine-forests and is highly gregarious.

Genus HÆMATOSPIZA, Blyth, 1844.

The genus *Hæmatospiza* is closely allied to *Loxia*, the male being red and the female green, but the bill is of a normal shape and very stout and strong. The wing is of considerable length, reaching beyond the middle of the tail. A curious feature of this species is the white colour of the bases to the feathers of the head and neck; these white bases show up when the feathers are disarranged and are hardly ever completely hidden.

751. Hæmatospiza sipahi. *The Scarlet Finch.*

Corythus sipahi, *Hodgs. As. Res.* xix, p. 151 (1836).
Hæmatospiza boetonensis (*Lath.*), *Blyth, J. A. S. B.* xiii, p. 951 (1844); *id. Cat.* p. 122.
Hæmatospiza sipahi (*Hodgs.*), *Blyth, Cat.* p. 342; *Horsf. & M. Cat.* ii, p. 454; *Jerd. B. I.* ii, p. 394; *Godw.-Aust. J. A. S. B.* xxxix, pt. ii, p. 110; *Hume, Cat. no.* 735.
Carpodacus sipahi (*Hodgs.*), *Sharpe, Cat. B. M.* xii, p. 397.

The Scarlet Grosbeak, Jerd.; *Phanying-pho biu*, Lepch.; *Labbia maphoo*, Bhut.

Fig. 57.—Head of *H. sipahi.*

Coloration. Male. The whole head and body brilliant scarlet, the concealed bases of the feathers ashy; wings black, every feather margined with scarlet; tail black with narrow crimson margins; thighs black; under tail-coverts black, with scarlet tips; axillaries and under wing-coverts ashy, with very small scarlet tips..

Female. Rump bright yellow; with this exception the whole upper plumage, sides of neck, lesser and median wing-coverts are dark brown, each feather with a large and well-defined greenish-yellow margin; greater coverts and quills dark brown, margined with greenish yellow on the outer webs; tail dark brown, with a greenish-yellow tinge on the outer webs; region of the eye and cheeks ochraceous yellow; ear-coverts greenish yellow with pale shafts; the whole lower plumage pale ochraceous, each feather with a subterminal black mark showing more or less clearly and sometimes concealed.

The young bird resembles the adult female; the male moults the first autumn into the female plumage again, but immediately after commences to assume the adult male plumage by a change of colour in the feathers, and probably attains the full plumage by the first breeding-season.

Bill yellow; legs brown; iris hazel-brown (*Jerdon*).

Length about 7·5; tail 2·8; wing 4; tarsus ·8; bill from gape ·8.

Distribution. Nepal and Sikhim from 5000 to 10,000 feet according to season; the Khási hills.

Habits, &c. According to Jerdon frequents both forest and bushy ground and has a loud whistling note. This species is captured in the Khási hills and kept in captivity.

Genus **PROPYRRHULA**, Hodgs., 1844.

The genus *Propyrrhula* connects *Loxia* and *Hæmatospiza* with the large group of the Rose-Finches. It has much the same bill as *Hæmatospiza* but a longer tail, and the plumage of the male is less brilliant, being much mixed with green above and the abdomen being brown. The female of *Propyrrhula*, however, is not very unlike the female of *Hæmatospiza*.

752. Propyrrhula subhimalayensis. *The Red-headed Rose-Finch.*

Corythus subhimachalus, *Hodgs. As. Res.* xix, p. 152 (1836).
Propyrrhula subhemachalana, *Hodgs.*, *Blyth, J. A. S. B.* xiii, p. 952 (1844); *Horsf. & M. Cat.* ii, p. 454.
Propyrrhula subhimachala (*Hodgs.*), *Blyth, Cat.* p. 123; *Jerd. B. I.* ii, p. 396; *Godw.-Aust. J. A. S. B.* xlv, pt. ii, p. 200; *Hume, Cat.* no. 736.
Propyrrhula subhimalayensis* (*Hodgs.*), *Sharpe, Cat. B. M.* xii, p. 462.

Coloration. *Male.* Forehead, supercilium, cheeks, chin, and throat crimson; fore neck and upper breast duller crimson, with the feathers subterminally paler, causing a mottled appearance; remainder of lower plumage greyish brown, paler on the abdomen; crown, nape, hind neck, sides of head and neck, back, scapulars,

---

* As pointed out by Sharpe, it will be convenient to employ this name for the present species rather than *subhimachalus* and *subhemachalana*.

and wing-coverts dull crimson, each feather with a brown centre; rump and upper tail-coverts bright crimson; quills and tail brown, margined with reddish.

Female. Forehead and an indistinct supercilium rather bright yellow; anterior part of crown yellow, the feathers with dusky centres, becoming ashy on the posterior portion of the crown, and the nape with narrow yellowish margins; back, scapulars, and wing-coverts bright olive-yellow, the feathers with dusky centres; rump and upper tail-coverts nearly pure olive-yellow; quills and tail brown, edged with olive-yellow; sides of the head ashy; lores, cheeks, and chin pale grey mottled with ashy; throat and sides of neck ashy; breast yellow, with ashy bases to the feathers; remainder of lower parts ashy, paler on the abdomen.

The young birds resemble the adult female.

Bill fleshy brown; legs pale brown; iris hazel-brown (*Jerdon*). Length about 8; tail 3·2; wing 3·8; tarsus ·8; bill from gape ·6.

Distribution. Nepal; Sikhim; Manipur.

Habits, &c. Jerdon states that he found this species near Darjeeling, frequenting the more open parts of the woods in small parties.

Genus PYRRHOSPIZA, Hodgs., 1844.

The genus *Pyrrhospiza* approaches the Rose-Finches still more than the preceding genus does, but differs in the male having the abdomen brown, streaked with black. The female is absolutely like the females of the Rose-Finches and has no green whatever in the plumage. In structure *Pyrrhospiza* resembles *Propyrrhula*, but in the former the bill is rather more slender and longer.

753. Pyrrhospiza punicea. *The Red-breasted Rose-Finch.*

Pyrrhospiza punicea, *Hodgs. Blyth, J. A. S. B.* xiii, p. 953 (1844); *Blyth, Cat.* p. 124; *Horsf. & M. Cat.* ii, p. 461; *Jerd. B. I.* ii, p. 406; *Stoliczka, J. A. S. B.* xxxvii, pt. ii, p. 60; *Blanf. J. A. S. B.* xli, pt. ii, p. 66; *Hume, Cat.* no. 747; *Sharpe, Cat. B. M.* xii, p. 431; *Oates in Hume's N. & E.* 2nd ed. ii, p. 152.

? Pyrrhospiza humii, *Sharpe, Cat. B. M.* xii, p. 433.

The Large Red-breasted Finch, Jerd.

Fig. 58.—Head of *P. punicea.*

Coloration. Male. The forehead and a broad but short supercilium crimson, each feather tipped with black or dusky; crown, nape, back, and scapulars black, each feather margined with light brown; rump rosy red with dusky tips; upper tail-coverts brown, with black shafts; wing-coverts dark brown with pale brown margins, the lesser series washed with rosy; quills and tail dark brown, very narrowly margined paler; a streak behind the eye and the sides of the neck and of the body pale brown, streaked with dark brown; cheeks, ear-coverts, chin, throat, and breast crimson, most of the feathers with white terminal shaft-streaks; abdomen ashy brown, sparingly streaked with black; under tail-coverts brown, margined with pink.

Female. The whole upper plumage, wings and tail, sides of the head and neck dark brown, each feather margined with pale brown and those of the rump with dull greenish; lower plumage pale fulvous with narrow black streaks, the breast more or less suffused with buff.

Bill horny brown.

Length 7·5; tail 3·2; wing 4·4; tarsus ·9; bill from gape ·6.

Sharpe has separated as a subspecies, under the name of *P. humii*, a pale race of this bird with the red parts of the head and breast rosy, not crimson, and the brown of the back quite pale. The frontal band is also much broader, extending back as far as the middle of the eye. The female has the rump-feathers broadly margined with olive-yellow. In the British Museum there is a pair of these birds procured in Kansu; one bird from Tibet; another from the "Borenda Pass;" and a fifth from Kotgarh. Altogether I am not satisfied that this race, as found in the Himalayas, is worthy of separation from *P. punicea.*

Distribution. The Himalayas from Kashmir to Sikhim, at elevations of from 10,000 to 17,000 feet, according to season, and extending into Tibet and Western China.

Habits, &c. Stoliczka found this Finch in Spiti and Ladák searching after food at the camping-grounds, and he also records the finding of a nest made of coarse grass and placed in a furze bush. The eggs were dirty white or greenish with some dark brown spots.

Genus PROPASSER, Hodgs., 1844.

The genus *Propasser* belongs to the Rose-Finches, the males of which are characterized by rose-coloured plumage, and the females by streaked brown plumage. The birds of this genus may be separated from those of the next genus *Carpodacus*, by the presence of a supercilium in both sexes and by the bluntness of the wing, the secondaries falling short of the tip of the wing by a distance less than the length of the tarsus. The bill of *Propasser* is of much the same shape as that of *Hæmatospiza*, but smaller in comparison to the size of the head.

Key to the Species.

a. Supercilium red.
  a'. Crown of head with coarse black
    streaks.
    a''. Feathers of forehead, supercilium,
      throat, and cheeks with shining
      white shaft-streaks .... *P. thura* ♂, p. 213.
    b''. Feathers of above parts without
      white shaft-streaks.
      a'''. Red parts of the head rosy.... *P. pulcherrimus* ♂, p. 215.
      b'''. Red parts of the head crimson. *P. ambiguus* ♂, p. 215.
  b'. Crown of head unstreaked or with
    mere black shaft-lines.
    c''. Rump red.
      c'''. Wing-coverts not tipped.
        a⁴. Feathers of supercilium and of
          sides of head and throat
          pointed and shining rosy:
          wing 3·5 .... *P. grandis* ♂, p. 216.
        b⁴. Feathers of supercilium and of
          sides of head and throat
          rosy red, not pointed; wing 3. *P. rhodochrous* ♂, p. 217.
      d'''. Wing-coverts broadly tipped
        with rosy .... *P. rhodopeplus* ♂, p. 217.
    d''. Rump coloured like back ...... *P. edwardsi* ♂, p. 218.
b. Supercilium buff or ochraceous.
  c'. Ground-colour of lower plumage not
    uniform; abdomen whitish ...... *P. thura* ♀, p. 213.
  d'. Ground-colour of lower plumage
    uniform throughout.
    e⁴. Lower plumage ashy white
      streaked with brown.
      e'''. Wing less than 3............ { *P. pulcherrimus* ♀, p. 215.
                                          { *P. ambiguus* ♀, p. 215.
      f'''. Wing more than 3·5 .... *P. grandis* ♀, p. 216.
    f'. Lower plumage ochraceous buff,
      streaked with brown.
      g'''. Wing under 3; culmen not
        exceeding ·35 .... *P. rhodochrous* ♀, p. 217.
      h'''. Wing over 3; culmen over ·4.
        c⁴. Lores and ear-coverts nearly
          uniform black; tail 2·8 .... *P. rhodopeplus* ♀, p. 217.
        d⁴. Lores and ear-coverts brown,
          mottled with buff; tail 2·6.. *P. edwardsi* ♀, p. 218.

754. Propasser thura. *The White-browed Rose-Finch.*

Carpodacus thura, *Bonap. & Schleg. Mon. Lox.* p. 21, pl. 23 (1850);
  Sharpe, Cat. B. M. xii, p. 425.
Propasser thura (*Bonap. & Schleg.*) *Horsf. & M. Cat.* ii, p.459; *Moore,
  P. Z. S.* 1855, p. 215, pl. 113; *Jerd. B. I.* ii, p. 400; *Blanf.
  J. A. S. B.* xli, pt. ii, p. 65; *Hume, Cat. no.* 710; *Oates in Hume's
  N. & E. 2nd ed.* ii, p. 152.
Propasser frontalis, *Blyth, Ibis,* 1862, p. 390; *id. J. A. S. B.* xxxii,

p. 458 ; *Jerd. B. I.* ii, pp. 403, 874 ; *Hume, Cat.* no 744 ; *G. F. L. Marshall, Ibis,* 1881, p. 84 ; *Hume, S. F.* ix, p. 349, note.
Carpodacus dubius, *Prjev. in Rowley's Orn. Misc.* ii, p. 201, pl. liv, v (1877).
Propasser blythi, *Biddulph, Ibis,* 1882, p. 283, pl. ix.

Coloration. Male. Lores and front part of face crimson ; forehead, supercilium, cheeks, ear-coverts, chin, and throat pale shining pink, with white shaft-streaks ; the end of the supercilium white ; a broad band behind the eye nearly black ; sides of neck plain brown ; crown, nape, back, and scapulars brown, broadly streaked with black ; rump rosy pink ; upper tail-coverts black, margined with rosy pink ; lesser wing-coverts black, edged with rosy ; median coverts black, broadly tipped with pink ; greater coverts black, narrowly margined with dull pink, and tipped on the outer web with pale buff ; quills, primary-coverts, and tail blackish, margined with dull rosy, the tertiaries with pale buff ; lower plumage from the throat downwards uniform rosy pink ; under tail-coverts black, edged with rosy ; axillaries and under wing-coverts whitish.

Female. Upper plumage dark brown, streaked with black ; the feathers of the rump and upper tail-coverts edged with golden yellow ; a broad supercilium and the feathers round the upper mandible buff, more or less streaked and mottled with black ; a broad band behind the eye black ; cheeks and ear-coverts pale rufous, streaked with black ; chin, throat, breast, and sides of the body rufous, streaked with black ; abdomen buffy white ; under tail-coverts black, margined with pale buffy white ; under wing-coverts and axillaries whitish.

Length about 6·5 ; tail 3·1 ; wing 3·4 ; tarsus ·9 ; bill from gape ·6.

Distribution. Sikhim ; Nepal ; Gilgit ; extending to Tibet, Alachan, and Kansu. This species has been recorded at 12,000 feet and upwards both by Blanford and Mandelli.

Habits, &c. This bird was observed by Blanford on rhododendron bushes, and sometimes on grassy hill-sides. Beavan observed it in flocks. Mandelli obtained the nest at Dolaka in Nepal in August. It was built in a thorny bush, was cup-shaped, and composed of fine grass coated exteriorly with brown moss, and was lined with white fur. The eggs, three in number, are dull greenish blue, sparingly marked with brownish grey, and measure about ·87 by ·65.

I have not been able to examine a specimen of *P. blythi* from Gilgit, but, judging from the description, this race of *P. thura* is identical with *P. dubius*, which is characterized by a pale brown upper plumage, and differs from *P. thura* in no other respect. There are two specimens of this pale race in the British Museum, marked as having been received from the N.W. Himalayas, but the labels are unsatisfactory and the locality doubtful. *P. thura* may be looked upon as a dark race from Sikhim and Nepal, and *P. dubius* (or *P. blythi*) as a pale race from the drier regions of

Gilgit and Tibet. I do not consider these races worthy of separation.

755. Propasser pulcherrimus. *The Beautiful Rose-Finch.*

Propasser pulcherrima, *Hodgs. in Gray's Zool. Misc.* p. 85 (1844, sine descr.*).
Propasser pulcherrimus, *Moore, P. Z. S.* 1855, p. 216; *Horsf. & M. Cat.* ii, p. 460; *Jerd. B. I.* ii, p. 402; *Hume, N. & E.* p. 471; *id. Cat.* no. 743; *Oates in Hume's N. & E.* 2nd ed. ii, p. 153.
Carpodacus davidianus, *Milne-Edwards, Nouv. Arch. Mus.* i, *Bull.* p. 19, pl. ii, fig. 2 (1864).
Carpodacus pulcherrimus (*Moore*), *Sharpe, Cat. B. M.* xii, p. 429.

Coloration. Male. Upper plumage ashy brown streaked with dark brown, the crown tinged with rosy; the rump rosy red; the upper tail-coverts rosy brown with dark shaft-streaks; wing-coverts blackish, broadly edged with ashy rufous; quills and tail black, narrowly edged with ruddy brown, the tertiaries more broadly with ashy; a very broad supercilium, cheeks, ear-coverts, chin, throat, and a small portion of the forehead pale rosy, with the black bases of the feathers showing more or less; lores and a band behind the eye rosy brown; breast and abdomen rosy red, with black shafts; sides of the body brown, streaked darker; under tail-coverts rosy red, with large black centres.

Female. The whole upper plumage fulvous-brown, streaked with black; wing-coverts, quills, and tail dark brown or black, margined with fulvous-brown; a very indistinct supercilium fulvous, mottled with brown; sides of the head and neck and the whole lower plumage ashy white, tinged with fulvous and densely streaked with brown.

Iris reddish brown; bill horny brown with the lower mandible greyish; legs rosy grey (*David*).

Length about 6; tail 2·6; wing 3; tarsus ·75; bill from gape ·45.

Distribution. The Himalayas from Kumaun to Sikhim, extending to Western China.

756. Propasser ambiguus. *Hume's Rose-Finch.*

Propasser ambiguus, *Hume, S. F.* ii, p. 326 (1874); *Brooks, S. F.* iii, p. 255; *Hume, Cat.* no. 743 bis.
Carpodacus ambiguus (*Hume*), *Sharpe, Cat. B. M.* xii, p. 428, pl. x.

Coloration. Male. The only male of this Rose-Finch that I have been able to examine is a carbolized specimen in the Hume Collection and the type of the species. It is now in very bad order, and I shall therefore quote the original description made when the specimen was fresher :—" Forehead, crown, occiput, back, and scapulars dark hair-brown, most of the feathers narrowly and inconspicuously margined with pale brown; a broad line from the nostrils over the

---

* Hodgson confounded this species with *P. rhodochrous*, but fortunately Moore, in redescribing the present species, retained Hodgson's name.

eye, the lores, cheeks, chin, and throat dull dark crimson, the feathers dusky at their bases ; the ear-coverts and sides of the neck like the back, but more broadly margined with very pale brown ; the wings, tail, and upper tail-coverts hair-brown ; the feathers with an excessively narrow pale brown margin, and the median coverts rather more broadly tipped with pale brownish pink ; the rump pale rose-colour ; breast, abdomen, vent, and under tail-coverts pale rose-colour, paling towards the lower tail-coverts, each feather dusky at the base and with brown shafts or narrow brown shaft-stripes."

Female. The specimens in the British Museum labelled as females of this species are absolutely inseparable from the females of *P. pulcherrimus* both as regards size and colour.

Length 5·5 to 6 ; tail 2·2 ; wing 3 ; tarsus ·8 ; bill from gape ·5.

Distribution. The only male of this species known was obtained in the valley of the Bhágbirati river in Garhwál. Females, presumably of this species, have been obtained at Suki and Darali in the same valley. Sharpe identifies with this species two females obtained by Hodgson in Nepal, but I cannot separate them from other females ascribed to *P. pulcherrimus* and also obtained by Hodgson in Nepal.

757. **Propasser grandis.** *The Red-mantled Rose-Finch.*

Carpodacus grandis, *Blyth, J. A. S. B.* xviii, p. 810 (1849) : *id. Cat.* p. 342 ; *Sharpe, Cat. B. M.* xii, p. 404.
Propasser rhodochlamys (*Brandt*), *apud Horsf. & M. Cat.* ii, p. 458 ; *Jerd. B. I.* ii, p. 401 ; *Stoliczka, J. A. S. B.* xxxvii, pt. ii, p. 60 ; *Hume & Henders. Lah. to Yark.* p. 259 ; *Hume, Cat.* no. 741 : *Biddulph, Ibis,* 1881, p. 84 ; *Scully, Ibis,* 1881, p. 578.

Coloration. Male. The whole upper plumage and the visible portions of the closed wings and tail rosy brown, becoming pure rosy on the rump ; the feathers of the head and back with dark brown streaks ; supercilium, sides of head, chin, and throat pale shining rosy, the feathers all pointed ; lores and a band behind the eye reddish brown ; lower plumage rosy red ; under wing-coverts and axillaries rosy white.

Female. Upper plumage ashy brown, streaked darker everywhere ; the lower plumage ashy white, streaked with dark brown ; wings and tail brown, the feathers with paler margins ; an indistinct supercilium pale buff mottled with brown.

Iris light brown : bill greyish brown, the lower mandible albescent ; legs pinkish carneous brown (*Wardlaw Ramsay*).

Length about 7 ; tail 3·1 ; wing 3·6 : tarsus ·8 ; bill from gape ·65.

This species differs from the true *P. rhodochlamys*, Brandt, in having no rosy plumes on the forehead.

Distribution. The whole Himalayas from Afghanistan and Gilgit eastwards to Garhwál and Kumaun. In the British Museum there is also a single female said to have been procured in Sikhim by Mandelli, but there is no original label attached to this specimen and I fear that some mistake may have been made regarding the locality. This Rose-Finch is found from about 4000 to 9000 feet according to season.

758. **Propasser rhodochrous.** *The Pink-browed Rose-Finch.*

Fringilla rhodochroa, *Vigors, P. Z. S.* 1831, p. 23; *Gould, Cent.* pl. 31, fig. 2.
Propasser rhodochrous (*Vig.*), *Horsf. & M. Cat.* ii, p. 459; *Jerd. B. I.* ii, p. 402; *Stoliczka, J. A. S. B.* xxxvii, pt. ii, p. 60; *Brooks, S. F.* iii, p. 255; *Hume, Cat.* no. 742.
Carpodacus rhodochrous (*Vig.*), *Blyth, Cat.* p. 122; *Sharpe, Cat. B. M.* xii, p. 415.

Gulabi tuti, Nep.: *Cheerya*, Nep. plains.

Coloration. Male. Crown and nape dusky crimson, with faint shaft-streaks; back and scapulars ruddy brown, streaked darker; rump rosy red; upper tail-coverts dull crimson; wing-coverts, quills, and tail dark brown, margined with ruddy brown; lores and a broad band behind the eye crimson-brown; a supercilium, cheeks, ear-coverts, and entire lower plumage rosy red, the plumage of the head with a pearly tinge; under wing-coverts and axillaries ashy rosy.

Female. Whole upper plumage olive-brown, broadly streaked with dark brown; wing-coverts, quills, and tail dark brown margined with olive-brown; a broad conspicuous ochraceous supercilium; lores and a band behind the eye dark brown; cheeks, chin, and upper throat ashy, streaked brown; the whole lower plumage ochraceous buff, streaked with dark brown.

The young resemble the adult female.

Upper mandible, legs and feet dark brown; lower mandible a lighter brown, fleshy brown towards the base; iris red-brown (*Hume*).

Length 6; tail 2·6; wing 2·8; tarsus ·75; bill from gape ·45.

Distribution. The Himalayas from Dharmsála to Nepal, not occurring in summer much below 7000 feet according to Stoliczka. Royle, as quoted by Jerdon, asserts that this Rose-Finch occurs in the plains near Sahāranpur.

Godwin-Austen (J. A. S. B. xxxix, pt. ii, p. 110) refers to this species, with some doubt, a female specimen of a Rose-Finch procured in the Khāsi hills by him.

759. **Propasser rhodopeplus.** *The Spotted-winged Rose-Finch.*

Fringilla rhodopepla, *Vigors, P. Z. S.* 1831, p. 23; *Gould, Cent.* pl. 31, fig. 1.
Propasser rhodopeplus (*Vig.*), *Horsf. & M. Cat.* ii, p. 458; *Jerd. B. I.* ii, p. 400; *Hume, Cat.* no. 739.
Carpodacus rhodopeplus (*Vig.*), *Blyth, Cat.* p. 121; *Sharpe, Cat. B. M.* xii, p. 417.

Gulabi tuti, Nep.

Coloration. Male. Upper plumage dark crimson-brown, the
centres of the feathers darker; lower back streaked with rosy;
rump-feathers broadly tipped rosy; wings and tail dark brown,
every feather margined with crimson-brown; median coverts with
a rosy tip; greater coverts and tertiaries with a rosy tip to the
outer web only; a very broad supercilium pale shining rosy; sides
of the head and neck crimson-brown; lower plumage rosy red, the
crimson-brown bases of the feathers everywhere visible, the
feathers of the throat pointed and tipped with pale shining rosy.

Female. The whole upper plumage olive-brown, tinged with
ochraceous and closely streaked with blackish; wing and tail dark
brown, the median and greater coverts tipped with ochraceous, the
tertiaries with a long oblique patch of ochraceous on the outer
web; all the quills and the tail margined with olive-brown; a
broad ochraceous-buff supercilium; lores and ear-coverts blackish;
the whole lower plumage ochraceous buff, streaked with dark brown.

Bill horny-brown; legs pale brown; iris brown (*Jerdon*).

Length nearly 7; tail 2·9; wing 3·3; tarsus ·9; bill from
gape ·6.

Distribution. The Himalayas from Garhwál to Sikhim.

760. Propasser edwardsi.　*Edwards's Rose-Finch.*

Carpodacus edwardsii, *Verr. Nouv. Arch. Mus.* vi, *Bull.* p. 39 (1870),
　　vii, *Bull.* p. 58, viii, *Bull.* pl. 3, fig. 4; *Sharpe, Cat. B. M.* xii, p. 418.
Propasser saturatus, *Blanf. J. A. S. B.* xli, pt. ii, p. 168, pl. viii
　　(1872); *id. Ibis,* 1873, p. 218; *Hume, S. F.* i, p. 418.
Propasser edwardsi (*Verr.*), *Hume, S. F.* vii, p. 415; *id. Cat. no.*
　　744 bis.

Coloration. Male. Forehead, crown, and nape dull crimson, with
black shaft-streaks; a broad supercilium, cheeks, chin, and throat
rosy pink, with dashes of the same on the forehead; lores and sides
of the head dull crimson; back, rump, scapulars, and upper tail-
coverts brown washed with crimson, the back and scapulars
with broad black streaks; wing-coverts, quills, and tail blackish
margined with reddish brown, the coverts and tertiaries also tipped
with pale rosy; breast dark rosy red, with black shaft-streaks;
remainder of lower parts pink, with black shaft-streaks.

Female. Similar to the female of *C. rhodopeplus,* from which it
can only be separated by its slightly shorter tail, measuring about
2·6 (whereas in the other species the tail measures quite 2·8); by
the colour of the lores and ear-coverts, which are brown mottled
with buff (not black); and by the less distinct supercilium.

Length about 6·5; tail 2·7; wing 3·1; tarsus ·9; bill from
gape ·58.

Distribution. The Himalayas from Nepal to Bhutan, extending to
the mountains of Western China. This species occurs at high
levels, Mandelli having procured it at 10,000 feet.

Genus CARPODACUS, Kaup, 1829.

The birds of the genus *Carpodacus* differ from those of the genus *Propasser* in having a longer and more pointed wing, the secondaries falling short of the tip of the wing by a distance greater than the length of the tarsus, and in having no supercilium.

Key to the Species.

a. Wing under 3·5 .... *C. erythrinus*, p. 219.
b. Wing considerably over 4 .... *C. severtzovi*, p. 220.

761. Carpodacus erythrinus. *The Common Rose-Finch.*

Loxia erythrina, *Pall. Nor. Comm. Petrop.* xiv, p. 587, pl. 23, fig. i (1770).
Carpodacus erythrinus (*Pall.*), *Blyth, Cat.* p. 122 ; *Horsf. & M. Cat.* ii, p. 456 ; *Jerd. B. I.* ii, p. 398 ; *Hume & Henders. Lah. to Yark.* p. 259 ; *Scully, S. F.* iv, p. 170 ; *Hume, Cat.* no. 738 ; *Scully, S. F.* viii, p. 335 ; *Biddulph, Ibis,* 1881, p. 83 ; *Davison, S. F.* x, p. 403 ; *Oates, B. B.* i, p. 345 ; *Barnes, Birds Bom.* p. 274 ; *Sharpe, Cat. B. M.* xii, p. 391 ; *Hume, S. F.* xi, p. 286 ; *Oates in Hume's N. & E.* 2nd ed. ii, p. 153.

Tuti, Hind.; Amonga tuti, Nep.; Chota tuti, Sylhet; Phulin-pho, Lepch.; Yedru-pichike, Yedru-jinawayi, Tel.

Coloration. Male. After the autumn moult the forehead, crown, and nape are dull crimson; back and scapulars crimson-brown, each feather margined with olivaceous; lower back and rump nearly uniform rosy red; upper tail-coverts ruddy brown, edged with olive; lesser wing-coverts crimson-brown; median coverts dark brown, broadly tipped with rufous; greater coverts brown, broadly edged with rufous; primary-coverts, winglet, quills, and tail brown, edged with ruddy brown tinged with olivaceous; lores and a band behind the eye dusky rufous; cheeks, chin, throat, and upper breast a beautiful rose-colour; lower breast paler rose, becoming albescent on the abdomen and under tail-coverts; sides of the neck and sides of the body olive-brown; axillaries and under wing-coverts ashy rufous.

After some months, owing to the wearing away of the margins and also to an increase of colour, the whole head and neck, chin, throat, and upper breast become bright crimson; the back, scapulars, rump, upper tail-coverts, and lesser wing-coverts very dark crimson.

Female. The whole plumage olive-brown, streaked with brown, the median and greater wing-coverts broadly tipped with ochraceous and the quills and tail margined with the same; under wing-coverts and axillaries pale ochraceous.

Young birds of both sexes resemble the adult female in general appearance. The young male appears to retain this female plumage during the first summer.

Iris dark brown; legs and feet dusky brown; bill dark horny brown, paling at base of upper mandible.

Length about 6·5; tail 2·6; wing 3·3; tarsus ·8; bill from gape ·5.

Distribution. A winter visitor to the whole of India as far south as the Nilgiri hills, and to the provinces to the east as far south as Arrakan and Northern Pegu. In summer the range of this species extends to Northern Europe and Northern Asia. Considerable numbers of these birds appear to summer in the higher and more remote portions of the Himalayas, breeding at 10,000 feet.

Habits, &c. Breeds in July and August, in low bushes near the ground, constructing a cup-shaped nest of grass. The eggs are described as being blue, marked with a few chocolate spots, and measuring about ·81 by ·6.

762. Carpodacus severtzovi. *Severtzoff's Rose-Finch.*

Propasser rubicillus (*Güld.*), *Horsf. & M. Cat.* ii, p. 457.
Carpodacus rubicilla (*Güld.*), *Jerd. B. I.* ii, p. 397; *Hume & Henders. Lah. to Yark.* p. 258; *Hume, N. & E.* p. 471; *id. Cat. no.* 737; *Biddulph, Ibis,* 1881, p. 82.
Carpodacus severtzovi, *Sharpe, P. Z. S.* 1886, p. 354; *id. Cat. B. M.* xii, p. 400; *Oates in Hume's N. & E.* 2nd ed. ii, p. 154.

The Caucasian Rose-Finch, Jerd.

Coloration. Male. Forehead and whole crown, except a patch on the hinder part, white, each feather margined with crimson; lores and feathers near bill deep crimson; hinder part of crown, hind neck, sides of neck, back, and scapulars rosy ashy; all the coverts and quills of the wing and the tail brown, edged with the colour of the back; rump and upper tail-coverts rose-colour; ear-coverts pale rose-colour; cheeks, chin, throat, and fore neck white and crimson, like the crown; remainder of lower surface rosy mottled with ashy, the feathers of the breast and middle abdomen with small subterminal white shaft-streaks.

Female. Upper plumage ashy brown, streaked with dark brown; wing-coverts and quills brown, margined with ashy; tail brown, edged with ashy, and the outer web of the outermost feather white; sides of the head and entire lower plumage paler than the upper plumage and everywhere streaked with brown.

Bill grey horny, with the upper mandible brownish above, the lower yellowish horny at base; iris brown; legs and feet brown or dusky brown; claws dusky brown and dusky blackish (*Scully*).

Length nearly 8; tail 3·5; wing 4·5; tarsus 1; bill from gape ·7.

C. rubicilla from the Caucasus differs from the present species in being everywhere of a much deeper red colour.

Distribution. Gilgit and Ladák, extending through Central Asia to Eastern Siberia. In summer this species is found in Turkestan from 10,000 to 12,000 feet and in winter it descends to the level of Gilgit.

Habits, &c. Stoliczka seems to have found the nest of this Rose-Finch in Western Tibet in July, but its authenticity is very doubtful.

C. stoliczkæ, Hume, from Yarkand, is a much smaller species, with the plumage pale ashy and the red parts of the head without the conspicuous white spangles which characterize *C. albicilla*. The female is plain unstreaked ashy throughout.

Genus **ERYTHROSPIZA**, Bonap., 1831.

The genus *Erythrospiza* contains the palest forms of the Rose-Finches, birds of the desert. In this genus the general colour of the males is brown or grey suffused with pale pink. The bill is short but extremely tumid, the lower mandible being as much curved as the upper. The wings are very long and reach much beyond the middle of the tail. The sexes do not differ much in colour.

Key to the Species.

a. In fresh plumage, upper parts bluish grey;
   greatest depth of closed bill ·4 .... *E. githaginea*, p. 221.
b. In fresh plumage, upper parts sandy brown;
   greatest depth of closed bill ·3 .... *E. mongolica*, p. 222.

763. **Erythrospiza githaginea.** *The Desert-Finch.*

Pyrrhula githaginea, *Temm. Pl. Col.* iii, pl. 400 (1826).
Bucanetes githaginea (*Temm.*), *Hume, S. F.* i, p. 210, vii, pp. 64,
   454; *Barnes, Birds Bom.* p. 273.
Erythrospiza githaginea (*Temm.*), *Hume, Cat.* no. 732 bis; *Sharpe,
   Cat. B. M.* xii, p. 284.

Coloration. Male. After the autumn moult the forehead, crown, sides of the head, and entire lower plumage are bluish grey, suffused with rosy on the lower parts; upper plumage and sides of the neck greyish brown, with a faint tinge of rosy on the rump; wings and

Fig. 59.—Head of *E. githaginea*.

tail brown, edged with vinous grey, the quills subterminally black.

The above plumage is retained by the male for only a short time after the moult, and at this period the whole plumage has a decided purplish tinge throughout, and hardly any pink is visible. A slight abrasion of the feathers soon causes a change, and as early as December the feathers round the bill, the cheeks, the whole lower plumage, the rump, and the margins of the quills and coverts become a beautiful rose-pink, becoming still brighter as the plumage gets more worn away.

Female. Resembles the male, but never becomes so rosy in tint at any time of the year.

Iris brown; legs and feet fleshy brown; claws dusky; soles whitish; bill orange-yellow, sometimes pale yellow, brownish on upper mandible (*Hume*).

Length nearly 6; tail 2·2; wing 3·5; tarsus ·7; bill from gape ·5.

Young birds appear to be characterized by the presence of some dark streaks on the breast and abdomen.

Distribution. The whole of Sind and a considerable portion of Rajputana, extending east as far as the Gurgaon district in the Punjab. This Finch is probably a resident, and it is found westwards throughout Afghanistan and Baluchistan to Europe.

Habits, &c. Hume observed this species in Sind, feeding in desert places in patches of mustard and other cultivation, and running about a good deal on the ground like Sparrows.

764. Erythrospiza mongolica. *The Mongolian Desert-Finch.*

Carpodacus mongolicus, *Swinh. P. Z. S.* 1870, p. 447, 1871, p. 387 ; *Scully, S. F.* iv, p. 169 ; *id. Ibis,* 1881, p. 577.
Erythrospiza incarnata, *Severtz. Turkest. Jevoln.* pp. 64, 117 (1873) ; *Biddulph, Ibis,* 1881, p. 82.
Erythrospiza mongolica (*Swinh.*), *Hume, S. F.* ix, p. 347 note; *Biddulph, Ibis,* 1882, p. 282 ; *Sharpe, Cat. B. M.* xii, p. 287.

Coloration. *Male.* After the autumn moult the whole upper plumage and lesser wing-coverts sandy brown, the centres of the feathers almost everywhere darker brown, the lower rump suffused with rosy; tail brown, broadly edged with pale buff; median wing-coverts brown, edged with rosy buff ; greater coverts brown, subterminally darker and broadly edged with rosy red; winglet very dark brown, edged with buff; primary-coverts paler brown, edged with buff; quills brown, the outer webs whity brown, and most of them suffused with rosy red on the greater part of the web ; tertiaries pale buff, with the middle portion brown : ear-coverts and sides of neck brown ; lores, round the eye, the cheeks, and the whole lower plumage, except the abdomen and under tail-coverts, pale rosy pink ; the two latter parts pale isabelline.

In the spring and autumn the rosy tinge on the plumage everywhere is much brighter owing to the wearing away of the margins of the feathers, and the outer webs of the quills and coverts become crimson in many birds.

Female. Very similar to the male in autumn, but with the rosy tinge much paler and entirely absent on the rump.

Bill and legs pale fleshy ; iris dark brown (*Murray*).

Length about 5·5 ; tail 2·3 ; wing 3·6 ; tarsus ·7 ; bill from gape ·45.

Distribution. Resident in Gilgit from 5000 to 10,000 feet according to season. To the west this Finch extends to Afghanistan. It is spread over a considerable portion of Central Asia and ranges to China.

Habits, &c. Occurs in large flocks throughout the winter.

The following two birds are likely to be found in India proper, but are not yet known to occur there :—

Rhodopechys sanguinea (Gould), entered by Hume in his Catalogue with a note of doubt. A fine Finch with a wing over four inches, the bases of all the quills largely white and the outer webs of the quills and coverts a beautiful rosy red, paler in the female. Occurs in Yárkand.

Rhodospiza obsoleta (Licht.). Sandy brown throughout, with the outer webs of the primaries pure white and those of the secondaries and greater coverts pink. Occurs in Afghanistan and Yárkand.

Genus PROCARDUELIS, Hodgs., 1844.

The genus *Procarduelis* contains certain species which are Rose-Finches in plumage, but differ remarkably from those birds in the shape of the bill, which is slender and pointed, with the culmen straight. The females differ from those of the Rose-Finches in being unstreaked.

Key to the Species.

a. Lower plumage red.
  a'. Forehead and supercilium of a different
    shade of red to the crown . . . . *P. nepalensis* ♂, p. 223.
  b'. Forehead and crown of the same red and
    no supercilium evident . . . . *P. rubescens* ♂, p. 224.
b. Lower plumage without a trace of red.
  c'. Upper plumage without a trace of red . . *P. nepalensis* ♀, p. 223.
  d'. Upper plumage suffused with red . . . . . . *P. rubescens* ♀, p. 224.

765. Procarduelis nepalensis. *The Dark Rose-Finch.*

Carduelis nipalensis, *Hodgs. As. Res.* xix, p. 157 (1836).
Procarduelis nipalensis (*Hodgs.*), *Blyth, Cat.* p. 121 ; *Horsf. & M. Cat.* ii, p. 492; *Jerd. B. I.* ii, p. 405; *Blanford, J. A. S. B.* xli, pt. ii, p. 66; *Hume, Cat. no.* 746; *Scully, S. F.* viii, p. 336; *Sharpe, Cat. B. M.* xii, p. 182.

Ka-biya, Lepch.

Coloration. Male. Forehead, anterior part of crown, broad supercilia, cheeks, chin, and throat rosy red, the forehead and anterior part of crown of a richer colour ; lores and a broad band through the eye over the ear-coverts black, tinged with red; upper plumage, scapulars, lesser wing-coverts, sides of the neck, and the whole breast dusky, tinged with vinaceous, and each feather margined with sanguineous; remaining wing-coverts, quills, and tail dark brown, edged with dusky red; abdomen rosy red; sides of the body brown, suffused with rufous; under tail-coverts brown, margined with pink; under wing-coverts and axillaries dark brown.

Fig. 60.—Head of *P. nipalensis.*

Female. The whole upper plumage brown, each feather more or less margined with ochraceous, most distinctly so on the back; wings dark brown, the coverts and the tertiaries very broadly tipped and margined near the tip with ochraceous; remainder of the wing and the tail brown, narrowly margined with ochraceous; sides of the head and neck and the whole lower plumage uniform ochraceous brown, the under tail-coverts lighter and with dusky centres.

This species does not appear to undergo any appreciable change of plumage according to season.

The young resemble the adult female closely.

Bill dusky, paler below; iris dark brown; legs fleshy brown (*Scully*).

Length about 6·3; tail 2·6; wing 3·6; tarsus ·85; bill from gape ·55.

Distribution. The Himalayas from Kashmir to Bhutan at elevations of from 6000 to 14,000 feet, according to season. This species extends into the mountains of Western China.

Habits, &c. This bird appears to be found in small flocks feeding on the ground, and is said not to be at all shy.

766. Procarduelis rubescens. *Blanford's Rose-Finch.*

Procarduelis rubescens, *Blanf. P. Z. S.* 1871, p. 694, pl. 74; *Hume, Cat. no. 746 bis; Sharpe, Cat. B. M.* xii, p. 184.
Procarduelis mandellii, *Hume, S. F.* i, pp. 14, 318 (1873).

Coloration. Male. The whole upper plumage, scapulars, lesser and median wing-coverts crimson, brightest on the crown, the bases of the feathers showing through more or less everywhere and imparting a brownish hue to the plumage; greater coverts brown, edged with red in some specimens, with rufous or fulvous brown in others; tertiaries the same; remaining quills and tail-feathers dark brown, narrowly margined with reddish; lores and a band through the eye over the ear-coverts brown stippled with red; the whole lower plumage rosy red, the lower part of the abdomen whitish; the under tail-coverts pale brown, edged with white; under wing-coverts and axillaries greyish brown tinged with fulvous.

Female. In general appearance similar to the female of *P. nepalensis*, but the whole upper plumage, except the back, and the margins of the wings and tail suffused with crimson, of which there is not a trace in the other species; the lower plumage much paler, becoming albescent on the abdomen.

Length about 5·5; tail 2·2; wing 3·3; tarsus ·75; bill from gape ·55. The bill is thicker in this species than in *P. nepalensis*.

Distribution. Sikhim and the eastern portion of Nepal, probably at high elevations.

Genus **CARDUELIS**, Briss., 1760.

The genus *Carduelis* contains the Goldfinches, of which two species are known, one inhabiting Europe and Western Asia, and the other Central Asia down to the Himalayas. The Goldfinches are characterized by a long, slender, straight and sharply pointed bill, long wings, the bright red colour of the face, and the bright yellow on the wing. The sexes are very closely similar.

767. Carduelis caniceps. *The Himalayan Goldfinch.*

Carduelis caniceps, *Vigors, P. Z. S.* 1831, p. 23; *Gould, Cent.* pl. 33, fig. 1; *Blyth, Cat.* p. 124; *Horsf. & M. Cat.* ii, p. 433; *Jerd. B. I.* ii, p. 408; *Stoliczka, J. A. S. B.* xxxvii, pt. ii, p. 61; *Hume, Cat.* no. 749; *Biddulph, Ibis,* 1881, p. 85; *Scully, Ibis,* 1881, p. 578; *Sharpe, Cat. B. M.* xii, p. 189.

Shira, Hind.; *Saira,* Kashm.

Fig. 64.—Head of *C. caniceps.*

Coloration. Male. Forehead, chin, and the cheeks next the bill crimson; lores black; upper plumage ashy brown, becoming whitish on the rump; upper tail-coverts white; lesser, median, and primary coverts with the winglet black, sometimes with ashy margins; greater coverts chiefly bright yellow; primaries and secondaries black, with a considerable portion of the outer webs of all but the first primary bright yellow, the inner webs margined with white; the tertiaries each with a large oval white mark on the outer web; tail black, the two outer pairs of feathers largely white on the inner webs, the two middle pairs tipped white; throat ashy white; sides

of the head and neck and the breast ashy brown ; abdomen and
under tail-coverts white; sides of the body fulvous ashy ; under
wing-coverts and axillaries whitish.

Female. Very similar to the male, but having the crimson on the
head paler and the yellow on the greater wing-coverts less extensive.

Bill carneous with a dusky tip; legs pale brown ; iris brown
(*Jerdon*).

Length about 5·5; tail 2·1 ; wing 3·2; tarsus ·55; bill from
gape ·6.

This species differs from the English Goldfinch, *C. elegans*,
chiefly in having no black on the head. Where the two species meet
they appear to interbreed, and every intermediate form between
the two may be found, as is well shown in the fine mounted series
of these birds in the Central Hall of the British Museum of
Natural History.

Distribution. The Himalayas from the Hazára country and Gilgit to Kumaun, from 5000 to 9000 or 10,000 feet according to
season. This species extends to Afghanistan on the west and
through Central Asia to Siberia on the north.

Habits, &c. This Goldfinch, like its European ally, affects open
country, feeding chiefly on the seeds of the thistle. Nothing is
known regarding its nidification on the Himalayas.

Genus CALLACANTHIS, Reichenb., 1850.

The genus *Callacanthis* contains one species which appears to
have considerable affinities for *Carduelis.* Both sexes are characterized by a large amount of white on the wings and tail ; the male
is rosy red, and the female brown, but they both preserve the same
pattern of colour. The bill is large and thick, but straight and
pointed, and the wings are very long.

This Finch probably resembles the Goldfinch in its habits.

768. Callacanthis burtoni. *The Red-browed Finch.*

Carduelis burtoni, *Gould, P. Z. S.* 1837, p. 90.
Fringilla burtoni (*Gld.*), *Blyth, Cat.* p. 337.
Callacanthis burtoni (*Gld.*). *Jerd. B. I.* ii, p. 407 ; *Stoliczka, J. A.
S. B.* xxxvii, pt. ii, p. 61 ; *Hume, N. & E.* p. 471 ; *Brooks, S. F.*
iii, p. 255 ; *Hume, Cat.* no. 748; *Sharpe, Cat. B. M.* xii, p. 232 ;
Oates in Hume's N. & E. 2nd ed. ii, p. 154.

Coloration. Male. Forehead, lores, and a large ring round the
eye crimson ; crown and nape black ; cheeks and ear-coverts black
with pale shafts ; upper plumage and scapulars brown suffused with
rose-colour ; lesser and median wing-coverts black, margined
with red ; the remaining wing-coverts and winglet black, tipped
with white, the tips of the greater coverts tinged with rosy ; quills
black, tipped with white; tail black, the middle pair of feathers
merely tipped with white, the others with an increasing amount of
white, the outermost feather having nearly the whole inner web

white; chin and throat blackish, tipped with red; lower plumage brown suffused with rosy red; under wing-coverts and axillaries white with ashy bases.

Female. The forehead, round the eye, and supercilium buff; crown and nape dusky brown; upper plumage ochraceous brown; lesser and median wing-coverts ochraceous brown, tipped paler; greater

Fig. 62.—Head of *C. burtoni.*

wing-coverts, primary-coverts, winglet, and quills black, tipped with white; tail as in male; lores and ear-coverts brown with pale shafts; entire lower plumage ochraceous brown; under wing-coverts and axillaries white with ashy bases.

In the dry state the bill is yellow and the legs fleshy brown.

Length 6·5; tail 2·6; wing 4; tarsus ·8; bill from gape ·7.

Distribution. The Himalayas from Murree to Garhwál and Kumaun. According to Stoliczka, this species is found occasionally in winter on the lesser ranges, about Kotgarh and Simla, between 4000 and 7000 feet; in summer it lives in the highest cedar-forests on the central range of the N.W. Himalayas.

Habits, &c. This Finch is said to make a large nest of moss in a pine-tree in dark forest situations. The eggs do not appear to be known.

Genus **ACANTHIS**, Bechst., 1802.

The genus *Acanthis* contains the Linnets, of which two species are found on the Himalayas. One of them is little more than a race of the common English Linnet, but it varies in certain constant particulars which I think entitle it to separation from the European form. The Linnets are brown, but the males have portions of the plumage suffused with red. The bill is short, straight and pointed. The sexes do not differ very much from each other except with regard to the rosy parts of the plumage.

Key to the Species.

a. Throat streaked .... *A. fringillirostris*, p. 228.
b. Throat unstreaked .... *A. brevirostris*, p. 229.

769. **Acanthis fringillirostris.** *The Eastern Linnet.*

Linota fringillirostris, *Bonap. & Schleg. Monog. Loc.* p. 45, pl. 49 (1850).

Linaria cannabina (*Linn.*), *Hume, S. F.* vii, p. 122; *Butler, S. F.* vii, p. 184; *Hume, Cat.* no. 751 ter; *Biddulph, Ibis,* 1881, p. 86; *Scully, Ibis,* 1881, p. 579; *Biddulph, Ibis,* 1882, p. 285.

Acanthis fringillirostris (*Bp. & Schleg.*), *Sharpe, Cat. B. M.* xii, p. 244.

Fig. 65. – Head of *A. fringillirostris.*

Coloration. Male. After the autumn moult the forehead, crown, nape, and hind neck are ashy brown, with dark brown streaks, the centres of the feathers of the forehead and front part of crown being red, but entirely concealed ; back, rump, scapulars, and wing-coverts dark brown, with broad chestnut-brown margins to all the feathers ; upper tail-coverts black, margined with white ; tail-feathers black, the inner webs broadly, the outer more narrowly, margined with pure white ; primaries black, margined and tipped with white ; secondaries dark brown, edged with reddish brown, which colour occupies nearly the whole of the tertiaries ; a broad band above and below the eye fulvous ; sides of the head pale brown ; chin and throat fulvous, the middle portions streaked with dark brown ; breast dull red, the feathers with very broad fulvous margins which nearly conceal the red ; sides of the breast and of the body fulvous, streaked with brown ; abdomen and under tail-coverts whitish suffused with fulvous.

In spring and summer the margins of the feathers of the forehead and front part of crown are worn down and the red centres become very evident, causing those parts to have a general red appearance ; the hinder crown, nape, and hind neck become more uniformly brown ; the breast becomes a deep rosy pink, with very narrow whitish margins ; a tinge of red is frequently observable on the rump.

Female. Resembles the male, but has no red whatever on the forehead, front part of crown, or breast, these parts being streaked with brown like the other parts of the plumage.

The young bird appears to resemble the adult female closely.

The colours of the bill &c. of this race have not been recorded ; in the Common Linnet the bill is horn-colour, the under mandible brown at base, legs pale reddish brown, iris brown.

Length 5·5 to 6 ; tail 2·4 ; wing 3·2 ; tarsus ·65 ; bill from gape ·45.

This race of Linnet differs from *A. cannabina* in being larger, and, as regards the males, in the colour of the forehead and breast in the full worn plumage of summer. In *A. cannabina* these parts are a deep carmine-red ; in *A. fringillirostris* a bright pomegranate-red. Other differences alleged to exist between the two birds as regards the amount of white on the wing and tail are, I find, of no service in distinguishing them.

Distribution. Occurs in Gilgit from November to February at an elevation of 5000 feet. In the Hume Collection there is a specimen said to have been procured at Daulatpur in Sind in November, and Butler is under the impression that he observed a Linnet, probably of this species, at Karáchi. This Linnet extends westwards to Asia Minor and it is found in Central Asia.

770. Acanthis brevirostris. *The Eastern Twite.*

Linota brevirostris, *Gould, Bonap. Comp. List B. Eur. & N. Am.* p. 34 (1838) ; *Horsf. & M. Cat.* ii, p. 493 ; *Stoliczka, J. A. S. B.* xxxvii, pt. ii, p. 62 ; *Hume & Henders. Lah. to Yark.* p. 260, pl. 26 ; *Scully, S. F.* iv, p. 170.

Linaria brevirostris (*Gould*), *Hume, S. F.* xii, p. 417 ; *id. Cat. no.* 751 bis ; *Biddulph, Ibis,* 1881, p. 86, 1882, p. 284 ; *Scully, Ibis,* 1881, p. 575.

Acanthis brevirostris (*Gould*), *Sharpe, Cat. B. M.* xii, p. 238 ; *Oates in Hume's N. & E.* 2nd ed. ii, p. 155.

Coloration. Male. Forehead, crown, nape, hind neck, back, and scapulars pale sandy brown, each feather with a dark brown streak down the middle ; rump rosy pink ; upper tail-coverts dark brown with broad whitish edges ; tail-feathers blackish, edged with white on both webs ; wing-coverts dark brown, suffused with rufous towards the edges and tipped with sandy white ; primaries and secondaries blackish, the fourth to the eighth primaries broadly edged with white, the others more narrowly with sandy white ; tertiaries brown, broadly edged with fulvous ; sides of head, chin, and throat sandy brown ; breast and sides of the body sandy brown, streaked with dark brown ; abdomen, under tail-coverts, under wing-coverts, and axillaries white.

Female. Differs from the male in having the rump of the same colour as the back.

In the summer the upper plumage is somewhat darker and the rump brighter pink than in the winter.

The young bird appears to resemble the adult female, but to be more fulvous.

Bill yellowish horny, brown on the culmen ; legs and feet brown, claws dusky or black with yellowish tips ; iris brown (*Scully*).

Length about 5·5 ; tail 2·6 ; wing 3·1 ; tarsus ·6 ; bill from gape ·4.

Distribution. Occurs in Gilgit apparently throughout the year from 5000 feet upwards, according to season. Stoliczka observes that this species is " rare in Ladák and visiting Kulu and the Sutlej valley in winter ; it is also in winter caught near Chini, and sometimes caged." No one has since confirmed this account of this

Linnet's distribution in the Himalayas. This species extends westwards to Asia Minor, and is found throughout a considerable portion of Central Asia.

Habits, &c. An egg of this species, said to have been found in Native Sikhim, is described as being white with a faint bluish tinge and mottled all over with reddish brown, and to have measured ·72 by ·55.

Genus **METOPONIA**, Bonap., 1853.

The genus *Metoponia* contains one Finch which has a considerable amount of yellow in its plumage, and connects the Linnets with the Siskins. The sexes are almost alike. The bill is very small but somewhat thick, with the culmen curved. In both sexes the front part of the crown is red.

771. **Metoponia pusilla.** *The Gold-fronted Finch.*

Passer pusillus, *Pallas, Zoogr. Ross.-Asiat.* ii, p. 28 (1811).
Serinus aurifrons, *Blyth, Cat.* p. 125 (1849).
Metoponia pusilla (*Pall.*), *Horsf. & M. Cat.* ii, p. 494; *Jerd. B. I.* ii,
    p. 410; *Stoliczka, J. A. S. B.* xxxvii, pt. ii, p. 61; *Hume & Henders. Lah. to Yark.* p. 259; *Hume, N. & E.* p. 473; *Brooks, J. A.
    S. B.* xli, pt. ii, p. 84; *Hume, Cat.* no. 751; *Wardlaw Ramsay,
    Ibis,* 1880, p. 67; *Biddulph, Ibis,* 1881, p. 86, 1882, p. 284;
    Scully, Ibis, 1881, p. 578; *Oates in Hume's N. & E.* 2nd ed. ii,
    p. 155.
Serinus pusillus (*Pall.*), *Sharpe, Cat. B. M.* xii, p. 373.
The Gold-headed Finch, Jerd.

Fig. 64. - Head of *M. pusilla.*

Coloration. Male. After the autumn moult the forehead and anterior portion of crown are crimson; remainder of crown, nape, and sides of the neck black, with broad grey fringes; sides of the head, chin, throat, and upper breast black, with narrow grey fringes; back, scapulars, and rump black, with broad yellowish margins; the shorter upper tail-coverts golden yellow, with a subterminal black mark and white tip; longer coverts black, margined with white; tail black, the outer webs of feathers basally yellow, margined with white elsewhere; lesser and median wing-coverts yellow, the longer ones tipped white, and the feathers more or less black in the middle; greater coverts black, broadly tipped yellowish white; primaries black, edged with yellow; the remaining quills black, broadly edged with dull white on the terminal half of outer web; lower plumage from the breast downwards yellowish streaked with black; under wing-coverts and axillaries bright yellow. In spring and autumn the fringes and margins are everywhere reduced or cast off, and the plumage becomes darker.

Female. Closely resembles the male.

The nestling is fulvous-brown streaked with brown above; below fulvous, with a few streaks, and tinged with yellow on the throat; the wings and tail are margined with paler yellow than in the adult. After the first autumn moult the plumage resembles that of the adult, but there is no black on the head, throat, and breast, this colour being gradually acquired during the winter. At the end of the first winter the young resemble the adult in all respects, but have no red on the forehead, this red being acquired at the second autumn moult.

Bill dull black; gape whitish; iris very dark brown; legs and feet black (*Hume*).

Length 5; tail 2·4; wing 3·1; tarsus ·55; bill from gape ·35.

Distribution. Kashmir and the Himalayas from Afghanistan to Garhwál, at heights from 5000 to 10,000 feet according to season. This Finch extends westwards to the Caucasus and Asia Minor and northwards to Turkestan.

Habits, &c. Breeds in Kashmir and Afghanistan in June and July, constructing a cup-shaped nest of grass and fibres and lined with wool, feathers, and hair. The eggs are described as being dull stone-white, marked with red-brown spots about the larger end, and one egg measured ·65 by ·49.

Genus **HYPACANTHIS**, Cabanis, 1851.

The genus *Hypacanthis* contains one species of Finch which is closely allied to the Common Greenfinch of Europe, but has a much more slender bill and a darker style of coloration, the upper plumage being a dark brown. The bill is of much the same shape as that of *Acanthis* (and is not therefore figured), but considerably deeper and broader.

772. Hypacanthis spinoides. *The Himalayan Greenfinch.*

Carduelis spinoides, *Vigors, P. Z. S.* 1831, p. 44; *Gould, Cent.* pl. 33, fig. 2.

Chrysomitris spinoides (*Vig.*), *Blyth, Cat.* p. 123; *Horsf. & M. Cat.* ii, p. 495; *Jerd. B. I.* ii, p. 409; *Stoliczka, J. A. S. B.* xxxvii, pt. ii, p.61; *Blanf. J. A. S. B.* xli, pt. ii, p.66; *Brooks, J. A. S. B.* xli, pt. ii, p. 84; *Godw.-Aust. J. A. S. B.* xlv, pt. ii, p. 200; *Sharpe, Cat. B. M.* xii, p. 201.

Hypacanthis spinoides (*Vig.*), *Hume, N. & E.* p. 472; *id. Cat.* no. 750; *Scully, S. F.* viii, p. 336; *Oates in Hume's N. & E.* 2nd ed. ii, p. 156.

Coloration. Male. Forehead either black or yellow or a mixture of both, the differences probably dependent on age; a very broad eyebrow, the sides of the neck meeting behind and forming a more or less distinct collar, the lores, a patch under the eye, a narrow band between the ear-coverts and the cheeks, the rump, and whole lower plumage bright yellow; crown, nape, ear-coverts, and a portion of the cheeks black tinged with green ; back and scapulars dark greenish brown, sometimes suffused with yellow ; upper tail-coverts greenish brown ; tail with the two middle pairs of feathers dark brown, the others brown largely mixed with yellow, the outermost pair being yellow with brown tips and brown shaft-streaks ; lesser and median coverts yellow ; greater coverts black tipped with yellow ; primary-coverts and quills black, all the quills except the first primary with a patch of yellow on the outer web ; the later secondaries and tertiaries broadly tipped and margined with white.

Female. Resembles the male, but has the yellow of the plumage paler and the dark upper plumage more tinged with green ; the forehead appears to be always black or brown.

The young bird has the lower plumage pale yellow streaked with brown ; the upper plumage dull greenish brown streaked with dark brown, the rump being merely tinged with yellow ; wing-coverts greenish brown, tipped with pale yellow ; quills and tail as in the adult but with less yellow ; sides of head brown where the adult is black.

The amount of yellow on the forehead of the adult male appears to depend more on age than on season ; some birds in January have a great deal of yellow on this part and a few summer birds fail to have any.

Bill fleshy, brownish on culmen and dusky at tip ; iris light or dark brown ; feet brownish fleshy ; claws dusky (*Scully*).

Length about 5 ; tail 1·9 ; wing 3·1 ; tarsus ·65 ; bill from gape ·5.

Distribution. The Himalayas from the Pir Panjal range in Kashmir to Sikhim ; Manipur. This species is found up to about 9000 feet.

Habits, &c. Breeds in July and August, constructing a cup-shaped nest of fine grass, hair, and moss in a branch of a tree and laying three eggs, which are pale green speckled with black, and measure about ·69 by ·52.

Genus CHRYSOMITRIS, Boie, 1828.

The genus *Chrysomitris* contains the Siskins, small birds of green plumage closely allied in form to the Linnets. In this genus the bill is very slender and pointed, but resembles that of the Linnets so closely in general shape that it is unnecessary to figure it. In the Siskins the sexes differ considerably in colour, the female being streaked.

773. Chrysomitris tibetana. *The Sikhim Siskin.*

Chrysomitris thibetana, *Hume, Ibis,* 1872, p. 107 ; *Brooks, Ibis,* 1872, p. 469 ; *Hume, S. F.* vii. p. 116 ; *id. Cat.* no. 750 bis ; *Sharpe, Cat. B. M.* xii, p. 226, pl. iii.

Coloration. Male. Upper plumage olive-green, the back and scapulars streaked with blackish ; an indistinct patch on the nape,

a broad supercilium, a streak under the eye and ear-coverts joining
the supercilium behind the ear-coverts, the chin, throat, breast,
abdomen, and under tail-coverts deep dull yellow, the sides of the
body greener and more or less streaked with brown; ear-coverts
and a broad moustachial streak olive-green; tail dark brown,
margined with olive-yellow; lesser and median wing-coverts olive-
green; larger coverts dark brown, tipped and edged with olive-
green; quills dark brown, margined on the outer web with olive-
green.

Female. Not very dissimilar to the male, but with the whole upper
plumage and most of the lower streaked with brown, the lower
plumage being also very pale yellow; wings, tail, and marks on side
of the head as in male.

Bill and legs (in dry state) dusky flesh-colour.

Length about 4·2; tail 1·6; wing 2·7; tarsus ·5; bill from
gape ·4.

Distribution. The interior of Sikhim, at high elevations, bordering
on Tibet*.

Genus FRINGILLA, Linn., 1766.

The genus *Fringilla* contains the typical Finches such as the
Brambling and Chaffinches. In this genus the plumage is much
variegated and in the Brambling, the only species found in India,
the summer plumage differs considerably from that of the winter,
owing to the margins of the feathers wearing away. The bill is
long and the culmen straight except near the tip, where it is slightly
deflected.

774. Fringilla montifringilla. *The Brambling.*

Fringilla montifringilla, *Linn. Syst. Nat.* i, p. 318 (1766); *Blyth,
Cat.* p. 121; *Horsf. & M. Cat.* ii, p. 491; *Jerd. B. I.* ii, p. 412;
Hume, S. F. vii, p. 465; *id. Cat.* no. 752; *Biddulph, Ibis,* 1881,
p. 87; *Scully, Ibis,* 1881, p. 579; *Sharpe, Cat. B. M.* xii, p. 178.

Colouration. Male. After the autumn moult the forehead, crown,
nape, hind neck, and back are black with broad rufous margins;
rump and the upper tail-coverts white, the sides of these black,

* SERINUS PICTORALIS, Murray, Vert. Zool. Sind, p. 190 (1884), is the
Crithagra chrysopyga of Swainson (Birds West Afr. i, p. 206, pl. 17) or
Serinus icterus, Bonn. et Vieill. apud Sharpe, Cat. B. M. xii, p. 356. This
species inhabits a considerable portion of Africa and is a very common cage-
bird, and it was doubtless a bird escaped from confinement that came under
Mr. Murray's observation. Both Mr. Murray's description and a coloured
sketch of the bird sent to me by Mr. Hume agree with the African bird in
every particular.
Forehead, supercilium, and cheeks bright yellow; a broad ashy band from
the lores through the eye to the ear-coverts; a dark brown moustachial streak;
upper plumage ashy green streaked with dark brown; lower plumage yellow,
the sides of the breast ashy; wings and tail brown, margined with yellow;
rump bright yellow. Tail 1·7; wing 2·5.

some of the longer tail-coverts mingled black and ashy; tail black, the feathers narrowly margined with white and the outermost feather with a good deal of white on it; lesser wing-coverts and scapulars orange-rufous; median coverts chiefly white; greater coverts black, tipped with pale rufous, the innermost feather or two with the inner web white; tertiaries black, edged with rufous;

Fig. 65.—Head of *F. montifringilla*.

remainder of quills black edged with pale yellow, and many of the primaries with a basal white patch on the outer web and a broad white margin on the inner; sides of the head and neck black, streaked and mottled with rufous; chin, throat, and breast orange-rufous; abdomen white; under tail-coverts pale buff; flanks buff spotted with black; axillaries primrose-yellow; under wing-coverts white suffused with yellow. In the late spring and summer the margins of the feathers of the head and back are cast or get worn away, leaving those parts deep black and the longer upper tail-coverts are also entirely black; the margins of the wing- and tail-feathers become reduced and in some cases entirely disappear.

Female. Not very different from the male in winter plumage, but the dark parts of the plumage are paler and the rufous margins broader; the lesser and greater wing-coverts and the scapulars are dark brown, fringed with rufous, and the median coverts are broadly tipped with white; the chin, throat, and breast are much paler rufous. Many specimens have an ashen patch on the nape, and this colour suffuses the sides of the neck.

Young birds resemble the female in general appearance, but are suffused with yellow.

Bill light grey at base, dusky at tip; iris brown; legs and feet fleshy brown (*Oates*). In summer the bill becomes black.

Length rather more than 6; tail 2·6; wing 3·5; tarsus ·75; bill from gape ·6.

Distribution. Occurs in Gilgit and N.W. Punjab at the spring and autumn migrations. This species summers and breeds in the more northerly portions of Europe and Asia and in winter migrates southwards, being found at that season in Southern Europe, South-western Asia from Asia Minor to Afghanistan, and in China.

Habits, &c. The Brambling is found in flocks and frequents forest country, but, like many other Finches, feeds on the ground both on seeds and insects as well as on trees.

Genus GYMNORHIS, Hodgs., 1844.

The genus *Gymnorhis* contains one Indian species of Finch which is generally termed a Sparrow, but its affinities for the Sparrows are not very great. In this genus the bill is long and slender with the culmen gently curved throughout, and the chief characteristic of the plumage is the presence of a yellow patch on the throat in both sexes.

Fig. 66. Head of *G. flavicollis*.

This Finch or Sparrow is found in all descriptions of jungle and frequently near houses, and it has much the same habits as the House-Sparrow.

775. Gymnorhis flavicollis. *The Yellow-throated Sparrow.*

Fringilla flavicollis, *Frankl. P. Z. S.* 1831, p. 120.
Petronia flavicollis (*Frankl.*), *Blyth, Cat.* p. 120; *Sharpe, Cat. B. M.* xii, p. 283.
Gymnorhis flavicollis (*Frankl.*), *Horsf. & M. Cat.* ii, p 497; *Hume, N. & E.* p. 461; *Ball, S. F.* vii, p. 223; *Hume, Cat.* no. 711; *Oates in Hume's N. & E.* 2nd ed. ii, p. 157.
Passer flavicollis (*Frankl.*), *Jerd. B. I.* ii, p. 368; *Legge, Birds Ceyl.* p. 695; *Barnes, Birds Bom.* p. 267.

The Yellow-necked Sparrow, Jerd.; *Raji, Jangli-churi*, Hind.; *Adari pichike, Konde pichike, Cheruka pichike*, Tel.

Coloration. Male. The whole upper plumage and scapulars ashy brown; tail brown, narrowly edged with whity brown; lesser wing-coverts chestnut; median coverts brown, tipped with white; greater coverts and tertiaries brown, margined and tipped with pale buff; the other quills dark brown, very narrowly margined with buff; primary-coverts and winglet black; chin dull white; throat yellow; sides of the head and neck and the breast pale ashy brown; remainder of lower plumage ashy white, the flanks darker.

Female. Resembles the male, but has the yellow throat-spot very pale and the lesser wing-coverts rufous-brown, not chestnut.

Iris dark brown; legs and feet greyish plumbeous; the male appears to have the bill black in winter, brown in summer; the female to have it always brown. The colour of the bill of the male is by no means constantly black in winter and brown in summer, but I cannot discover any reasons for the exceptions.

Length about 6; tail 2·1; wing 3·2; tarsus ·6; bill from gape ·6.

Distribution. The plains of India from the foot of the Himalayas down to Travancore, and from Sind eastwards to about the longitude of Midnapore in Bengal; also Ceylon. This species ascends the Himalayas in parts, and the hill-tracts of the south of India up to about 4000 feet. It extends westwards to Persia.

Habits, &c. Breeds from April to July, constructing a small nest of grass and feathers in holes of trees and more rarely in houses. The eggs, three or four in number, are greenish white, densely blotched all over with brown, and measure ·74 by ·55.

Genus **PASSER**, Briss., 1760.

The genus *Passer* contains the true Sparrows, which are well represented over the greater part of the Old World. In this genus, as restricted by me, both sexes agree in exhibiting a peculiar pattern on the outer webs of the earlier primaries, caused by the varying width of the margins of the feathers. The bill is short and stout and the culmen slightly curved.

The Sparrows are mostly well-known birds which frequent the neighbourhood of towns and villages. A few species, however, are only found in the open country away from houses.

Key to the Species.

a. Back streaked with black.
  a'. No supercilium.
    a''. Crown of head ashy grey.
      a'''. Chin, throat, and whole breast
        black .... *P. domesticus* ♂, p. 236.
      b'''. Chin and upper throat only
        black .... *P. pyrrhonotus* ♂, p. 238.
    b''. Crown of head red or rufous.
      c'''. Chin, throat, and breast black. *P. hispaniolensis* ♂, p. 239.
      d'''. Chin and throat only black.
        a⁴. A black patch on ear-coverts. *P. montanus*, p. 240.
        b⁴. No black patch on ear-coverts. *P. cinnamomeus* ♂, p. 240.
  b'. A supercilium.
    c''. Lower plumage more or less
      streaked .... *P. hispaniolensis* ♀, p. 239.
    d''. Lower plumage unstreaked.
      e'''. No yellow in lower plumage.
        c⁴. Wing about 3 .... *P. domesticus* ♀, p. 236.
        d⁴. Wing about 2·5 .... *P. pyrrhonotus* ♀, p. 238.
      f'''. Lower plumage decidedly
        yellow .... *P. cinnamomeus* ♀, p. 240.
b. No black streaks on back .... *P. flaveolus*, p. 242.

776. **Passer domesticus.** *The House-Sparrow.*

Fringilla domestica, *Linn. Syst. Nat.* i, p. 323 (1766).
Passer indicus, *Jard. & Selby, Ill. Orn.* iii, pl. 118 (1835?): *Blyth, Cat.* p. 119: *Horsf. & M. Cat.* ii, p. 490; *Jerd. B. I.* ii, p. 362; *Hume, N. & E.* p. 457; *Oates, B. B.* i, p. 346.
Passer domesticus (*Linn.*), *Hume, Cat.* no. 706; *Legge, Birds Ceyl.*

p. 600; *Sharpe, Cat. B. M.* xii, p. 307; *Barnes, Birds Bom.*
p. 265; *Oates in Hume's N. & E.* 2nd ed. ii, p. 159.

The Indian House-Sparrow, Jerd.; *Gauriya,* Hind. in the North; *Chiri*
and *Khas churi,* Hind. in the South; *Charia* or *Chata,* Beng.; *Uri-pichike,*
Tel.; *Adiki lam kuravi,* Tam.

Fig. 67. Head of *P. domesticus.*

Coloration. Male. Head from forehead to nape ashy grey;
lores and round the eye blackish; cheeks, ear-coverts, and sides of
neck pure white; a broad streak from the eye over the ear-coverts,
and passing partially round the end of them, chestnut; chin, throat,
and the median portion of the breast black, some of the lowermost
feathers margined with ashy; remainder of lower plumage ashy
white; back and scapulars chestnut, the terminal two thirds of the
inner webs black; rump and upper tail-coverts ashy grey; tail
brown, margined paler; lesser wing-coverts chestnut; median
coverts blackish, broadly tipped with white; greater coverts
blackish, broadly margined with rufous and tipped paler; quills
dark brown, margined with pale rufous.

Female. The head from the forehead to the nape and the
extreme upper back with the rump and upper tail-coverts brown;
the back and scapulars pale rufous, with the inner webs chiefly
black; tail brown, edged paler; a rather broad supercilium pale
rufous-white; sides of the head ashy brown; the whole lower
plumage ashy white, darker on the breast; lesser wing-coverts
brown; median coverts blackish, broadly tipped with rufous-white;
greater coverts and wings dark brown, edged with pale rufous.

In fresh or autumn plumage the male has the feathers of the
back and breast margined with ashy; but these margins soon
wear off.

In summer the bill of the male is usually black, but this is not
always the case; in winter the colour is a light horn-colour but
occasionally black; the female has the bill always brown; in both
sexes the iris is brown, the legs pale brown.

Length 6; tail 2·2; wing 3; tarsus ·75; bill from gape ·55.

The House-Sparrow of the East differs from the House-Sparrow
of the West in being much whiter about the sides of the head, and
in having more black below the eye and at the base of the cheeks,
but these characters vary considerably and it is not advisable to
keep the two birds distinct.

Distribution. The entire Empire and Ceylon, except the
Andamans and Nicobars and the portion of Tenasserim south of

Moulmein. This species ascends the Himalayas to moderate elevations. It is capricious in its distribution, being rare in some parts of the Empire and extremely common in others.

The House-Sparrow to the eastward is found in Cochin-China, and on the west it extends to Europe.

Habits, &c. Breeds usually from February to May, but also at other times of the year, making a shapeless nest of grass and various materials in holes about houses, in walls, in wells, and occasionally in some thick tree or shrub. The eggs, which are usually five in number, are white or greenish marked with various shades of brown, and measure about ·81 by ·6.

777. **Passer pyrrhonotus.** *The Rufous-backed Sparrow.*

Passer pyrrhonotus, *Blyth, J. A. S. B.* xiii, p. 946 (1844); *id. Cat.* p. 119; *Jerd. B. I.* ii, p. 365; *Hume, Cat.* no. 709; *Hume, S. F.* ix, pp. 232, 442; *Doig, S. F.* ix, p. 280; *Sharpe, Cat. B. M.* xii, p. 316, pl. v; *Barnes, Birds Bom.* p. 266; *Oates in Hume's N. & E.* 2nd ed. ii, p. 162.

Coloration. Male. After the autumn moult the forehead, crown, and hind neck are ashy grey; lores and under the eye blackish; a broad band behind the eye and ear-coverts chestnut, with ashy fringes; cheeks, ear-coverts, and sides of the neck ashy grey; chin and throat black, with whitish fringes, and bordered on both sides by a broad whitish band; lower plumage pale ashy, becoming whiter on the abdomen and under tail-coverts; back chestnut, fringed with fulvous, and the inner web of each feather with a black streak; lesser wing-coverts, scapulars, and rump chestnut with ashy fringes; median coverts almost entirely white; greater coverts blackish, edged with chestnut-brown; quills dark brown, edged with pale chestnut-brown; rump and upper tail-coverts ashy; tail brown, edged with dull fulvous. In the spring and summer all the fringes on the various parts of the plumage get worn away, and those parts become a dark uniform colour.

Female. The whole upper plumage ashy brown, the feathers of the back with black streaks on the inner web; median wing-coverts black, broadly tipped with pale buff; the greater coverts blackish, broadly edged with buff; the quills dark brown edged with buff, most broadly so at the base near the coverts; tail brown, narrowly edged with buff; a broad supercilium isabelline; sides of head ashy; the entire lower plumage pale ashy white.

Iris light brown; eyelids leaden slaty; legs and toes dusky fleshy brown (*Doig*). In winter the bill is dusky brown, in summer probably black.

Length about 5·5; tail 2·1; wing 2·6; tarsus ·65; bill from gape ·45.

Distribution. Bahawalpur; the Eastern Nara, Sind.

Habits, &c. Mr. Doig remarks that he never met with these Sparrows at any distance from water, and that they were usually seen in small flocks. Their food consists of seeds and insects.

He found three nests in August, built on the tops of acacia trees growing in the water. The nest resembles that of *P. domesticus*, and the eggs do not differ in colour from those of that species. They measure about ·69 by ·51.

778. Passer hispaniolensis. *The Spanish Sparrow.*

Fringilla hispaniolensis, *Temm. Mon. d'Orn.* éd. 2, p. 353 (1820).
Fringilla salicicola, *Vieill. Faune Franç.* p. 417 (1828).
Pyrgita salicaria, *Bonap. Comp. List B. Eur. & N. Am.* p. 30 (1838).
Passer salicaria (*Bonap.*), *Blyth, Cat.* p. 119.
Passer salicicola (*Vieill.*), *Horsf. & M. Cat.* ii, p. 501 ; *Jerd. B. I.*
   ii, p. 364 ; *Hume, S. F.* i, p. 209 ; *Scully, S. F.* iv, p. 164.
Passer hispaniolensis (*Temm.*), *Hume, Cat.* no. 707 ; *Sharpe, Cat.
   B. M.* xii, p. 317 ; *Barnes, Birds Bom.* p. 266.

The Willow Sparrow, Jerd.

Coloration. Male. After the autumn moult the forehead, crown, nape, and a band behind the ear-coverts are chestnut with creamy-white fringes; sides of the forehead and a small patch behind the eye, forming a disconnected supercilium, dull white; lores and under the eye black ; cheeks, ear-coverts, and sides of the neck white; chin, throat, and upper breast black with whitish fringes ; lower breast and sides of the body creamy white streaked with black ; abdomen and under tail-coverts creamy white unstreaked ; back black with broad fulvous margins, some of the lateral feathers with the whole outer web fulvous and the inner black; lesser wing-coverts chestnut; median coverts almost entirely white; greater coverts and quills dark brown, broadly edged with chestnut-brown ; rump and upper tail-coverts fulvous ashy with paler margins; tail brown, edged with fulvous. In spring and summer the fringes everywhere get worn away and the parts affected become uniform.

Female. Resembles the female of *P. domesticus* so closely as to require no separate description. It differs in having a much larger bill and the lower plumage faintly streaked throughout.

Legs and feet horny brown; soles yellow; iris brown; bill brown, yellow at gape (*Hume*, December); in summer the bill of the male is black.

Length about 6; tail 2·4; wing 3·2; tarsus ·8; bill from gape ·55.

Distribution. Sind ; the Punjab ; the northern part of Rajputana down to Sámbhar ; the N.W. Provinces and Oudh as far east as Mirzapore. This Sparrow is apparently found throughout the above tracts in the winter months only. It occurs in Kashmir, so far as I have been able to ascertain, both in summer and winter. It extends north to Yárkand, where it breeds. To the west it ranges to Europe and Northern Africa.

Habits, &c. This Sparrow is said by Dr. Scully to frequent reeds, poplar trees, and corn-fields. He states that it breeds in Turkestan in May and June, nesting in trees.

779. **Passer montanus.** *The Tree-Sparrow.*

Fringilla montana, *Linn. Syst. Nat.* i, p. 324 (1766).
Passer montanus (*Linn.*), *Blyth, Cat.* p. 120; *Horsf. & M. Cat.* ii,
p. 500; *Jerd. B. I.* ii, p. 365; *Hume, N. & E.* p. 460; *Anders.
Yunnan Exped., Aves,* p. 601; *Hume, Cat.* no. 710; *Oates, B. B.* i,
p. 348; *Sharpe, Cat. B. M.* xii, p. 304; *Oates in Hume's N. & E.
2nd ed.* ii, p. 162.

The Mountain Sparrow, Jerd.

Coloration. Male and female. The whole head from forehead to
nape vinous chestnut; lores, feathers under the eye, and a patch
under the ear-coverts and encroaching upon them black; with this
exception the sides of the face and neck are white; chin and
throat black; lower plumage ashy, whitish on the abdomen and
tinged with fulvous on the sides of the breast, flanks, and under
tail-coverts; back and scapulars pale chestnut, with the inner webs
of the feathers chiefly black; rump and upper tail-coverts yellowish
brown; tail brown, edged with fulvous; lesser wing-coverts chestnut; median coverts black, broadly tipped with white; greater
coverts blackish, edged with pale chestnut and tipped with whitish;
quills dark brown edged with rufous.

Bill black; iris brown; legs flesh-colour; claws brown.

Length 5·6; tail 2·3; wing 2·7; tarsus ·7; bill from gape ·55.

This species throughout its vast range remains very constant in
coloration. A slight exception occurs, however, in birds from
Yárkand and Central Asia, where the lower plumage of this
Sparrow becomes white.

Distribution. The whole of the Himalayas from Afghanistan to
Assam up to 7000 feet in summer, descending to lower levels in
winter; the whole of the countries from Assam southwards to the
extreme southern point of Tenasserim. In the British Museum
there is a skin of this Sparrow said to have been procured in the
Deccan by Sykes, but probably erroneously, as Horsfield and
Moore do not record the specimen in their Catalogue and the
locality is quite outside the range of this bird.

The Tree-Sparrow has a wide range over Europe, Africa, and
Asia, extending south to Java.

Habits, &c. The Tree-Sparrow nests in the east chiefly in holes
about houses and other buildings, but in Europe it nests generally
in trees. The nest is constructed of all sorts of materials and is a
shapeless mass, suited roughly to the cavity it occupies. The eggs
resemble those of the House-Sparrow and measure about ·73 by
·54. The nest may be found at most times of the year, but more
commonly from February to May.

780. **Passer cinnamomeus.** *The Cinnamon Tree-Sparrow.*

Pyrgita cinnamomea, *Gould, P. Z. S.* 1835, p. 185.
Passer cinnamomeus (*Gould*), *Blyth, Cat.* p. 119; *Horsf. & M. Cat.*

ii, p. 500 : *Jerd. B. I.* ii, p. 365; *Hume & Henders. Lah. to Yark.*
p. 252, pl. 25; *Hume, N. & E.* p. 459; *Anders. Yunnan Exped.,*
Aves, p. 602; *Hume, Cat.* no. 708; *Sharpe, Cat. B. M.* xii, p. 325;
Hume, S. F. xi, p. 275; *Oates in Hume's N. & E.* 2nd ed. ii, p. 164.

The *Cinnamon-headed Sparrow,* Jerd.

Coloration. Male. The upper plumage from the forehead to the
rump, including the scapulars and lesser wing-coverts, bright
cinnamon-rufous, the feathers of the back with the inner web black,
wholly or partially, and all the feathers with very narrow pale
fringes; upper tail-coverts brown with ashy margins; tail brown
with greenish margins; median coverts black, broadly tipped
white; greater coverts and tertiaries black, edged with pale rufous;
primaries and secondaries black, edged with pale fulvous, more
broadly so at the base and just above the emarginations of the first
few primaries; lores and round the eye black; cheeks and ear-
coverts pale yellowish white; chin and throat black, fringed with
whitish; a large patch on each side of the throat bright yellow;
lower plumage greyish yellow, more yellow on the abdomen and
under-tail coverts. The difference between the summer and winter
plumage of this Sparrow is very slight, the colours in the former
season being slightly more intense owing to the narrow fringes
wearing away.

Female. The whole upper plumage ruddy brown, tinged with red
on the rump and with black and fulvous streaks on the back;
lesser wing-coverts ruddy brown; median coverts black, tipped
with white; greater coverts, quills, and tail dark brown edged with
fulvous; a broad fulvous supercilium, with a broad dusky band
below it; sides of the head and neck and the whole lower plumage
pale ashy yellow.

Iris reddish brown; legs and feet dark reddish brown; bill pale
brown in winter, black in summer.

Length about 5·5; tail 2·3; wing 2·9; tarsus ·65; bill from
gape ·55.

Distribution. The Himalayas from Murree to Bhutan up to 7000
feet; the Khásí hills; the Nága hills; Manipur; the hills east of
Bhámo; the Karen hills east of Toungngoo.

Habits, &c. Chiefly a jungle-sparrow. Breeds in May and June,
constructing its nest in holes of trees as a rule, but sometimes in
houses. Eggs, four to six, of the Sparrow type, and measuring
about ·76 by ·57.

Passer assimilis, Walden, A. M. N. H. (4) v, p. 218 (1870), is,
I now find after a re-examination of the type, to be referred to
P. rutilans, Temm., as already noted by Sharpe (Cat. B. M. xii,
p. 827). The type of *P. assimilis* is said to have been procured at
Toungngoo, but there may be probably some mistake about this, as
the specimen was not shot by Wardlaw Ramsay or other trustworthy collector. It appears to be a dealer's skin. It is also to
be noted that a pair of true *P. cinnamomeus* were procured by
Wardlaw Ramsay on the Karen hills near Toungngoo, and it is

unlikely that two distinct but closely allied Sparrows should be found together in Toungngoo and its neighbourhood. Under these circumstances I shall not include *P. rutilans* among the birds of the Indian Empire. In case, however, it should be met with, the male may be recognized by its similarity to the male of *P. cinnamomeus*, from which it differs in having the cheeks and ear-coverts pure white and the lower plumage ashy white without a trace of yellow. The females of the two species are undistinguishable from each other. *P. rutilans* is found in China.

781. Passer flaveolus. *The Pegu House-Sparrow.*

Passer flaveolus, *Blyth*, J. A. S. B. xiii, p. 946 (1844); *id. Cat.* p. 120; *Hume, N. & E.* p. 460; *Blyth & Wald. Birds Burm.* p. 94; *Hume, S. F.* iii, p. 156; *Anders. Yunnan Exped., Aves*, p. 602; *Hume, Cat.* no. 708 bis; *Oates, B. B.* i, p. 349; *Sharpe, Cat. B. M.* xii, p. 330; *Salvadori, Ann. Mus. Civ. Gen.* (2), vii, p. 419; *Oates in Hume's N. & E.* 2nd ed. ii, p. 165.
Passer jugiferus, *Temm., Bonap. Consp. Av.* i, p. 508 (1850).

Coloration. Male. After the autumn moult the lores, chin, and a stripe down the throat are black; a line over the lores from the nostrils to the eye yellow; cheeks and ear-coverts with the whole lower plumage and under wing-coverts rather bright yellow; a patch extending from the eye over the ear-coverts to the sides of the nape chestnut; forehead, top of head, nape, and hind neck greenish olive; back, scapulars, and lesser wing-coverts chestnut fringed with greenish; lower back and rump brown tinged with yellow; tail brown, the outer webs tinged with olive-yellow; median wing-coverts dark brown, broadly tipped with white; greater wing-coverts and all the quills dark brown, edged with yellowish white. In the summer the greenish fringes are worn away.

Female. The chin, throat, cheeks, and the whole lower plumage with the under wing-coverts pale yellow; a streak from the eye to the nape yellowish white; the upper plumage from the forehead to the upper tail-coverts, the scapulars and lesser wing-coverts hair-brown, the shafts of all the feathers darker; the median and greater wing-coverts and the quills dark brown, each feather edged with yellowish white; tail brown, the feathers edged with whitish on the outer webs.

Bill black in the male, flesh-colour in the female; iris dark hazel-brown; legs and claws plumbeous flesh-colour.

Length 5·5; tail 2·1; wing 2·7; tarsus ·6; bill from gape ·55.

Distribution. Mengoon on the Irrawaddy river in Upper Burma; Arrakan; the greater part of Pegu, but the species is more abundant in the northern portion; the Karen hills east of Toungngoo; Karennee; extending into Cochin China.

Habits, &c. Frequents habitations and jungle near houses, breeding in March and other months.

The following Sparrow, which occurs in Turkestan, may possibly hereafter be found within Indian limits:—

Passer ammodendri, Severtz. The male has the forehead, crown, and nape narrowly black; the sides of the crown and the sides of the nape clear rufous; chin and throat black; lower plumage ashy white; upper plumage ashy brown, streaked with black. The female has no rufous on the sides of the crown and nape, and the chin and throat are pale brown. Tail 2·7, wing 3.

Passer pyrrhopterus (Less.), mentioned by Jerdon (B. I. ii, p. 367), is probably referable to *P. domesticus*, but it is not easy to identify it. It is not, however, likely to be a Sparrow which remains to be rediscovered in Southern India.

Genus **PETRONIA**, Kaup, 1829.

The genus *Petronia* contains some species which resemble Sparrows in structure and habits, but differ from them in having a much stronger bill and longer wings. The only species found in India is characterized by the presence of a yellow patch on the throat in both sexes. This species has the same pattern on the earlier primaries as the Sparrows.

The Rock-Sparrows frequent open rocky land and are gregarious.

782. **Petronia stulta.** *The Rock-Sparrow.*

Fringilla petronia, *Linn. Syst. Nat.* i, p. 322 (1766).
Fringilla stulta, *Gmel. Syst. Nat.* i, p. 919 (1788).
Petronia stulta (*Gm.*) *Blyth, Cat.* p. 120; *Horsf. & M. Cat.* ii,
    p. 497; *Biddulph, Ibis*, 1881, p. 79; *Scully, Ibis*, 1881, p. 574;
    Hume, S. F. ix, p. 343 (note).
Petronia petronia (*Linn.*), *Sharpe, Cat. B. M.* xii, p. 289.

Fig. 68.—Head of *P. stulta.*

Coloration. Male. The forehead and crown dark brown, with a broad mesial buff band from the bill to the nape; a broad supercilium buff; back and scapulars with the inner webs of the feathers black, the outer buff; rump mingled brown and buff; upper tail-coverts brown edged with buff; middle tail-feathers ashy brown, becoming black towards the tip and margined whitish; the other feathers the same but with a large terminal white patch

n 2

on the inner web; quills and coverts brown edged with buff; a dark brown band from the eye under the supercilium and a spot near the gape; ear-coverts brown : lower plumage whity brown more or less streaked with dark brown, and with a yellow patch on the lower throat.

Female. Resembles the male closely.

Bill horn-brown above, light brown below with a dark tip; legs light brown; iris brown.

Length about 6·5; tail 2·3; wing 3·8; tarsus ·7; bill from gape ·7.

Distribution. A winter visitor to Gilgit, occurring there from November to March. This species has a wide range, being found over a considerable portion of Europe, in North Africa, Asia Minor, Persia, Afghanistan, Central Asia, Siberia, and Northern China.

Genus **MONTIFRINGILLA**, Brehm, 1828.

The genus *Montifringilla* is closely allied to the genus *Petronia* in structure, but the wings are rather longer and the bill less deep though otherwise similar in outline. The birds of this genus are characterized by a large amount of white on the wings and tail. The sexes are alike or nearly so. The claws are somewhat lengthened. The tail is perfectly square, the middle feathers being as long as the others.

The Mountain-Finches are found at high elevations feeding in flocks on the ground.

Key to the Species.

a. Sides of neck ferruginous; lores black.
    a'. Chin black : no moustachial streaks ...... M. *blanfordi*, p. 245.
    b'. Chin white ; black moustachial streaks .. M. *ruficollis*, p. 245.
b. Sides of neck pale fulvous; lores pale ; no
    black marks on the head .... M. *adamsi*, p. 246.

Montifringilla mandellii, Hume (S. F. iv. p. 488), was procured by Mandelli in Tibet, north of Sikhim. It does not appear to have yet been met with in British territory. This species resembles *M. ruficollis*, but has no ferruginous on the sides of the head and neck, nor has it any trace of the black moustachial streaks. The bill in the dry state is yellow with a black tip, not bluish black throughout as in *M. ruficollis*. Wing 4·1 ; tail 3.

M. alpicola (Pall.) occurs in Persia and Afghanistan. In the latter country this species was procured by Griffith " near Gurdan Dewar, on the Helmund, at an elevation of 11,500 feet " (Horsf. & M. Cat. ii, p. 491). The specimen referred to is now in the British Museum. This species has a very long bill, a wing measuring 4·7, and the wing-coverts and secondaries entirely white.

783. Montifringilla blanfordi. *Blanford's Mountain-Finch.*

Montifringilla blanfordi, *Hume, S. F.* iv, p. 487 (1876); *id. Cat.* no. 752 quint.; *Sharpe, Cat. B. M.* xii, p. 264, pl. iv.

Coloration. Forehead white, with a median black streak; a large patch between the eye and the bill black, extending down to the cheeks and up to the forehead; a short but broad white supercilium; cheeks and ear-coverts white; the whole upper plumage fulvous; wing-coverts and tertiaries brown, broadly edged with fulvous; quills dark brown, edged with fulvous, the last few primaries with a patch of white on the outer web; all the quills except the first two or three primaries also with a patch of white on the inner web; middle tail-feathers brown edged with fulvous, the others ashy brown, then white and broadly tipped brown; sides of the neck ferruginous, reaching forward to the sides of the throat and breast but not meeting in front; chin black; remainder of lower plumage pale fulvous-white.

None of the birds in the British Museum series of this species are sexed. The sexes are, however, probably alike. Some few birds with the chin white or pale brown are obviously young birds just fledged; they resemble the adult in other respects, but have the marks on the head paler, and some have the lower plumage suffused with yellow.

In the dry state the bill is bluish and the legs black.

Length about 6; tail 2·2; wing 3·8; tarsus ·7; bill from gape ·55.

Distribution. In the Hume Collection there are four specimens of this species which were procured " near Darjeeling " and numerous others from Tibet immediately north of Sikhim. Some of these latter are quite young birds.

784. Montifringilla ruficollis. *The Red-necked Mountain-Finch.*

Montifringilla ruficollis, *Blanf. J. A. S. B.* xli, pt. ii, p. 66; *Hume, S. F.* vii, p. 420; *id. Cat.* no. 752 quat.; *Sharpe, Cat. B. M.* xii, p. 263.

Coloration. Male. Forehead whitish, turning to ashy and becoming umber-brown on the crown, nape, and hind neck; a broad white supercilium; lores and a band under the eye and over the ear-coverts black, passing into rufous posteriorly; sides of the nape, sides of the neck, the ear-coverts, and the sides of the lower throat ferruginous; chin, throat, and cheeks pure white; a moustachial streak black; remainder of lower plumage white tinged with fulvous; back and scapulars umber-brown streaked with dark brown; rump and upper tail-coverts plain umber-brown; middle tail-feathers brown, the others chiefly ashy with broad brown tips, the portion of each feather immediately before the brown tip being white; lesser wing coverts brown; median and greater coverts chiefly white; winglet and primary-coverts dark brown; quills brown margined with fulvous; the outer web of the first primary entirely white; all the quills except the first four primaries and the last two or three secondaries with a basal patch of white on the inner web; tertiaries brown, broadly edged with fulvous.

Female. Apparently differs from the male in having the ferruginous on the sides of the neck and throat produced so as to form a continuous collar across the lower throat; and the white on the forehead less extensive.

The young bird resembles the adult closely, but is paler and has the marks on the head less distinct.

In the dry state the bill is bluish black and the legs black.

Length about 6; tail 2·3; wing 3·7; tarsus ·85; bill from gape ·55.

Distribution. Sikhim and Tibet, extending to Western China. This species is found at great altitudes, Blanford meeting with it at 15,000 and 16,000 feet. Mandelli procured many young birds, just able to fly, in Tibet, immediately north of Sikhim, in June.

785. **Montifringilla adamsi.** *Adams's Mountain-Finch.*

Montifringilla adamsi, *Moore, Adams, P. Z. S.* 1858, p. 482, 1859, p. 178, pl. 156; *Hume & Henders. Lah. to Yark.* p. 262; *Hume, N. & E.* p. 473; *Stoliczka, J. A. S. B.* xxxvii, pt. ii, p. 62; *id. S. F.* ii, p. 463, iii, p. 220; *Scully, S. F.* iv, p. 172; *Hume, Cat.* no. 752 ter; *Sharpe, Cat. B. M.* xii, p. 261; *Oates in Hume's N. & E.* 2nd ed. ii, p. 165.

Coloration. The whole upper plumage brown, the feathers of the back with a broad median darker brown streak; upper tail-coverts with the outer webs whitish, the inner brown; middle tail-feathers black with fulvous edges; the next black at base of inner web and at tip of both webs, white elsewhere; the others white with black tips; lesser and median wing-coverts and tertiaries brown; greater coverts dark brown, broadly tipped white; primary-coverts white, tipped brown; primaries blackish, edged with fulvous; secondaries brown, with a large amount of white near the tips; the sides of the head and neck and the whole lower plumage pale fulvous-white; under wing-coverts and axillaries pure white.

In spring and summer the plumage is much worn down and consequently duller, but no other change takes place. The sexes appear to be alike.

Legs, feet, and claws black; iris brown; bill black in summer, orange-yellow, dusky on culmen and brown at tip, in winter (*Hume*).

Length nearly 7; tail 2·7; wing 4·3; tarsus ·8; bill from gape ·65.

Distribution. The higher regions of the Himalayas beyond the first snowy range. On the east this species occurs as far as Sikhim according to Hume, and on the west it is met with over the greater part of the north and east of Kashmir, Ladák, Kulu, &c. It has also been met with near Gilgit, and it extends to Káshgarh and Tibet. This species appears to be found between 11,000 and 14,000 feet in summer.

According to Adams this species breeds in Ladák and Little Tibet, constructing its nest in the long dykes built by the Tartars over their dead.

Genus **FRINGILLAUDA**, Hodgs., 1836.

The genus *Fringillauda* resembles *Montifringilla*, but the bill is more slender and the tail is forked, the middle feathers being considerably shorter than the outer. There is moreover no white in the wings and tail except on the margins.

The birds of this genus, like those of the last, are found at considerable altitudes and feed on the ground in flocks.

Key to the Species.

a. Wing nearly 4 ; no rose-colour on rump or
    wing-coverts.
  a'. Axillaries yellow .... *F. nemoricola*, p. 247.
  b'. Axillaries pale ashy .... *F. sordida*, p. 248.
b. Wing nearly 5; rump and wing-coverts
    suffused with rose-colour .... *F. brandti*, p. 248.

786. **Fringillauda nemoricola.** *Hodgson's Mountain-Finch.*

Fringillauda nemoricola, *Hodgs. As. Res.* xix, p. 158 (1836); *Horsf. & M. Cat.* ii, p. 491 (part.); *Jerd. B. I.* ii, p. 414 (part.); *Blanf. J. A. S. B.* xli, pt. ii, p. 66; *Hume, S. F.* i, p. 41; *id. Cat.* no. 753.
Montifringilla nemoricola (*Hodgs.*), *Blyth. Cat.* p. 121 (part.); *Sharpe, Cat. B. M.* xii, p. 268.
The Himalayan Lark-Finch, Jerd. (part.).

Fig. 69.—Head of *F. nemoricola.*

Coloration. Forehead, crown, nape, hind neck, back, scapulars, and lesser wing-coverts dark brown, with rufous margins ; rump ashy grey ; upper tail-coverts black, with distinct white margins ; tail brown, with narrow pale margins ; median wing-coverts ashy brown, with white margins ; greater coverts brown, mottled with black and tipped white ; primary-coverts tipped with dark brown and edged with ashy ; quills brown, margined with rufous, and the inner webs of the tertiaries black ; a broad but indistinct supercilium ashy white streaked with brown ; sides of the head and neck rufous-brown, the cheeks and the part under the eye streaked with brown ; lower plumage plain brown, the sides of the breast streaked with darker brown, most of the feathers edged paler, the

middle of the abdomen whitish, the flanks a darker brown, the
under tail-coverts very broadly edged with white ; axillaries yellow ;
under wing-coverts ashy white.

The sexes appear to be alike.

The nestling bird has the whole crown uniform deep rufous, the
upper plumage darker than in the adult and with deeper rufous
margins, and the lower plumage uniform rufous, a little paler than
the crown. The change to the adult plumage is difficult to trace,
but it is probable that the young bird retains the rufous head till
the second autumn.

Bill and legs fleshy-brown ; iris red-brown (*Hodgson*) ; iris clear
nut-brown (*A. David*).

Length about 6·5 ; tail 2·8 ; wing 3·9 ; tarsus ·8 ; bill from
gape ·55.

Distribution. The Himalayas from Nepal to Bhutan, extending
to Moupin and Western China.

787. Fringillauda sordida. *Stoliczka's Mountain-Finch.*

Montifringilla nemoricola (*Hodgs.*), *Blyth, Cat.* p. 121 (part.)
Fringillauda nemoricola, *Hodgs., Horsf. & M. Cat.* ii, p. 491 (part.) ;
  Jerd. B. I. ii, p. 414 (part.) ; *Hume & Henders. Lah. to Yark.*
  p. 264 ; *Brooks, J. A. S. B.* xli, pt. ii, p. 84.
Fringillauda sordida, *Stoliczka, J. A. S. B.* xxxvii, pt. ii, p. 63
  (1868) ; *Hume, S. F.* i, p. 41 ; *id. Cat.* no. 753 bis ; *Biddulph,*
  Ibis, 1881, p. 88 ; *Scully, Ibis,* 1881, p. 579.
Montifringilla sordida (*Stol.*), *Sharpe, Cat. B. M.* xii, p. 266.

The Himalayan Lark-Finch, Jerd. (part.)

Coloration. Resembles *F. nemoricola,* but has the axillaries pale
ashy and the tips to the median and greater wing-coverts less
distinct and tinged with rufous, not pure white.

Iris cinnabar-red ; bill brown, a spot of brownish fleshy at base
of forehead between nostrils, and base of lower mandible brownish
fleshy ; legs, feet, and claws blackish brown (*Hume*).

Of much the same size as *F. nemoricola.*

Distribution. The Himalayas from Afghanistan and Gilgit to
Kumaun, extending to parts of Central Asia. Common at about
5000 feet in winter about Gilgit, and in summer at 9000 or
10,000 feet.

788. Fringillauda brandti. *Brandt's Mountain-Finch.*

Leucosticte brandti, *Bonap. Consp. Av.* i, p. 537 (1850) ; *Biddulph,*
  Ibis, 1881, p. 88.
Montifringilla hæmatopygia, *Gould, P. Z. S.* 1851, p. 115 ; *Stoliczka,*
  J. A. S. B. xxxvii, pt. ii, p. 62 ; *Hume & Henders. Lah. to Yark.*
  p. 261 ; *Scully, S. F.* iv, p. 171.
Leucosticte hæmatopygia (*Gould*), *Blanf. J. A. S. B.* xli, pt. ii,
  p. 66 ; *Hume, Cat.* no. 752 bis.
Montifringilla brandti (*Bonap.*), *Sharpe, Cat. B. M.* xii, p. 269.

Coloration. After the autumn moult the forehead, anterior part of crown, and lores are black, margined with sandy brown; the remainder of the crown, nape, hind neck, and sides of neck brown, margined with sandy brown; back and scapulars ashy brown with fulvous fringes and dark shafts; rump ashy, each feather delicately tipped with rosy red, the shafts dark; upper tail-coverts ashy, broadly tipped and margined with white, and the shafts dark; wing-coverts pale ashy with darker shafts, the lesser and median coverts fringed with rosy red, the greater coverts with pale fulvous white; winglet, primary-coverts, primaries, and secondaries black edged with white; tertiaries ashy brown; tail black edged with white; sides of the head, chin, throat, and breast ashy, the feathers edged with pale fulvous; remainder of lower plumage pale ashy with darker shaft-streaks; under wing-coverts and axillaries pale ashy white. The sexes appear to be alike.

As the winter progresses, the margins of all the feathers get worn away, and the whole head and mantle become dark blackish brown; the other parts are also much darker, and while the red on the rump becomes more intense in colour, the red on the wing-coverts disappears by abrasion; the wings and tail become nearly uniform brown.

The young bird has very broad fulvous margins to the feathers of the head, and the margins everywhere more fulvous; there is no red on the wing-coverts, but the rump-feathers are rather broadly margined with red or reddish yellow from the earliest period.

Bill, legs, and feet black; iris brown (*Hume*).

Length about 7·5; tail 3·1; wing 4·8; tarsus ·85; bill from gape ·55.

Distribution. The Himalayas from Gilgit to Sikhim, extending into Turkestan and Tibet. This species is found at high elevations from 12,000 to 19,000 feet, but descends occasionally in winter to the level of Gilgit.

Subfamily EMBERIZINÆ.

The *Emberizinæ* comprise the Buntings, a very large group, of which fifteen species are found in India, the majority visiting that country in the winter and retiring north in the summer. A few remain to breed, but chiefly in the Himalayas.

The Buntings have a conical and sharply pointed bill, with the culmen straight or nearly so; the edges of the two mandibles, however, unlike those of the other *Fringillidæ*, are not in contact throughout their length, but form a gap or angle about midway between the gape and the tip of the bill. The upper mandible, moreover, has the palate furnished with a small hard process or knob. With this exception the Buntings conform in structure to the Finches. Like them also they have a double plumage, caused in most cases by the abrasion or dropping off of the margins of the feathers in spring, while a few Buntings have in addition a partial spring moult.

The young of Buntings resemble the female, but are characterized, where striation is present, by a greater amount of streaking.
At the autumn moult of the first year the young assume the
plumage of the adult female, and then the males gradually put on
the plumage of the adult male, the process taking several months.

The Buntings frequent cornfields, waste lands, and grassy tracts
of country. They are more or less gregarious. They devour grain
in large quantities and also feed on seeds of all sorts. Their nests
are cup-shaped and placed on or near the ground in grass and
bushes, and sometimes in crevices of rocks and walls. The eggs, as
a rule, are richly marked with spots and lines of red and purple.

With the exception of one Bunting which is crested, all the
Indian species appear to me to be sufficiently similar in structure
to be congeneric, and I have accordingly placed them all in the
genus *Emberiza*.

Key to the Genera.

```
a. No crest  ..............................  EMBERIZA, p. 250.
b. A well-developed crest  ....................  MELOPHUS, p. 265.
```

Genus **EMBERIZA**, Briss., 1760.

Fig. 70.—Head of *E. aureola*.

The genus *Emberiza* contains the typical Buntings, which are
crestless and have a slightly forked tail.

Key to the Species *.

```
a. A large distinct white patch on the
     outermost tail-feather.
   a'. Sides of body streaked or differ-
        ently coloured to abdomen.
     a". No trace of yellow on lower
          plumage.
       a'''. Chin and throat black.
         a³. Breast white ............  E. schœniclus ♂, p. 251.
         b³. Breast chestnut ..........  E. stewarti ♂, p. 256.
       b'''. Chin and throat chestnut ..  E. leucocephala ♂, p. 254.
```

---

* This key applies only to fully adult birds, and the margins of the feathers
in winter plumage are disregarded, the colours noted being those which are
most fully developed at the breeding-season.

789. Emberiza schœniclus. *The Reed-Bunting.*

Emberiza schœniclus, *Linn. Syst. Nat.* i, p. 311 (1766); *Hume, S. F.*
vii, p. 412; *Biddulph, Ibis,* 1881, p. 81; *Scully, Ibis,* 1881, p. 575;
Hume, Cat. no. 720 ter; *Sharpe, Cat. B. M.* xii, p. 480.

Coloration. Male. After the autumn moult the forehead,
crown, and nape are black, each feather broadly margined with
fulvous; a broad collar round the hind neck white, the feathers
with broad fulvous-ashy tips, which conceal the white bases of the
feathers; the whole upper plumage and wings dark brown, each
feather broadly margined with rufous or chestnut, the rump and
upper tail-coverts strongly tinged with ashy: the four middle
pairs of tail-feathers very dark brown, margined with rufous; the

two outer pairs mostly white, the bases and a longitudinal streak along the shaft being brown or black; sides of the head rufous with concealed black bases; a broad white moustachial band more or less dimmed by rufous tips; chin, throat, and fore neck black, with broad white edges; sides of the neck and remaining lower plumage dull white, the sides of the body streaked with ochraceous brown.

Soon after the moult the margins and tips of the feathers begin to wear away, and in full breeding-plumage the moustachial band and the collar round the hind neck become pure white; the forehead, crown, nape, sides of the head, chin, throat, and fore neck become deep black; the margins of the feathers of the upper plumage decrease in size, causing the plumage to become much blacker.

Female. Closely resembles the male after the moult, but has no concealed black bases to the feathers of the head, chin, throat, and fore neck, these parts being rufous or fulvous, more or less streaked and mottled with black; the moustachial band, which is pale fulvous, is bordered below by another blackish band, and the breast and sides of the body are boldly streaked with rufous.

The young resemble the adult female closely.

In winter the bill is dark brown, the lower mandible paler and whitish; legs and feet dark bronze-brown; claws black; iris brown (*Hume Coll.*); in the summer the bill of the male becomes nearly black.

Length about 6; tail 2·7; wing 3·1; tarsus ·75; bill from gape ·4.

Distribution. A winter visitor to the north-west of the Empire. This species occurs in Gilgit from December to March, and a specimen in the Hume collection was obtained at Rohtak in the Punjab in December.

This Bunting has a very wide range, being found over the whole of Europe and Central and Northern Asia.

790. **Emberiza fucata.** *The Grey-headed Bunting.*

Emberiza fucata, *Pall. Reis. Russ. Reichs,* iii, p. 698 (1776); *Jerd. B. I.* ii, p. 375; *Anders. Yunnan Exped., Aves,* p. 603; *Hume, Cat.* no. 719; *Oates, B. B.* i, p. 551; *Sharpe, Cat. B. M.* xii, p. 493; *Barnes, Birds Bom.* p. 280; *Hume, S. F.* xi, p. 279; *Oates in Hume's N. & E.* 2nd ed. ii, p. 166.
Euspiza fucata (*Pall.*), *Blyth, Cat.* p. 129; *Horsf. & M. Cat.* ii, p. 488.
Citrinella fucata (*Pall.*), *Hume, N. & E.* p. 465.
Emberiza arcuata, *Sharpe, Cat. B. M.* xii, p. 494 (1888).
Putthur-chirta, Hind.

Coloration. Male. After the moult the forehead, crown, nape, and sides of the neck are ashy streaked with black; back and scapulars reddish brown, with broad black streaks; rump reddish brown, with obsolete brown streaks; upper tail-coverts fawn-brown, streaked with brown; lesser and median wing-coverts chestnut with concealed dark bases; greater coverts and tertiaries black,

with broad rufous edges; quills dark brown edged with rufous;
lores and round the eye fulvous mottled with ashy; ear-coverts
chestnut; cheeks fulvous, continued as a band under the ear-
coverts; a moustachial band black, gradually widening and reaching
to the lower throat, where it meets the other moustachial streak,
thus forming a gorget which on the fore neck is more or less
interrupted by fulvous streaks; chin and throat fulvous; a band
of chestnut across the upper breast; remaining lower plumage
pale fulvous, the sides of the breast and of the body streaked with
dark brown; tail dark brown edged with rufous, the penultimate
feather with a triangular patch of white at the tip, the outer
feather with the outer web almost entirely white and half of the
inner, next the shaft, also white.

In spring the chin, throat, and streaks on the gorget become pure
white, the rufous pectoral band becomes broader and brighter by
the wearing away of the tips of the feathers which partially overlie
the band, and the sides of the body become uniform bright
chestnut; the ashy parts of the head become purer ashy, and the
upper plumage in general becomes a richer rufous. The difference
between breeding and non-breeding plumage in this species is not
very marked or noteworthy.

Female. Resembles at all seasons the male after the moult; is
perhaps a trifle duller in colour.

The young bird resembles the female.

Bill dark fleshy brown, the lower mandible paler; iris brown;
feet and claws pinkish.

Length about 6; tail 2·7; wing 2·8; tarsus ·8; bill from
gape ·6.

The Himalayan Buntings of this type differ from the true
E. fucata of Siberia, and from those which visit the Eastern portion
of the Empire in having the scapulars and lesser wing-coverts and
the whole of the sides of the body uniform chestnut. Should this
form be distinct, it will bear Sharpe's name *E. arcuata.*

Distribution. A resident species in the whole of the Himalayas
from Kashmir to Assam, and a fairly common winter visitor to all
parts of the Eastern portion of the Empire from Assam down to
about the latitude of Moulmein.

According to Jerdon, this species is spread sparingly through
Northern and Central India and it has been found near Calcutta,
Jálna in the Deccan, Mhow, Saugor, and Nágpur. Barnes re-
cords it from Neemuch. There is, however, no specimen of this
Bunting from any part of the plains of India proper either in the
Hume or Tweeddale collections, nor have I ever seen a specimen
from those parts.

This species, if the same as *E. fucata,* ranges over the greater
part of Northern and Eastern Asia.

Habits, &c. Breeds from 6000 to 8000 feet in May, June, and
July, constructing a saucer-shaped nest of dry grass on the ground
under shelter of a bush or stone. The eggs, four in number, are
pale greenish grey speckled all over with dull reddish or purplish
brown, and measure about ·83 by ·6.

791. Emberiza pusilla. *The Dwarf Bunting.*

Emberiza pusilla, *Pall. Reis. Russ. Reichs.* iii, p. 697 (1776); *Jerd. B. I.* ii, p. 376; *Anders. Yunnan Exped., Aves,* p. 603; *Oates, B. B.* i, p. 353; *Sharpe, Cat. B. M.* xii, p. 487 : *Hume, Cat.* no. 720 : *Hume, S. F.* xi, p. 280.

Euspiza pusilla (*Pall.*), *Blyth, Cat.* p. 130.

Oeyris oinopus, *Hodgs., Horsf. & M. Cat.* ii, p. 488.

Coloration. Male. After the moult a broad rufous band over the crown from the forehead to the nape, some of the feathers with a brown mesial streak ; a broad dark brown band on either side of the coronal band, all the feathers broadly margined with rufous : a distinct pale rufous supercilium : lores and ear-coverts chestnut ; upper plumage and wings dark brown or blackish, each feather margined with rufous ; tail dark brown margined paler, the penultimate feather with a streak of white near the tip, the outer feather largely white on both webs ; cheeks pale fulvous, produced as a band under the ear-coverts : chin and throat white : sides of the throat, the whole breast, and the sides of the body white, sullied with fulvous and densely streaked with dark brown ; abdomen, vent, and under tail-coverts white without streaks.

In spring, the broad coronal band becomes richer rufous, and the broad lateral bands pure deep black, in consequence of the rufous margins getting worn away, and the supercilium becomes very well defined and somewhat broader.

Female. Resembles the male in winter plumage very closely, but apparently never acquires the deep black coronal bands.

The young resemble the adults in winter but are paler.

Bill horny ; legs pale fleshy brown ; iris brown (*Jerdon*).

Length about 5·5 ; tail 2·4 ; wing 2·8 ; tarsus ·7 ; bill from gape ·45.

Distribution. The Himalayas from the Sutlej valley to Assam. This species has been observed at numerous localities in the Eastern part of the Empire from Assam through the hill-ranges to Bhámo. It has also been obtained in Karennee and on Muleyit mountain in Tenasserim. This Bunting does not appear to be found in the plains of the Indian Peninsula, but Jerdon records it from the Purneah district. A specimen from the Andamans is in the Tweeddale collection.

The Dwarf Bunting visits the Empire in the winter only. In the summer it is found throughout Northern Asia and China.

792. Emberiza leucocephala. *The Pine-Bunting.*

Emberiza leucocephala, *S. G. Gm. N. Comm. Acad. Sc. Imp. Petrov.* xv, p. 480, tab. xxiii, fig. 3 (1770); *Hume & Henders. Lah. to Yark.* p. 254 ; *Hume, Cat.* no. 712 ; *Biddulph, Ibis,* 1881, p. 79, 1882, p. 282 ; *Scully, Ibis,* 1881, p. 574 ; *Sharpe, Cat. B. M.* xii, p. 549.

Emberiza pithyornis, *Gm. Syst. Nat.* i, p. 875 (1788) ; *Horsf. & M. Cat.* ii, p. 482 ; *Jerd. B. I.* ii, p. 370.

Emberiza albida, *Blyth, J. A. S. B.* xviii, p. 811 (1849); *id. Cat.*
p. 128.
The White-crowned Bunting, Jerd.

Coloration. Male. After the moult the forehead and crown are
ashy, streaked with brown, the base of the feathers white, but not
showing at first; lores, round the eye, and a short but broad
supercilium, cheeks, chin, throat, and sides of the neck chestnut,
each feather margined with white; ear-coverts brown, divided down
the middle by a band of white which extends under the eye to the
gape; hind neck ashy, turning to rufous on the back and scapulars,
the feathers of which are streaked with black; rump and upper tail-
coverts chestnut, margined with white; tail dark brown, narrowly
margined with pale rufous, the two outer pairs of feathers with
the terminal two-thirds of the inner web, and a margin on the
outer web, white; lesser coverts pale rufous; median and greater
coverts and tertiaries black, with broad rufous margins; quills
brown, narrowly margined with pale rufous; a large patch of white
on the lower throat; breast chestnut, margined with white; abdomen white; sides of the body white, streaked with chestnut.

In the spring the crown and nape become pure white, bounded
on each side and on the forehead by a black band; the chestnut
on the sides of the head and on the chin and throat becomes pure,
owing to the white margins wearing off; the breast and sides of
the body also become purer chestnut, but the white on these parts
never entirely disappears.

Between the two extreme plumages described every intermediate
stage is to be found.

Female. The forehead, crown, and nape ashy streaked with
brown, and without any white at the base of the feathers; the
remaining upper plumage, wings, and tail as in the male, but
duller; the white band on the side of the head and all chestnut on
this part and the chin and throat are wanting the former part
being more or less plain brown, and the two latter whitish streaked
with brown; breast and sides of the body rufous, streaked with
brown and varied with white; abdomen white.

The young bird appears to resemble the male.

Upper mandible very dark brown, the base from gape to nostril
yellowish; lower mandible very pale horny bluish; legs yellowish
fleshy, feet slightly tinged brown; iris dark brown (*Hume
Coll.*).

Length about 7; tail 3·4; wing 3·6; tarsus ·75; bill from
gape ·5.

Distribution. A winter visitor to Gilgit, Kashmir, and the Himalayas down to Garhwál. At this season the Pine-Bunting is also
found in Afghanistan and Europe, but in the summer it is confined to Northern Asia.

793. Emberiza stewarti. *The White-capped Bunting.*

Emberiza stewarti, *Blyth, J. A. S. B.* xxiii, p. 215 (1854); *Horsf. & M. Cat.* ii, p. 475; *Jerd. B. I.* ii, p. 374; *Stoliczka, J. A. S. B.* xxxvii, pt. ii, p. 58; *Hume, Cat. no.* 718; *Biddulph, Ibis,* 1881, p. 81, 1882, p. 282; *Scully, Ibis,* 1881, p. 575; *Sharpe, Cat. B. M.* xii, p. 547; *Barnes, Birds Bom.* p. 269; *Oates in Hume's N. & E.* 2nd ed. ii, p. 167.
Citrinella stewarti (*Blyth*), *Hume, N. & E.* p. 465.

Coloration. Male. After the autumn moult the forehead, crown, nape, and ear-coverts are grey, with brownish tips to the feathers; a broad black supercilium, each feather tipped with grey; the whole upper plumage and scapulars chestnut with pale fulvous margins to the feathers; lesser and median wing-coverts dull chestnut, edged paler; greater coverts and quills dark brown, edged with rufous; tail brown margined with rufous, the two outer pairs of feathers almost entirely white, the bases and the shafts with a narrow portion of the outer webs only being brown; chin and upper throat, produced laterally down the sides of the lower throat, black, each feather margined with whitish; lower throat and fore neck white; breast chestnut, margined with white; remaining lower plumage pale fulvous, the sides of the head streaked or blotched with rufous.

In spring the margins and tips to the feathers of the crown, supercilium, upper plumage in general, chin, throat, and lower breast disappear in part or wholly by abrasion, leaving each part entirely of one colour or nearly so.

Female. Forehead, crown, nape, hind neck, back, and scapulars ashy brown, streaked with blackish, the scapulars tinged with chestnut; rump and upper tail-coverts chestnut, with paler edges and blackish shafts; tail as in the male, but with rather less white on the two outer pairs of feathers; wings brown, each feather margined with pale rufous or fulvous; lores and round the eye fulvous; ear-coverts and sides of the neck brown; lower plumage pale fulvous streaked with brown.

The young resemble the female.

Bill brown, paler on lower mandible; iris slightly reddish brown; legs and feet pinkish fleshy; claws pale brown (*Hume*).

Length about 6·5; tail 3; wing 3·3; tarsus ·75; bill from gape ·4.

Distribution. The Himalayas from the Hazára country, Gilgit, and Kashmir to about Almora; the Punjab, Sind, Rajputana, and the N.W. Provinces as far east as Etáwah. This species is found in the plains on the lower ranges of the Himalayas in winter only and on the higher parts of the latter (up to 6000 or 8000 feet) in summer. It extends into Afghanistan.

Habits, &c. Breeds in June and July, constructing a deep cup-shaped nest, of fibres and grass, in low bushes, or on the ground by the side of a road or bank. The eggs, usually four in number, are white mottled with purple, and measure about ·78 by ·59.

791. **Emberiza stracheyi**. *The Eastern Meadow-Bunting.*

Euspiziza cia (*Linn.*), *apud Blyth, Cat.* p. 130.

Emberiza stracheyi, *Moore, P. Z. S.* 1855, p. 215, pl. 112 ; *Horsf. & M. Cat.* ii, p. 483; *Jerd. B. I.* ii, p. 372; *Hume, Cat.* no. 714; *Biddulph, Ibis,* 1881, p. 79; *Scully, Ibis,* 1881, p. 574; *Sharpe, Cat. B. M.* xii, p. 539; *Oates in Hume's N. & E.* 2nd ed. ii, p. 168.

Emberiza cia, *Linn. apud Jerd. B. I.* ii, p. 371; *Stoliczka, J. A. S. B.* xxxvii, pt. ii, p. 57; *Hume & Henders. Lah. to Yark.* p. 256.

Citrinella cia (*Linn.*), *Hume, N. & E.* p. 461.

Coloration. Male. After the autumn moult a longitudinal broad coronal band from the bill to the nape is bluish grey with a few black streaks ; a broad lateral band on each side of the crown black with rufous tips, succeeded by a distinct pale fulvous eyebrow reaching from the nostrils to the nape ; lores and band through the eye black ; cheeks and ear-coverts pale fulvous ; a narrow black moustachial streak passing under and behind the ear-coverts and joining the eye-band ; back and scapulars chestnut-brown, streaked with black ; lesser wing-coverts bluish grey ; median and greater coverts, secondaries and tertiaries black, broadly margined with chestnut-brown ; primaries brown, narrowly edged with rufous ; rump chestnut with paler edges ; upper tail-coverts chestnut-brown, with black shaft-streaks ; middle pair of tail-feathers black, broadly edged with chestnut-brown, the next two pairs entirely black, with narrow pale margins ; the next pair black with a white tip ; the outer two pairs white on the terminal half with black shafts ; chin whitish ; throat and breast bluish grey, each feather with an indistinct triangular brownish tip ; remainder of lower plumage and the sides of the neck plain chestnut-brown.

In the spring the tips and margins of the feathers become abraded, and the mesial coronal band becomes pure bluish grey ; the lateral bands, the eye-band, and the moustachial streak deep black ; the eyebrows, cheeks, and ear-coverts pure white ; the throat and breast lose all traces of the triangular tips to the feathers.

Female. Resembles the male in every respect, but is perhaps a little paler ; undergoes the same seasonal change of colour.

The young bird is rufous-brown above, everywhere densely streaked with black, and the lower plumage is pale rufous, deepening on the abdomen and densely streaked with black on the throat, breast, and sides of the body.

Bill dark plumbeous above, light plumbeous below ; iris dark brown ; legs and feet fleshy yellow (*Hume*).

Length 6·5 ; tail 3·4 ; wing 3·2 ; tarsus ·7 ; bill from gape ·5.

This Bunting may be looked upon as a race of *E. cia* of Europe. *E. cia* differs in wanting the pure black and white marks on the head which are so conspicuous in *E. stracheyi*, the white in *E. cia* being always tinged with grey and the black obscured by rufous tints. In *E. cia* the median and greater wing-coverts are tipped with a more or less pure white, whereas in the Indian bird the tippings to these parts are chestnut-brown of the same colour as the back. Lastly, in *E. cia* the rufous of the lower parts is much paler.

Distribution. The Himalayas, from the Hazára country and Gilgit to Kumaun. This species is resident on the Himalayas, moving vertically according to season. A few Buntings of this species appear to visit the plains of the Punjab in the winter. The range extends into Baluchistan.

Habits, &c. Breeds in the Himalayas from 4000 to 9000 feet, making a loose cup-like nest of grass on the ground. The eggs are pale greenish white or grey mottled with purplish, and covered by a series of delicate lines and scrawls which are dark brown or black. They measure about ·83 by ·65.

795. Emberiza buchanani. *The Grey-necked Bunting.*

Emberiza buchanani, *Blyth, J. A. S. B.* xiii, p. 957 (1844), xvi, p. 780; *Hume, S. F.* vii, p. 150; *id. Cat.* no. 716; *Sharpe, Cat. B. M.* xii, p. 535; *Barnes, Birds Bom.* p. 268.

Euspiza huttoni, *Blyth, J. A. S. B.* xviii, p. 811 (1849).

Emberiza huttoni (*Blyth*), *Horsf. & M. Cat.* ii, p. 485; *Jerd. B. I.* ii, p. 373; *Hume, J. A. S. B.* xxxix, pt. ii, p. 121.

Citrinella huttoni (*Bl.*), *Stoliczka, J. A. S. B.* xli, pt. ii, p. 247.

Jamjohara, Hind.

Coloration. Male. After the autumn moult the whole upper plumage is ashy brown, each feather with a dark brown shaft-streak, the back slightly tinged with rufous; lesser wing-coverts ashy brown; the remaining coverts and quills dark brown, broadly margined with rufous; tail brown, the middle pair broadly, the next three pairs narrowly, margined with rufous; the two outer pairs with the terminal half of the inner web white, as also a small portion of the outer web of the outermost feather; lores and a ring round the eye fulvous; sides of the head and neck ashy brown; an indistinct black moustachial streak; entire lower plumage rufous, palest on the chin and darkest on the breast, most of the feathers with pale fulvous margins.

In the spring the fulvous margins on the lower plumage get worn away.

Female. Hardly distinguishable from the male, but slightly paler.

Iris dark brown; legs and feet yellowish brown; bill fleshy brown (*Butler*).

Length about 6·5; tail 2·8; wing 3·3; tarsus ·75; bill from gape ·45.

Distribution. A winter visitor to the whole of the North-western portion of the plains of India, extending south as far as Khandála and Chánda and east as far as Etáwah. This species migrates through Kashmir and has been observed in Gilgit in September, and our Indian birds probably summer in Turkestan and Persia. Its range extends westwards to the Caucasus.

796. **Emberiza hortulana.** *The Ortolan Bunting.*

Emberiza hortulana, *Linn. Syst. Nat.* i, p. 309 (1766); *Horsf. & M.
Cat.* ii, p. 484; *Jerd. B. I.* ii, p. 372; *Hume, Cat.* no. 715; *Bid-
dulph, Ibis,* 1881, p. 80; *Scully, Ibis,* 1881, p. 574; *Sharpe, Cat.
B. M.* xii, p. 530.
Euspiza hortulana (*Linn.*), *Blyth, Cat.* p. 129.
Emberiza shah, *Bonap. Consp. Av.* i, p. 465 (1850).

Coloration. Male. Forehead, crown, and nape dusky olive-green;
back and scapulars pale rufous, with broad black streaks: rump
and upper tail-coverts pale rufous, with less distinct black streaks;
tail dark brown edged with rufous, the two outer pairs of feathers
white on the terminal half of the inner web; lesser wing-coverts
ashy; remaining coverts and quills brown with rufous margins;
feathers on the eyelids, lores, cheeks, chin, and throat yellow; sides
of the head and neck dusky olive; a moustachial streak pale brown;
upper breast dull olive-yellow; remainder of lower plumage
cinnamon-rufous.

The above is the full breeding-plumage. I have not been able
to examine freshly moulted autumn birds, but these are said to have
pale margins to the feathers of the head and breast as in the other
species of Buntings.

Female. Resembles the male very closely but is much paler on
chin and throat, and the upper breast is frequently streaked with
brown, which may, however, be only remains of the immature
plumage.

The young bird is pale rufous throughout, densely streaked with
dark brown both above and below.

Bill dark flesh-colour, rather darker above than below; iris
brown; legs pale fleshy red (*Dresser*).

Length about 6·5; tail 2·8; wing 3·6; tarsus ·75; bill from
gape ·55.

Distribution. A rare visitor to Gilgit, where this species has
been obtained in May. This Bunting is found in Afghanistan
and Turkestan and extends westwards throughout Europe.

797. **Emberiza aureola.** *The Yellow-breasted Bunting.*

Emberiza aureola, *Pall. Reis. Russ. Reichs,* ii, p. 711 (1773);
Anders. Yunnan Exped., Aves, p. 602; *Oates, B. B.* i, p. 355;
Sharpe, Cat. B. M. xii, p. 509.
Mirafra flavicollis, *McClell. P. Z. S.* 1839, p. 164.
Euspiza aureola (*Pall.*), *Blyth, Cat.* p. 129; *Horsf. & M. Cat.* ii.
p. 487; *Jerd. B. I.* ii, p. 380; *Hume, Cat.* no. 723; *id. S. F.* xi, p. 282.
Euspiza flavogularis, *Blyth, J. A. S. B.* xviii, pp. 86, 811 (1849);
id. Cat. p. 129.

Coloration. Male. After the autumn moult the whole upper
plumage is a dull chestnut, each feather margined with ashy; tail
brown, margined with dull rufous, the outermost feather with a broad
diagonal white band across the inner web, the penultimate with a
narrow white tip; lesser wing-coverts chestnut edged with ashy;

s 2

median coverts almost entirely white ; greater coverts and quills dark brown, margined with chestnut-brown ; a distinct supercilium, lores, cheeks, and ear-coverts ashy yellow ; a band above the ear-coverts and another below chestnut-brown ; the whole lower plumage yellow, with a chestnut band across the breast, and the sides of the body streaked with chestnut-brown ; the under tail-coverts paler than the other parts.

In the spring the margins on the upper plumage become worn away and the general colour becomes rich maroon chestnut, and in a similar manner the lower plumage becomes rich yellow and the pectoral band broader and deeper chestnut. A change takes place in the colour of the head, but this is effected by a complete moult of the feathers of the parts affected : these parts are the forehead, anterior part of crown, lores, ear-coverts, cheeks, chin, and a small portion of the throat, which become deep black.

The plumage of the males varies a good deal, as does also the time at which the black on the head is assumed.

Female. Head chestnut-brown, with dark brown streaks ; nape and back of the neck olive-brown, with indistinct brown streaks and the centres of the feathers tinged with chestnut ; back and scapulars olive-brown, with broad distinct dark brown streaks ; rump pale chestnut, edged with grey : upper tail-coverts brown, centred darker ; lesser wing-coverts brown ; median ones brown, very broadly tipped with white : greater coverts and all the quills brown, edged on the outer webs with pale rufous-brown ; tail as in the male : a broad supercilium reaching to the nape yellowish white ; sides of the head mixed brown and yellowish white ; chin and throat whitish ; breast, sides of neck, and abdomen bright yellow, tinged with brown across the breast, which is also faintly streaked with brown ; flanks faint yellow, streaked with brown ; vent and under tail-coverts pale yellow, the latter indistinctly streaked.

The young bird is very similar to the female, but has no chestnut on the head and rump and the whole breast is boldly streaked with brown.

Iris rich brown ; upper mandible dark brown, lower fleshy brown ; feet and claws pinkish brown.

Length 6·2 ; tail 2·4 ; wing 3 ; tarsus ·85 ; bill from gape ·55.

Distribution. A common winter visitor to the Himalayas from Nepal to Assam and to the whole of the eastern portion of the Empire from Assam southwards to Tenasserim, and also to the Nicobar Islands. This species occurs from October to May, and at this season it ranges to the southern extremity of the Malay peninsula and to China and Siam. In summer it is found chiefly in Northern Asia.

798. Emberiza spodocephala. *The Black-faced Bunting.*

Emberiza spodocephala, *Pall. Reis. Russ. Reichs,* iii, p. 698 (1776) ;
    Jerd. B. I. ii, p. 374 ; *Hume, Cat. no.* 717 ; *id. S. F.* xi, p. 275 ;
    Sharpe, Cat. B. M. xii, p. 522.

Emberiza melanops, *Blyth, J. A. S. B.* xiv, p. 554 (1845).
Euspiza melanops (*Blyth*), *Blyth, Cat.* p. 129.

Coloration. Male. After the autumn moult the lores, the region
of the gape, and the point of the chin are black : the whole head,
whole neck, and breast dull olive-green, some of the feathers of the
hind neck with dull rufous-brown tips and the feathers of the
crown with indistinct dark shaft-streaks ; back and scapulars dull
rufous-brown streaked with black ; rump and upper tail-coverts
olive-brown ; tail dark brown, edged with olive-brown, the outermost feather with the basal portion of the outer web and the
terminal half of the inner web white ; the penultimate feather
with a large triangular white tip to the inner web ; lesser coverts
rufous-brown ; remaining coverts and quills dark brown, broadly
edged with rufous-brown ; lower plumage from the breast downwards yellow, the sides of the body tinged with rufous and streaked
with brown.

The change that takes place in the plumage in spring is very
trifling, the rufous tips to the feathers of the hind neck wearing
away and the plumage in general becoming brighter.

Female. The whole upper plumage, wings, and tail as in the male,
but the head and hind neck less green and the shaft-streaks well-
developed ; a supercilium, lores, cheeks, chin, and throat pale
yellowish ; ear-coverts brown ; a series of brown spots on each
side of the throat extending to the breast, which is dull yellowish
streaked with brown ; remainder of lower plumage yellow, the
sides of the body streaked with brown.

Young birds resemble the female, but have the head more
streaked and the throat spotted with brown. Young males have
assumed the adult plumage by February or March, the last signs
of immaturity left at that time being small triangular tips to the
feathers of the crown.

Legs and feet pale brownish fleshy ; upper mandible dark
brown ; lower mandible and gape horny pinkish white ; iris brown
(*Hume*).

Length about 6 ; tail 2·6 ; wing 2·8 ; tarsus ·75 ; bill from gape ·5.

Distribution. A winter visitor to the Himalayas from Nepal to
Assam and to the eastern portion of the Empire from Assam down
to Manipur. In the winter this species extends to China and it
summers in Eastern Siberia.

Habits, &c. According to Hume, this species is very partial to
long grass and watery localities.

799. Emberiza melanocephala. *The Black-headed Bunting.*

Emberiza melanocephala, *Scop. Ann.* i, p. 142 (1769) ; *Sharpe, Cat.
B. M.* xii, p. 503 ; *Oates in Hume's N. & E.* 2nd ed, ii, p. 170.
Euspiza simillima, *Blyth, J. A. S. B.* xviii, p. 811 (1849): *id. Cat.*
p. 128 ; *Horsf. & M. Cat.* ii, p. 486 ; *Hume, N. & E.* p. 465.
Euspiza melanocephala (*Scop.*), *Blyth, Cat.* p. 128 ; *Jerd. B. I.* ii,
p. 378 ; *Blanf. J. A. S. B.* xxxviii, pt. ii, p. 186 ; *Hume, Cat. no.*
721 ; *Barnes, Birds Bom.* p. 271.

Gondana, Hind.

Coloration. Male. After the autumn moult the forehead, crown, and nape are black almost concealed by ashy margins ; a concealed yellow collar on the hind neck ; back, rump, and scapulars orange-chestnut with ashy margins ; upper tail-coverts brown, edged with ashy ; tail brown, margined with fulvous ; lesser wing-coverts orange-chestnut, margined with ashy ; remaining coverts and quills dark brown, edged with fulvous ashy ; lores and under the eye deep black ; ear-coverts black tipped with yellow ; cheeks, side of the neck, and the whole lower plumage deep yellow with pale lilac margins.

In spring the margins everywhere get worn away ; the forehead, crown, nape, lores, under the eye, and the ear-coverts become deep black : the upper plumage and lesser coverts become uniform deep orange-chestnut, and the whole lower plumage a deep yellow.

Female. The whole upper plumage and lesser wing-coverts fulvous brown, streaked with dark brown, the streaks almost obsolete on the rump and upper tail-coverts ; coverts, quills, and tail as in the male ; the entire lower plumage is a delicate fulvous, washed with ochraceous on the breast and with yellow on the abdomen ; under tail-coverts bright yellow. The difference between the summer and the winter plumage of the female is slight.

Young birds resemble the female closely ; young males not quite adult have brown ear-coverts.

Iris dark brown ; legs and feet fleshy brown ; bill pale greenish horn, brown on culmen (*Butler*).

Length about 7·5 ; tail 3·1 ; wing 3·8 ; tarsus ·85 ; bill from gape ·6.

Distribution. A winter visitor to the plains of India as far east as Delhi, Nágpur, and Chánda, and as far south as Belgaum. This species passes through Baluchistan, and, in smaller numbers, through Gilgit on migration, and the Indian birds probably breed in Persia. This Bunting extends westwards to South-western Europe.

Habits, &c. This Bunting is usually found in India in large flocks, which commit great devastation in corn-fields. It breeds about May in Western Asia and South-western Europe ; the nest, a cup of straw or grass lined with hair or roots, is usually placed in a bush, vine, or low tree, and the eggs, four to six in number, are pale greenish-blue, spotted throughout, more profusely round the larger end, and measure about 0·87 by 0·62.

---

800. **Emberiza luteola.** *The Red-headed Bunting.*

Emberiza luteola, *Sparrm. Mus. Carls.* fasc. iv. Taf. 93 (1788) ; *Sharpe, Cat. B. M.* xii, p. 506.

Euspiza luteola (*Sparrm.*), *Blyth, Cat.* p. 128 ; *Horsf. & M. Cat.* ii, p. 486 ; *Jerd. B. I.* ii, p. 378 ; *Hume, S. F.* iii, p. 408 ; *Scully, S. F.* iv. p. 167 ; *Wardlaw Ramsay, Ibis,* 1880, p. 66 ; *Hume, Cat.* no. 722 ; *Barnes, Birds Bom.* p. 271.

Gándam, Hind. ; *Dalchidi,* Sind ; *Pacha jinnwayi,* Tel.

Coloration. Male. After the autumn moult the forehead, crown, and nape are rich golden brown, the feathers tipped with ashy; hind neck and sides of neck olive-yellow; back, scapulars, and lesser coverts olive-yellow streaked with brown; rump yellow; upper tail-coverts olive-brown, margined with olive-yellow; tail dark brown, edged with fulvous; middle and greater coverts and quills dark brown, margined with fulvous; lores, sides of the head, chin, throat, and middle portion of breast chestnut, the feathers margined with ashy; sides of the breast and remainder of lower plumage deep yellow.

In spring the forehead, crown, and nape become deep golden brown, and the lores, sides of the head, chin, throat, and breast pure chestnut. This change is caused by the abrasion of the ashy margins on those parts.

Occasionally the golden brown of the crown suffuses the entire upper plumage. This occurs probably in very old males only.

Female. The whole upper plumage ashy brown, the back and scapulars streaked with dark brown and the rump tinged with olive-yellow; tail and wings as in the male; lores pale ashy white; sides of the head and neck dark fulvous; the whole lower plumage pale fulvous, the abdomen washed with yellow, and the under tail-coverts pure yellow.

The young bird resembles the female, but has the whole upper plumage, sides of the throat, and the whole breast thickly streaked with brown.

Iris dark brown; legs and feet brown; bill greyish brown above, darkest on the culmen and greenish horn below (*Butler*).

Length about 7; tail 2·8; wing 3·5; tarsus ·8; bill from gape ·6.

Distribution. A winter visitor to the plains of India from the foot of the Himalayas down to the Nilgiris and from Sind to Chutia Nagpur. This species passes through Gilgit on migration and breeds in Turkestan and Northern Asia. It extends to Afghanistan, Turkestan, and Persia *.

Habits, &c. Not so commonly found in flocks, and not associating in as large numbers as the last species, and less confined to well-cultivated tracts. The nest and eggs are very similar to those of *E. melanocephala*, and have been taken in Eastern Turkestan by Stoliczka and Scully in May and June, and by Wardlaw Ramsay in the Hariab valley, Western Afghanistan.

801. Emberiza rutila. *The Chestnut Bunting.*

Emberiza rutila, *Pall. Reis. Russ. Reichs,* iii, p. 698 (1776); *Oates, B. B.* i, p. 554; *Sharpe, Cat. B. M.* xii, p. 514.
Euspiza rutila (*Pall.*), *Blanf. J. A. S. B.* xli, pt. ii, p. 168; *Hume, Cat.* no. 722 bis; *id. S. F.* xi, p. 282.
Citrinella rutila (*Pall.*), *Hume, S. F.* iii, p. 157.

* Dr. Scully (Ibis, 1881, p. 575) indicates an undescribed species of Bunting allied to *E. luteola.*

Coloration. *Male.* After the autumn moult the whole upper plumage, lesser and median wing-coverts, sides of the head and neck, chin, throat, and fore neck are deep chestnut, each feather fringed with ashy yellow; greater coverts and tertiaries dark brown, margined with chestnut; primaries, secondaries, and tail-feathers dark brown, narrowly margined with ochraceous; lower plumage from the fore neck downwards yellow, the sides of the body and the under tail-coverts streaked with dusky green.

In the spring the ashy-yellow margins get worn away on all the chestnut parts of the plumage, and these become deep chestnut.

Female. Forehead, crown, nape, hind neck, back, and lesser wing-coverts ashy brown streaked with black; rump chestnut, edged with ashy; upper tail-coverts rufous-brown, edged with ashy; median and greater wing-coverts and tertiaries dark brown, broadly margined with fulvous or rufous; quills and tail brown, edged with ashy or fulvous; lores, an indistinct supercilium, cheeks, chin, and throat fulvous; a brown moustachial streak; remaining lower plumage oil-yellow, tinged with brown on the breast, which is also obsoletely streaked; sides of the body boldly streaked with dusky green.

The young bird is brown above and pale yellowish below, everywhere streaked with dark brown.

Iris bright brown; bill above dark horny, below pale; legs pale yellowish brown (*Wardlaw Ramsay*).

Length about 6; tail 2·5; wing 2·9; tarsus ·7; bill from gape ·5.

Distribution. The Eastern portion of the Empire. This species has been recorded from Sikhim, the Bhutan Doars, Manipur, Karennee, and various localities in Pegu and Northern and Central Tenasserim. It visits India only in the winter, at which season it is also found in China and Siam. It summers in North China and Eastern Siberia.

802. **Emberiza striolata.** *The Striolated Bunting.*

Fringillaria striolata, *Licht. Verz. Doubl.* p. 24 (1823); *Hume, N. & E.* p. 465; *Sharpe, Cat. B. M.* xii, p. 561.
Emberiza striolata (*Licht.*), *Hume, Ibis,* 1870, p. 339; *id. S. F.* vii, p. 410; *id. Cat.* no. 720 bis; *Oates in Hume's N. & E.* 2nd ed. ii, p. 170.

Coloration. *Male.* Forehead, crown, and nape greyish white, thickly streaked with black; upper plumage, scapulars, and lesser coverts pale rufous with dark streaks; wings and tail dark brown, each feather margined with rufous; lores and a distinct supercilium white; a black eye-band followed below by a broad whitish band from the gape to the middle of the ear-coverts and by another black band meeting the eye-band behind the ear-coverts; a white moustachial streak; chin, throat, and upper breast ashy, with black streaks to the feathers; remainder of lower plumage, under wing-coverts, and a portion of the inner webs of the quills rufous.

The difference in plumage in winter and summer is in this
species very trifling.

Female. Resembles the male closely but is somewhat duller in
coloration.

Iris brown; legs and feet yellowish fleshy; claws pale brown;
upper mandible dark brown, lower fleshy brown (*Hume Coll.*).

Length nearly 6; tail 2·6; wing 3·1; tarsus ·6; bill from
gape ·45.

Distribution. A permanent resident in a great portion of the
plains of the north-west portion of the Empire from Sind to
Etáwah in the N.W. Provinces and from the Punjab down to
Cutch. This species also occurs as far as Saugor in the Central
Provinces. It extends westwards to Arabia and Palestine.

Habits, &c. Breeds in November and December, and probably
also a second time in June or July, constructing a nest of grass
under or between blocks of stone. The eggs are marked with
brown of various shades and measure about ·76 by ·56.

Genus MELOPHUS, Swains., 1837.

In the genus *Melophus* both sexes are crested, the crest of the
female being shorter than that of the male, and the tail is more
even or square at the tip than in *Emberiza*. The sexes are very

Fig. 71.—Head of *M. melanicterus.*

differently coloured, but both have a considerable amount of red on
the wings and tail. The only Bunting of this genus affects rocky
hill-sides and the banks of streams, and is solitary in its habits.

803. **Melophus melanicterus.** *The Crested Bunting.*

Fringilla melanictera, *Gm. Syst. Nat.* i, p. 910 (1788).
Emberiza cristata, *Vigors, P. Z. S.* 1831, p. 35.
Euspiza lathami (*J. E. Gr.*), *Blyth, Cat.* p. 129.
Melophus melanicterus (*Gm.*), *Horsf. & M. Cat.* ii, p. 489; *Jerd.
B. I.* ii, p. 384; *Hume, N. & E.* p. 467; *Anders. Yunnan Exped.,
Aves,* p. 604; *Hume, S. F.* vii, p. 517; *id. Cat.* no. 724; *Scully,
S. F.* viii, p. 334; *Oates, B. B.* i, p. 357; *Barnes, Birds Bom.*

p. 272; *Sharpe, Cat. B. M.* xii, p. 568; *Oates in Hume's N. & E.* 2nd ed. ii, p. 173.

Pothar chirta, Hind.

Coloration. Male. After the autumn moult the whole head, neck, back, rump, scapulars, lesser wing-coverts, and the whole lower plumage except the thighs and under tail-coverts are black, each feather with a broad ashy margin; wings, tail, under tail-coverts, and thighs chestnut, the quills, tail, and some of the wing-coverts tipped with black; upper tail-coverts chestnut, margined with black, sometimes wholly black.

In the spring the ashy margins disappear wholly or in part.

Female. The crest shorter and not very apparent in some specimens; the whole upper plumage dark brown, each feather edged with olive-brown, frequently with a tinge of rufous; lesser wing-coverts dark brown, narrowly edged with pale rufous; median and greater coverts dark brown, very broadly edged with cinnamon; primaries and secondaries with the outer webs blackish, edged exteriorly with pale cinnamon; inner webs cinnamon, broadly tipped with dark brown; tertiaries dark brown, edged with pale cinnamon on the outer webs; outer tail-feathers cinnamon, with a broad band of brown on the inner web; the other feathers brown, with a narrow margin of pale cinnamon on the outer webs, and the fifth pair from the middle also with a streak of cinnamon on the inner webs; ear-coverts and cheeks dark brown, the former tipped with olive-brown; lower plumage dull buff to yellowish brown, streaked and mottled, especially on the throat and breast, with dark brown; vent and under tail-coverts brighter and sometimes tinged with rufescent.

The young bird resembles the female very closely, and the young male commences in the autumn to put on the chestnut body-plumage of the adult.

Bill dusky, blackish above and fleshy at base of lower mandible; irides dark brown; feet fleshy brown, the toes darker; claws blackish, pale at tips (*Scully*).

Length 6·5; tail 2·8; wing 3·2; tarsus ·75; bill from gape ·6; crest in male about ·8, in female ·5.

Distribution. The Himalayas from Kashmir to Bhutan; the plains of India from Sind to Bengal as far south as about the latitude of Mahableshwar; thence through the Assam hills, Manipur, and Upper Burma to Arrakan on the one hand, and to Karennee and Northern Tenasserim on the other. This species is somewhat capricious in its choice of localities, and it is absent throughout large tracts of country. It extends into China. It is everywhere apparently a resident species.

Habits, &c. Breeds from April to August, making a saucer-like nest of grass on the ground or in holes of banks and walls. The eggs are thickly marked with red or purple and measure about ·79 by ·63.

Fig. 72.—*Hirundo erythropygia.*

Family HIRUNDINIDÆ.

The intrinsic muscles of the syrinx fixed to the ends of the bronchial semi-rings; the edges of both mandibles perfectly smooth, with a single notch or the indication of one in the upper; the hinder part of the tarsus longitudinally bilaminated, the laminæ entire and smooth; wing with nine primaries, the first and second nearly equal; bill flat, broad and triangular when viewed from above; gape very wide; the longest secondaries reaching to about the middle of the wing; front of tarsus smooth; rectrices twelve; sexes alike; a moult in the spring only; young very similar to the adult; rictal bristles weak.

The Swallows form a well-defined group of birds remarkable for their great powers of flight, the whole of their food, which consists of small insects, being caught on the wing. Many of the Swallows migrate vast distances, others are resident, and some species are confined to small areas.

The Swallows resemble each other closely in structure, and the only point in which they vary is the shape of the tail. As the shape of the tail is, however, different in almost every species, it cannot very well be utilized as a generic character. The Indian Swallows may be divided into four genera by characters which are of considerable value, such as the feathered or bare condition of the leg, the colour of the plumage and of the tail, the mode of nidification, and the colour of the eggs.

Key to the Genera.

a. Tarsus and toes feathered ... . CHELIDON, p. 268.
b. Tarsus and toes bare.
   a'. Upper plumage brown and without
      gloss.
     a''. Tail-feathers uniform .... COTILE, p. 271.
     b''. Tail-feathers with white spots .... PTYONOPROGNE, p. 273.
   b'. Upper plumage, or the greater portion
      of it, black and highly glossy ...... HIRUNDO, p. 276.

Genus **CHELIDON**, Forster, 1817.

The genus *Chelidon* contains the Martins, which are distinguished from all the other birds of this group by their feathered tarsus and toes. The rump in all the species is white, and forms a conspicuous feature of the Martins when flying. The shape of the tail varies in the different species, *C. urbica* having the tail somewhat deeply forked, and *C. nepalensis* having it quite square.

The Martins build nests of mud lined with feathers, and lay four to six pure white eggs.

Figs. 73, 74, 75.—Foot, head, and bill of *C. urbica.*

Key to the Species.

a. Under tail-coverts white.
   a'. The longer upper tail-coverts black.
     a''. Lower plumage pure white; fork
      of tail half to three-quarters inch
      deep .... *C. urbica,* p. 269.
     b''. Lower plumage pale grey; fork of
      tail one-quarter inch, or less, deep.. *C. kashmiriensis,* p. 269.
   b'. The longer upper tail-coverts white .. *C. lagopus,* p. 270.
b. Under tail-coverts black .... *C. nepalensis,* p. 271.

804. **Chelidon urbica.** *The Martin.*

Hirundo urbica, *Linn. Syst. Nat.* i, p. 344 (1786); *Blyth, Cat.* p. 198.
Chelidon urbica (*Linn.*), *Horsf. & M. Cat.* i, p. 385; *Jerd. B. I.* i,
p. 166; *Hume, Cat.* no. 92; *Butler, S. F.* ix, p. 378; *Scully, Ibis,*
1881, p. 428; *Biddulph, Ibis,* 1882, p. 269; *Barnes, Birds Bom.*
p. 84; *Sharpe, Cat. B. M.* x, p. 87; *Oates in Hume's N. & E.* 2nd
ed. ii, p. 177.
The English House Martin, Jerd.

Coloration. Forehead, crown, nape, lores, a small space below
the eye, hind neck, back, and scapulars glossy bluish black; rump
and the shorter upper tail-coverts white, with the shafts very
narrowly brown; longer upper tail-coverts glossy black; tail
black with a slight gloss; coverts and quills dull black, some of
the smaller coverts margined with glossy bluish black; cheeks,
ear-coverts, and lower plumage white, washed with ashy on the
sides of the breast and body and on the axillaries.

The young have the chin, throat, fore neck, cheeks, and ear-
coverts dull smoky brown, and the quills next the body tipped
with white; the upper plumage is dull brown.

Bill black; feet pale flesh-colour; iris deep brown.

Length nearly 6; tail 2·5; wing 4·4; tarsus ·5; bill from gape
·5; bifurcation of tail from ·5 to ·75.

Distribution. The series of Indian-killed specimens of this
Martin in the Hume Collection is remarkably poor and the skins
are in almost all cases badly prepared. It is not therefore easy
to identify some of them with absolute certainty, especially the
younger birds, which are very close to *C. kashmiriensis.* There are
three nearly adult birds killed in April at Mussooree; one nearly
adult and four immature birds from Hazára, killed in September;
one nearly adult from Khandesh killed in November; another
quite adult, but moulting, from Shimoga, Mysore, obtained in
April, and four young January birds from Coimbatore. Until
well-preserved adult specimens are obtained, the distribution of
this Martin in India must remain in great doubt. Jerdon records
this species from the Nilgiris in March. Scully informs us that
it is very common in Gilgit in May and June, and Biddulph
obtained it at Gilgit in July.

This species is found in Europe, Africa, and the western half of
Asia.

Habits, &c. Has been found breeding in Mysore in May, constructing a nest of mud pellets lined with feathers under a large
rock in the bed of the river Tungabhadra. This Martin probably
breeds in other similar localities. The eggs, varying from two to
four in number, are pure white and measure about ·75 by ·54.

805. **Chelidon kashmiriensis.** *The Kashmir Martin.*

Chelidon cashmeriensis, *Gould, P. Z. S.* 1858, p. 356; *Jerd. B. I.* i,
p. 167; *Hume, N. & E.* p. 84; *id. Cat.* no. 93; *Biddulph, Ibis,*

1881, p. 47, 1882, p. 269; *Sharpe, Cat. B. M. x, p. 90; Oates in Hume's N. & E.* 2nd ed. ii, p. 177.

The Cashmere House Martin, Jerd.

Coloration. Very similar to *C. urbica,* but with the whole lower plumage pale smoky brown and the axillaries darker brown. The shaft-lines on the rump are generally coarser.

This species and the preceding are so closely allied that they can only be separated with certainty when full-grown and when the tail is perfect. In *C. urbica* the difference between the middle and the outermost pair of tail-feathers varies from half to three-quarters of an inch, whereas in *C. kashmiriensis,* it is never more than a quarter of an inch. *C. urbica* is a larger bird.

Length about 5; tail 2·1; wing 4; tarsus ·5; bill from gape ·45.

Distribution. The series of this bird in the Hume Collection is little better than that of *C. urbica.* There are three adult specimens from Sikhim (April); one nearly adult from the Sutlej valley; five from Kashmir; two from Gilgit (May and July); one from Hazára (November); one from Garhwál (December); and a solitary specimen from the plains, obtained by Blanford at Biláspur in the Central Provinces in April.

This species ascends the Himalayas up to 12,000 or 13,000 feet, and it appears to breed along the whole range, from Kashmir to Sikhim. Its range in the plains is quite unknown.

Habits, &c. Breeds in Kashmir in April and May, and probably a second time later on, constructing a mud nest, shallow and cup-shaped, in the hollows of rocks, many birds breeding together. The eggs are not known, but will undoubtedly prove to be pure white.

806. Chelidon lagopus. *The Siberian Martin.*

Hirundo lagopoda, *Pall. Zoogr. Rosso-Asiat.* i, p. 532 (1811).
Chelidon urbica (*Pall.*) *apud Tick. J. A. S. B.* xxiv, p. 277 note; *Blyth, Birds Burm.* p. 127; *Hume & Dav. S. F.* vi, p. 45.
Chelidon whiteleyi, *Swinhoe, P. Z. S.* 1862, p. 320; *id. Ibis,* 1874, p. 152, pl. vii, fig. 2.
Chelidon lagopoda (*Pall.*), *Oates, B. B.* i, p. 311; *Seebohm, Hist. Brit. Birds,* ii, p. 179 note.
Chelidon lagopus (*Pall.*), *Sharpe, Cat. B. M. x, p. 93.*

Coloration. Forehead, crown, nape, back, scapulars, and lesser wing-coverts glossy steel-black; rump and upper tail-coverts white, the shafts dark; tail, wings, and greater coverts brown; lores, the feathers under the eye and above the ear-coverts dull black; cheeks, lower ear-coverts, and all the lower plumage pure white; under wing-coverts and axillaries dark smoky brown.

Length nearly 5 inches; tail 2·3; wing about 4·5; bill from gape ·45.

Distribution. A House-Martin was observed in Burma by the late Colonel Tickell many years ago. He identified it with *C. urbica,* but his description and figure of it in his MS. work, now in the Library of the Zoological Society of London, show that he procured the present species. I have frequently seen large flocks of a Martin in Southern Pegu, but have failed to secure a specimen: they were most probably of this species.

The Siberian Martin summers in Northern and Central Asia, and visits Burma in the winter months. Colonel Tickell's specimen of this bird was obtained at Moulmein.

807. Chelidon nepalensis. *Hodgson's Martin.*

Delichon nepalensis, *Hodgs., Moore, P. Z. S.* 1854, p. 104, pl. lxiii; *Horsf. & M. Cat.* i, p. 384; *Hume, Cat.* no. 94; *id. S. F.* xi, p. 29.
Chelidon nepalensis (*Hodgs.*), *Jerd. B. I.* i, p. 168; *Blanf. J. A. S. B.* xli, pt. ii, p. 156; *Godw.-Aust. J. A. S. B.* xlv, pt. ii, pp. 68, 195; xlvii, pt. ii, p. 13; *Sharpe, Cat. B. M.* x, p. 95.

The Little Himalayan Martin, Jerd.

Coloration. Rump white, the feathers delicately fringed with black; with this exception the whole upper plumage is glossy bluish black; wings and tail dull black, some of the coverts margined and tipped with glossy bluish black; lores and sides of the head deep black with a very slight gloss; chin and throat black speckled with white; fore neck, breast, abdomen, vent, and legs white; under tail-coverts, axillaries, and under wing-coverts deep black.

Some birds, probably the young, have the point of the chin black, the remainder of the chin and the whole throat white; and in these birds the underparts are not quite so pure a white as in those with black throats.

Bill brown, paler at gape; legs and toes fleshy white.

Length about 4·5; tail 1·8; wing 3·7; tarsus ·45; bill from gape ·4: tail quite square at tip.

Distribution. The Himalayas from Naini Tal to the Daphla hills in Assam, and thence southwards through the hill-ranges to Manipur. This Martin appears to ascend the Himalayas up to at least 8000 feet, and it also appears to be found on those mountains throughout the year, as I have seen specimens procured in Sikhim in every month from June to January. It probably visits the lower valleys and plains in the winter months only.

Genus COTILE, Boie, 1822.

The genus *Cotile* comprises the Sand-Martins, which frequent large rivers and construct their nests in holes excavated in the banks. The eggs are white.

In *Cotile* the legs and toes are bare except in *C. riparia*, which has a small tuft of feathers at the base of the tarsus and behind it. The tail is forked to a very small extent, and the colour of the plumage is extremely plain and dull. The tail-feathers are not spotted with white as in the next genus. The Sand-Martins are highly gregarious and breed in large societies.

808. **Cotile riparia.** *The Sand-Martin.*

Hirundo riparia, *Linn. Syst. Nat.* i, p. 344 (1766); *Blyth, Cat.*
  p. 199.
Cotyle riparia (*Linn.*), *Horsf. & M. Cat.* i, p. 95; *Jerd. B. I.* i,
  p. 163; *Blanf. S. F.* iv, p. 507; *Hume, Cat. no.* 87; *id. S. F.*
  xi, p. 28; *Barnes, Birds Bom.* p. 82.
Cotile riparia (*Linn.*), *Oates, B. B.* i, p. 310; *Sharpe, Cat. B. M.* x,
  p. 96.

The European Sand-Martin, Jerd.

Fig. 76.—Foot of *C. riparia.*

Coloration. The whole upper plumage greyish brown, each
feather with a more or less distinct pale margin; a dark spot in
front of the eye; lores and ear-coverts brown; quills and coverts
dark brown; tail dark brown, narrowly margined with whitish;
a broad and distinct band across the breast brown; cheeks and
remainder of lower plumage pure white.

Young birds have all the feathers of the upper plumage and
the wings margined with rufous, the chin and throat fulvous, and
the breast broadly brown.

Bill black; iris brown; legs dark brown.

Length about 5; tail 2·3; wing 4; tarsus ·45; bill from gape
·5; bifurcation of tail ·4.

Distribution. The Sand-Martin is probably spread over the
whole of the northern portion of India proper as far south as the
latitude of Bombay. It appears to be rare in India, for the Hume
Collection contains but very few specimens, and these from Sind
only; specimens from the eastern part of the Empire are more
numerous. This species extends from Assam to Tenasserim, the
most southern locality from which I have seen a specimen being
Thatone near Moulmein. It appears to be a winter visitor to
India for the most part, but specimens procured in May and
June are contained in the Hume Collection. This Martin is found
over the whole northern hemisphere.

800. **Cotile sinensis.** *The Indian Sand-Martin.*

Hirundo chinensis, *J. E. Gray in Hardw. Ill. Ind. Orn.* i, pl. 35,
    f. 3 (1830-2).
Hirundo subsoccata, *Hodgs. in Gray's Zool. Misc.* p. 82 (1844, desc.
    null.).
Hirundo sinensis (*J. E. Gr.*), *Blyth, Cat.* p. 199.
Cotyle sinensis (*J. E. Gr.*), *Horsf. & M. Cat.* i, p. 96; *Jerd. B. I.* i,
    p. 164; *Anders. Yunnan Exped., Aves,* p. 651; *Hume, Cat. no.
    89; *Barnes, Birds Bom.* p. 82.
Cotyle subsoccata (*Hodgs.*), *Jerd. B. I.* i, p. 163; *Hume, Cat. no.* 88.
Cotile subsoccata (*Hodgs.*), *Hume, N. & E.* p. 82.
Cotile sinensis (*J. E. Gr.*), *Hume, N. & E.* p. 82; *Oates, B. B.* i,
    p. 300; *Sharpe, Cat. B. M.* x, p. 104; *Oates in Hume's N. & E.*
    2nd ed. ii, p. 178.

The Dusky Martin, The Indian Bank Martin, Jerd.; *Abali,* Hind.;
Nakuti, Beng.

Coloration. Upper plumage greyish brown, most of the feathers
margined with paler brown; wings and tail darker brown: chin,
throat, breast, sides of the head and neck pale grey; abdomen,
vent, and under tail-coverts white.

The young bird has all the feathers of the upper plumage and
wings broadly margined with rufous, and the chin, throat, and
breast are pale rufous.

Iris brown; bill black; legs dark brown.

Length about 4; tail 1·8; wing 3·4; tarsus ·35; bill from gape
·45; bifurcation of tail about ·2.

Distribution. A resident species over the whole of the northern
half of India down to about the latitude of Bombay, and probably
further south. This Martin ascends the Himalayas wherever
the streams are suitable to its habits. In the eastern part of
the Empire it extends from Assam to Northern Tenasserim. It
is found in Southern China, Siam, and the Philippine Islands.

Habits, &c. Breeds in large societies in the sandy banks of
rivers, constructing its nest, which consists of a few feathers and
a little grass, in a roundish chamber at the end of a narrow
tunnel, frequently three feet in length. The breeding-season lasts
from November to February in the greater part of India, but in
some parts these Martins breed in April and May. The eggs,
either four or five in number, are pure white, and measure about
·68 by ·48.

Genus **PTYONOPROGNE**, Reichenb., 1850.

The genus *Ptyonoprogne* comprises the Crag-Martins, which are
similar in appearance and structure to the Sand-Martins, but differ
from them in some important points.

The Crag-Martins have a white spot on the inner web of all
the tail-feathers except the middle and outermost pair; they

274   HIRUNDINIDÆ.

construct mud-nests amongst rocks, in caves or old buildings, and
they lay spotted eggs.

Key to the Species.

a. Chin and upper throat streaked or spotted.
   a'. Wing 5; under tail-coverts much darker
      than abdomen........................ *P. rupestris*, p. 274.
   b'. Wing little more than 4; under tail-coverts
      of same colour as abdomen .... *P. concolor*, p. 275.
b. Chin and upper throat unmarked .... *P. obsoleta*, p. 275.

810. **Ptyonoprogne rupestris.** *The Crag-Martin.*

Hirundo rupestris, *Scop. Ann.* i, *Hist. Nat.* p. 167 (1769); *Blyth,
  Cat.* p. 198.
Cotyle rupestris (*Scop.*), *Horsf. & M. Cat.* i, p. 95; *Jerd. B. I.* i,
  p. 166; *Hume & Henders. Lah. to Yark.* p. 177; *Blanf. J. A. S. B.*
  xxxviii, pt. ii, p. 173; *Hume, J. A. S. B.* xxxix, pt. ii, p. 116;
  Barnes, Birds Bom. p. 83.
Ptyonoprogne rupestris (*Scop.*), *Hume, N. & E.* p. 84; *id. Cat.* no.
  91; *Oates in Hume's N. & E.* 2nd ed. ii, p. 180.
Cotile rupestris (*Scop.*), *Sharpe, Cat. B. M.* x, p. 100.
The Mountain Crag Martin, Jerd.

Coloration. Upper plumage, sides of the head, wings, and tail
ashy brown; a large white spot on the inner web of all the tail-
feathers except the middle and outermost pair; chin, throat, and
breast white, tinged with pale rufous, the chin and upper throat
spotted with brown; abdomen and sides of the body rufous ashy;
under tail-coverts dark ashy brown.

The young has the whole upper plumage, wings, and under tail-
coverts margined with rufous, and the lower plumage uniform pale
rufous.

Bill black; legs and feet fleshy; claws dusky; iris dark brown
(*Scully*).

Length about 6; tail 2·4; wing up to 5·4; tarsus ·45; bill from
gape ·55.

Distribution. The whole Himalayas as far east as Bhutan and
the plains as far south as the Nilgiri hills. The range of this
Martin is probably much greater than above indicated, for
Davison observed a *Ptyonoprogne* in Tenasserim, which was
probably of this species. The Crag-Martin has an immense range
out of India, being found in Southern Europe, Northern Africa,
and a great portion of Asia.

Habits, &c. Breeds amongst precipitous rocks high up in the
Himalayas in April, constructing a saucer-shaped nest of mud,
attached to the rock. The eggs are described as being white,
speckled with red and purple.

811. Ptyonoprogne concolor. *The Dusky Crag-Martin.*

Hirundo concolor, *Sykes, P. Z. S.* 1832, p. 83; *Blyth, Cat.* p. 199.
Cotyle concolor (*Sykes*), *Horsf. & M. Cat.* i, p. 97; *Jerd. B. Ind.*
   i, p. 165; *Barnes, Birds Bom.* p. 83.
Ptyonoprogne concolor (*Sykes*), *Hume, N. & E.* p. 83; *id. Cat. no.*
   90; *Oates in Hume's N. & E.* 2nd ed. ii, p. 181.
Cotile concolor (*Sykes*), *Sharpe, Cat. B. M.* x, p. 108.

Coloration. The whole upper plumage, wings, and tail dark
sooty brown; a white spot on the inner web of all the tail-feathers
except the middle and outermost pairs, the spot on the pair
next the middle one being obsolete or frequently wanting; cheeks,
chin, throat, and fore neck rufescent, streaked with brown;
remainder of lower plumage sooty brown, many of the feathers
with fulvous margins and darker shafts.

The young bird has the upper plumage and wings margined
with rufous.

Iris dark brown; bill, legs, and claws brown (*Bingham*).

Length about 5; tail 2; wing 4·2; tarsus ·4; bill from gape ·5.

Distribution. The plains of India, from the foot of the Himalayas
to the Nilgiris, and extending eastwards to Behar and Western
Bengal. To the west this species is found throughout Rajputana,
but does not appear to be found in Sind.

Habits, &c. Breeds during many months of the year, the
time varying according to locality. The nest is cup-shaped,
constructed of mud and attached to a rock or wall in caves, old
buildings, and cliffs. The eggs are white, speckled with yellowish
and reddish brown, generally four in number, and measure about
·72 by ·52.

812. Ptyonoprogne obsoleta. *The Pale Crag-Martin.*

Cotyle obsoleta, *Cab. Mus. Hein.* i, p. 50 (1850); *Barnes, Birds*
   Bom. p. 83.
Ptyonoprogne pallida, *Hume, S. F.* i, pp. 1, 417 (1873).
Ptyonoprogne obsoleta (*Cab.*), *Hume, Cat. no.* 91 bis.
Cotile obsoleta (*Cab.*), *Sharpe, Cat. B. M.* x, p. 111.

Coloration. The whole upper plumage pale greyish brown;
wings and tail darker; all the tail-feathers with a white spot on
the inner web except the middle and outermost pair, but with
signs of a spot on the latter sometimes; a black spot in front of
the eye; sides of the head like the upper plumage; lower plumage
white, tinged with fulvous, and gradually turning to pale brown
on the lower abdomen and under tail-coverts, which latter are
margined paler.

Legs and feet dusky greyish brown; bill blackish brown; iris
blackish brown (*Butler*).

Length about 5·5; tail 2·2; wing 4·5; tarsus ·4; bill from
gape ·55.

Distribution. Sind, extending west to Arabia and Egypt. This species appears to be only a winter visitor to Sind.

Genus **HIRUNDO**, Linn., 1766.

The genus *Hirundo* comprises the true Swallows, which are for the most part familiar and well-known birds.

The Swallows have the upper plumage, or the greater portion of it, deep steel-blue and highly glossy. Many of them have the tail greatly forked, and a few have it nearly square. They all construct nests of mud lined with feathers, some making their nests cup-shaped, while others add a long tubular entrance. The eggs in some species are speckled, in others white without any marks.

Key to the Species.

a. Rump blue or brown.
  a'. White spots on tail.
    a''. A complete or broken band across the breast.
      a'''. Pectoral band complete.......... *H. rustica*, p. 277.
      b'''. Pectoral band more or less interrupted in the middle.
        a⁴. Chin and throat chestnut; abdomen white .... *H. gutturalis*, p. 277.
        b⁴. Chin, throat, and abdomen uniform deep chestnut.......... *H. tytleri*, p. 278.
        c⁴. Chin and throat much deeper chestnut than the abdomen .. *H. erythrogastra*, p. 279.
    b''. No trace of a pectoral band.
      c'''. Chin, throat, and fore neck chestnut; abdomen grey........... *H. javanica*, p. 279.
      d'''. Chin, throat, and fore neck white like abdomen .... *H. smithii*, p. 280.
  b'. No white spots on tail .... *H. fluvicola*, p. 280.
b. Rump chestnut.
  c'. Lower plumage pale rufous, much paler than ear-coverts.
    c''. Rump and upper tail-coverts of same colour throughout, or very slightly paler posteriorly.
      e'''. Wing 4·9 to 5·3.
        d⁴. Shaft-streaks on rump very distinct; lower plumage nearly white, with very coarse striations..................... *H. striolata*, p. 281.
        e⁴. Shaft-streaks on rump absent or obsolete; lower plumage decidedly rufous, with fine striations...................... *H. daurica*, p. 282.
      f'''. Wing 4·3 to 4·7.
        f⁴. Striations on lower plumage much broader than the shafts .. *H. nepalensis*, p. 282.

g'. Striations on lower plumage
    hardly broader than the shafts ..   *H. erythropygia*, p. 283.
d''. Rump paling posteriorly and be-
    coming creamy white .... *H. rufula*, p. 284.
d'. Lower plumage chestnut, quite as dark
    as the ear-coverts .... *H. hyperythra*, p. 284.

813. **Hirundo rustica.** *The Swallow.*

Hirundo rustica, *Linn. Syst. Nat.* i, p. 343 (1766); *Blyth, Cat.* p. 197; *Jerd. B. I.* i, p. 157; *Hume, N. & E.* p. 72; *Anders. Yunnan Exped., Aves*, p. 649; *Legge, Birds Ceyl.* p. 587; *Hume, Cat.* no. 82; *Oates, B. B.* i, p. 302; *Barnes, Birds Bom.* p. 79; *Sharpe, Cat. B. M.* x, p. 128; *Oates in Hume's N. & E.* 2nd ed. ii, p. 184.

The Common Swallow, Jerd.; *Ababil*, Hind.; *Talli-illedi kurari*, Tam.; *Wanna 'korela*, Tel.; *Paras pitta* of the Mharis and Gonds; *Tám pádi*, Tam.; *Pyun-hlwa*, Burm.; *Wahelaniya*, Cing.

Coloration. Forehead, chin, and throat chestnut; lores black; upper plumage and wing-coverts glossy purplish blue; quills and tail black suffused with glossy green, all the tail-feathers, except the middle pair, with a white patch on the inner web; sides of the head and neck and a very broad pectoral band glossy black, a few of the feathers of the latter part narrowly fringed with chestnut; lower plumage from the pectoral band downwards pale rufous, becoming rather darker on the under tail-coverts.

The young bird does not differ very much from the adult, but has the colour of its plumage very dull.

Bill black; feet black; iris dark brown.

Length up to 8; tail up to 4·5; wing 5; tarsus ·5; bill from gape ·6; bifurcation of tail about 2·7.

Distribution. Every portion of the Empire, breeding throughout the whole range of the Himalayas and being found in the plains during the winter. Young birds of this species are to be met with in the plains at nearly all times of the summer in small numbers.

The Swallow has an enormous range, being found over the whole of Europe and Africa and over a great part of Asia.

Habits, &c. Breeds throughout the Himalayas in April and May from 4000 to 7000 feet, constructing its nest of mud, lined with feathers, in outbuildings, verandahs of houses, and sheds. The eggs, four or five in number, are white or pale pink speckled with red and purple, and measure about ·76 by ·53.

814. **Hirundo gutturalis.** *The Eastern Swallow.*

Hirundo gutturalis, *Scop. Del. Flor. et Faun. Insubr.* ii, p. 96 (1786); *Hume & Dav. S. F.* vi, p. 41; *Hume, Cat.* no. 82 bis; *Sharpe, Cat. B. M.* x, p. 134.
Hirundo panayana, *Gm. Syst. Nat.* i, p. 1018 (1788); *Horsf. & M. Cat.* i, p. 91.
Hirundo andamanensis, *Tytler, Beavan, Ibis*, 1867, p. 316; *Ball, S. F.* i, p. 55; *Hume, Cat.* no. 82 quat.

Coloration. Resembles *H. rustica*, but the chestnut of the throat encroaches on the black pectoral band so as to nearly sever it down the middle of the breast: the lower plumage below the pectoral band is pure white.

Length about 6·5; tail 3·6; wing 4·6; tarsus ·4; bill from gape ·6; bifurcation of tail 1·7.

Typical examples of this Swallow from Japan and North-eastern Asia are very distinct from *H. rustica*, the lower plumage being pure white and the pectoral band severed in two by the encroachment of the chestnut of the throat. Many examples procured in Burma are sufficiently typical to be easily recognizable, but the majority of Swallows from the eastern portion of the Empire are quite intermediate between the two species.

Distribution. Common in winter over the whole of the Empire east of the Bay of Bengal and extending to Assam and Bengal. The western limits of this species cannot be determined with any accuracy, as many birds from the continent of India are quite intermediate between *H. rustica* and *H. gutturalis*, and I have seen no bird which could unhesitatingly be assigned to *H. gutturalis* from any point west of Calcutta. This species ranges from North-eastern Asia to the Malay islands and Singapore.

815. Hirundo tytleri. *Tytler's Swallow.*

Hirundo tytleri, *Jerd. B. I.* iii, App. p. 870 (1864); *Hume, S. F.* iii, p. 41; *Wardlaw Ramsay, Ibis,* 1877, p. 466; *Hume & Dav. S. F.* vi, p. 41; *Hume, Cat.* no. 82 ter; *Simson, Ibis,* 1882, p. 84; *Oates, B. B.* i, p. 304; *Sharpe, Cat. B. M.* x, p. 140.

Coloration. Forehead rufous; lores black; whole upper plumage glossy purplish blue; wing-coverts brown, margined with glossy purplish blue; quills black, suffused with glossy green; tail brown, all the feathers, except the middle pair, with a patch of white on the inner web; the whole lower plumage chestnut, the chin and throat very little if at all darker than the other parts; sides of the head and neck, continued to the sides of the breast and forming a pectoral band, interrupted in the middle, glossy purplish blue.

Length 6·5 to 7; tail up to 4; wing 4·8; tarsus ·45; bill from gape ·6.

Distribution. A common but uncertain visitor to the eastern parts of the Empire. This species has been observed in Sadiya, Dacca, Cachar, the Khási hills, Manipur, Pegu, and Tenasserim, and there are specimens from all these parts in the British Museum. It has been obtained during February, March, April, May, and June; and it may breed on or near the eastern borders of Assam and Burma. It occurs in Eastern Siberia and Kamtschatka and probably has a very wide range. Some specimens of Swallows from Peru and Brazil in the British Museum are perfectly undistinguishable from *H. tytleri.*

816. **Hirundo erythrogastra.** *The American Swallow.*

Hirundo erythrogaster, *Bodd. Tabl. Pl. Enl.* p. 45 (1783)
Hirundo horreorum, *Bart. Fragm. Nat. Hist.* p. 17 (1799); *Hume
& Dav. S. F.* vi, p. 42; *Hume, Cat. no.* 82 quint.; *Oates, B. B.* i,
p. 303.
Hirundo erythrogastra, *Bodd., Sharpe, Cat. B. M.* x, p. 137.

Coloration. Resembles *H. tytleri,* and like it has an interrupted
pectoral band, but the plumage below the band is pale rufous,
whereas the chin, throat, and middle of the breast are chestnut.
Of the same size as *H. tytleri.*
Distribution. The only specimens of this species that I have seen
from within Indian limits are two procured by myself in February,
one at Toungngoo and one at Pegu. The former is adult and the
latter young. There are two typical adult specimens from Cochin
China in the British Museum.
This Swallow is found in Eastern Asia and over the whole of the
continent of America as far south as Brazil.

817. **Hirundo javanica.** *The Nilgiri House-Swallow.*

Hirundo javanica, *Sparrm. Mus. Carls.* iv, pl. 100 (1789); *Legge,
Birds Ceyl.* p. 597; *Hume, Cat.* no. 83; *Sharpe, Cat. B. M.* x,
p. 142; *Oates in Hume's N. & E.* 2nd ed. ii, p. 186.
Hirundo domicola, *Jerd. Madr. Journ. L. S.* xiii, p. 173 (1844);
Blyth, Cat. p. 198; *Horsf. & M. Cat.* i, p. 384; *Jerd. B. I.* i, p. 158.
Hypurolepis domicola (*Jerd.*), *Hume, N. & E.* p. 73.
Hypurolepis javanica (*Sparrm.*), *Oates, B. B.* i. p. 308.

Coloration. A broad band on the forehead, the chin, throat, upper
breast, cheeks, and ear-coverts deep ferruginous; lores dusky;
upper plumage glossy black; wings and tail dark brown, slightly
glossy on the outer webs; the tail with an oval spot on all the inner
webs of the feathers except those of the median pair; lower plumage
pale ashy, albescent on the abdomen; under tail-coverts ashy, the
feathers with white tips and subterminal patches of black.
The young bird is without gloss above and has some of the
secondaries margined with rufous.
Bill black; legs and claws lighter black; iris dark brown.
Length about 5; tail 2·1; wing 4·2; tarsus ·4; bill from
gape ·65; bifurcation of tail ·3.
Distribution. Ceylon and Southern India as far north as the
Nilgiris, being resident and confined to the higher parts of the hills;
Tenasserim; the Andamans. This species extends down the Malay
peninsula and is found throughout the islands.
Habits, &c. Breeds from February to June, constructing a cup-
shaped mud nest in bungalows and outbuildings and laying three
eggs, which are white spotted with brown and purple and measure
about ·7 by ·5.

818. Hirundo smithii. *The Wire-tailed Swallow.*

Hirundo smithii, *Leach, App. to Tuckey's Voy. Congo*, p. 407 (1818);
Sharpe, Cat. B. M. x, p. 150; *Oates in Hume's N. & E.* 2nd ed. ii,
p. 188.
Hirundo filifera, *Steph. Gen. Zool.* xiii, p. 78 (1826); *Blyth, Cat.*
p. 197; *Horsf. & M. Cat.* i, p. 93; *Jerd. B. I.* i, p. 159; *Anders.
Yunnan Exped., Aves.* p. 650; *Hume, Cat.* no. 84; *Barnes, Birds
Bom.* p. 79.
Uromitrus filiferus (*Steph.*), *Hume, N. & E.* p. 75; *Oates, B. B.* i,
p. 307.

Leishra, Hind.

Coloration. Forehead, crown, and nape chestnut; sides of the
head and neck and the whole upper plumage with the wing-coverts
glossy steel-blue; quills and tail dark brown, margined with steel-
blue: all the tail-feathers except the two median pairs with a
white spot on the inner web; the whole lower plumage white.

The nestling has the chestnut of the head replaced by brown,
and the chin, throat, and breast are tinged with pale fulvous.

Bill, legs, and feet black; iris dark brown (*Bingham*).

Length to tip of ordinary feathers of the tail about 5; tail to
end of ordinary feathers 1·8; outer tail-feather with lengthened
shaft about 7 in male, somewhat less in female; wing 4·6; tarsus
·45; bill from gape ·55.

The Wire-tailed Swallow of India is quite identical with the
Wire-tailed Swallow of Africa.

Distribution. The whole peninsula of India as far south as Mysore
and the Nilgiris; this Swallow is apparently absent from, or rare in,
Bengal, Assam, and Upper Burma, but reappears in Pegu and Tenasserim. It ascends the Himalayas to a considerable height in the
summer, and appears to be a constant resident in the plains. It
extends to Africa and is found in various parts of that continent.

Habits, &c. Breeds from January to December, according to
locality, constructing a small cup-shaped nest under bridges and
culverts, and also under rocks in the immediate vicinity of water.
The eggs, three or four in number, are white marked with various
shades of brown and red, and measure about ·72 by ·53.

819. Hirundo fluvicola. *The Indian Cliff-Swallow.*

Hirundo fluvicola, *Jerd., Blyth, J. A. S. B.* xxiv, p. 470 (1855);
Jerd. B. I. i, p. 161; *Blanf. J. A. S. B.* xxxviii, pt. ii, p. 172;
Hume, J. A. S. B. xxxix. pt. ii, p. 115; *id. Cat.* no. 86; *Barnes,
Birds Bom.* p. 81; *Oates in Hume's N. & E.* 2nd ed. ii, p. 191.
Lagenoplastes fluvicola (*Jerd.*), *Hume, N. & E.* p. 80.
Petrochelidon fluvicola (*Jerd.*), *Sharpe, Cat. B. M.* x, p. 200.

Coloration. Forehead, crown, and nape dull chestnut, with black
shaft-streaks; back and scapulars glossy steel-blue; rump and
upper tail-coverts dull brown, with narrow pale margins; wings
and tail dull brown; lores and upper part of the ear-coverts dark

brown; remainder of the sides of the head, chin, throat, and upper breast white, tinged with very pale fulvous, and boldly streaked with brown; remainder of lower plumage white, the sides of the body slightly streaked with brown.

The young bird has the head brownish, the feathers of the back margined with rufous, those of the rump very broadly margined with fulvous, and the wings more or less margined with the same.

Legs dark brown; toes black; iris brown; bill black (*Hume*).

Length about 4·5; tail 1·75; wing 3·6; tarsus ·4; bill from gape ·5; bifurcation of tail ·15.

Distribution. A considerable portion of India proper. Towards the south this species has been found as far as Coimbatore; to the east as far as Etáwah; and on the north and west to the foot of the Himalayas and in Kashmir, extending into all parts of the north-west except Sind.

Habits, &c. Breeds in large societies from February to April and in July and August, constructing a mud nest, which is spherical or oval with a long neck or tubular entrance attached to it. Large numbers of nests are plastered together against the face of cliffs or under bridges. The eggs are sometimes spotless white, more frequently white speckled with yellowish or reddish brown, and usually three in number. They measure about ·76 by ·53.

820. **Hirundo striolata.** *The Japanese Striated Swallow.*

Hirundo striolata, *Temm., Temm. & Schleg. Faun. Jap., Aves*, p. 33 (1850); *Walden in Blyth's Birds Burm.* p. 127; *Sharpe, Cat. B. M.* x, p. 161.

Hirundo alpestris japonica, *Temm. & Schleg. op. cit.* p. 33, pl. ii (1850).

Lillia substriolata, *Hume, S. F.* v, p. 264 (1877).

Hirundo substriolata (*Hume*), *Hume, Cat.* no. 85 quat.

Hirundo japonica (*T. & S.*), *Oates, B. B.* i, p. 305; *Sharpe, Cat. B. M.* x, p. 162.

Coloration. Forehead, crown, nape, hind neck, back, scapulars, lesser and median wing-coverts, and tail-coverts glossy steel-blue; quills and tail black, slightly glossy on the outer webs; rump chestnut, with well-marked black shafts; lores black: a mark in front and above the eye, continued as a narrow line over the eye, and the ear-coverts chestnut, streaked with black; lower plumage white, with a very pale fulvous tinge throughout, and very coarsely streaked with black throughout: under tail-coverts with their terminal halves entirely black.

Iris dark brown; bill and legs dark brown (*Wardlaw Ramsay*).

Length nearly 8; tail up to 4·2; wing 4·9 to 5·3; tarsus ·6; bill from gape ·6; bifurcation of tail 2·1.

Distribution. The only specimens of this species killed within Indian limits that I have seen are: two specimens from Cachar, February (types of *Lillia substriolata*, Hume): a specimen from Karennee, 2600 feet, 29th March; and two from the Karen hills, east of Toungngoo, 3000 feet, 15th January.

This species is found from Japan down to Java and Flores, summering in the northern part of its range, and wintering in the south. It is common throughout China.

821. Hirundo daurica. *The Daurian Striated Swallow.*

Hirundo daurica, *Linn. Mantissa Plant.* p. 528 (1771); *Sharpe. Cat.*
  B. M. x, p. 159.
Hirundo alpestris, *Pall. Reis. Russ. Reichs,* ii, p. 709 (1773).
Lillia intermedia, *Hume, S. F.* v, p. 263 (1877).
Hirundo intermedia (*Hume*), *Hume, Cat.* no. 85 ter.

Coloration. Resembles *H. striolata* closely, but has the lower plumage distinctly rufous, and the striations less coarse : the shafts of the feathers of the rump are with few exceptions rufous, a very few only being very finely black.

The dimensions are much the same as those of *H. striolata.*

This species, when compared with the last, appears very distinct, and it has a very different geographical distribution.

Distribution. The only two specimens of this species killed within Indian limits that I have seen, are two in the Hume Collection from Sadiya in Assam, obtained in June. These are the types of *Lillia intermedia* (Hume). This species is found over a large portion of Northern and Central Asia, but does not extend to Japan or China. I have seen specimens from the Irtisch river, Dauria, and Mongolia, all killed in the height of summer. It is probably a resident species except in the more northern part of its range.

822. Hirundo nepalensis. *Hodgson's Striated Swallow.*

Hirundo nipalensis, *Hodgs. J. A. S. B.* v, p. 780 (1836); *Hume, Cat.*
  no. 85 bis; *Oates, B. B.* i, p. 306; *Sharpe, Cat. B. M.* x, p. 160:
  Oates in Hume's N. & E. 2nd ed. ii, p. 195.
Hirundo daurica, *Linn., Blyth, Cat.* p. 198 (pt.); *Horsf. & M. Cat.* i,
  p. 92 (pt.); *Jerd. B. I.* i, p. 160 (pt.).
Cecropis arctivitta, *Swinh. P. Z. S.* 1871, p. 346.
Lillia daurica (*Linn.*) *apud Hume, N. & E.* p. 78.
Lillia nipalensis (*Hodgs.*), *Hume, S. F.* v, p. 262.

The Red-rumped Swallow, Jerd.

Coloration. Merely a small form of *H. striolata,* the wing seldom exceeding 4·7, and being frequently under 4·5 ; the colour of the lower plumage is also more rufous and the striations rather less coarse ; very few of the feathers of the rump have black shafts, while many are fringed paler.

The expediency of separating this form from *H. striolata* may be questioned by many and with much justice. Having regard, however, to the fact that no Swallow of this type either from the Himalayas or from the plains of India ever attains the size of the true *H. striolata,* nor has the striations on the lower plumage as coarsely marked, nor those on the rump as numerous or distinct, I am inclined to consider the two races sufficiently well differentiated to be easily recognizable, notwithstanding that a few birds may occasionally be met with which it is difficult to assign to the one species or the other with certainty.

Distribution. Every portion of the Empire, breeding along the whole extent of the Himalayas, and visiting the plains in the winter months.

This species extends into China, where it appears to be common, and it ranges as far as Japan.

Habits, &c. Breeds in the Himalayas from April to August, constructing a retort-shaped nest like that of *H. fluvicola* in the verandahs of houses, and on rocks and cliffs. The eggs are pure white, and measure about ·85 by ·55.

823. Hirundo erythropygia. *Sykes's Striated Swallow.*

Hirundo erythropygia, *Sykes, P. Z. S.* 1832, p. 83; *Hume, Cat.* no. 85; *Legge, Birds Ceyl.* p. 594; *Sharpe, Cat. B. M.* x, p. 164; *Barnes, Birds Bom.* p. 80; *Oates in Hume's N. & E.* 2nd ed. ii, p. 197.
Hirundo daurica, *Linn., Blyth, Cat.* p. 198 (pt.); *Horsf. & M. Cat.* i, p. 92 (pt.); *Jerd. B. I.* i, p. 160 (pt.).
Lillia erythropygia (*Sykes*), *Hume, N. & E.* p. 76; *id. S. F.* v, p. 255.

The Red-rumped Swallow, Jerd.; *Masjid-ababil,* Hind.

Coloration. Forehead, crown, nape, back, scapulars, lesser and median wing-coverts glossy steel-blue; rump and the shorter upper tail-coverts plain chestnut; longer upper tail-coverts glossy black; greater wing-coverts, quills, and tail dark brown, glossed with blue on the outer webs; the outermost tail-feather with an obsolete white patch on the inner web; lores brown; under the eye mixed rufous and brown; ear-coverts and a narrow partially-interrupted collar on the hind neck chestnut; the whole lower plumage pale rufous, with delicate brown streaks, hardly anywhere broader than the shafts themselves; under tail-coverts tipped with black; sides of the neck glossy blue, extending to the sides of the breast.

The young bird has the striations on the lower plumage very faint, and the quills are tipped with rufous.

Bill and legs black; iris brown.

Length about 6·5; tail 3·2; wing 4·3; tarsus ·5; bill from gape ·45; bifurcation of tail 1·4.

Distribution. A resident in the plains of India from the foot of the Himalayas to the Nilgiris, and from Sind to about the longitude of Calcutta. This Swallow occasionally wanders to Ceylon.

Habits, &c. Breeds from April to August, constructing a retort-shaped nest of mud under arches, against walls and rocks, and sometimes in old buildings. The eggs are pure white, three in number, and measure about ·78 by ·55.

824. **Hirundo rufula.**   *The European Striated Swallow.*

Hirundo rufula, *Temm. Man.* 2ᵉ éd. iii, p. 298 (1835); *Wardlaw Ramsay, Ibis,* 1880, p. 48; *Scully, Ibis,* 1881, p. 427; *Sharpe, Cat. B. M.* x, p. 156.

Hirundo scullii, *Seebohm, Ibis,* 1883, p. 167; *Sharpe, Cat. B. M.* x, p. 158.

Coloration. Resembles *H. erythropygia,* but has the striations on the lower plumage even finer, and the chestnut on the rump paling off to creamy white posteriorly : much larger.

Bill blackish ; iris blackish brown : feet dark brown (*Dresser*).

Length about 7·5 ; tail 4 ; wing 4·8 : tarsus ·55 ; bill from gape ·6 ; bifurcation of tail 2·3.

Mr. Seebohm has separated a small form of this species from Nepal as *H. scullii.* As this only differs from *H. rufula* in being very slightly smaller, I do not propose to keep it distinct. On examining the type of *H. scullii,* kindly lent to me for the purpose by Mr. Seebohm, I have strong doubts as to whether it belongs to *H. rufula* or to *H. nepalensis.* The rump pales slightly posteriorly it is true, but not more so than in many specimens of undoubted *H. nepalensis* ; the striations, however, on the lower plumage appear to me to be too coarse for *H. rufula.* Fortunately, the occurrence of this latter species within Indian limits rests on other evidence than that of this dubious specimen collected in Nepal, Dr. Scully having obtained *H. rufula* in Gilgit, where its presence was not unexpected. Major Wardlaw Ramsay also procured an undoubted specimen of this species at Byan Khel in Afghanistan, and this is deposited in the British Museum.

Distribution. Found in Gilgit in summer, and probably extends along the Himalayas to Nepal. This Swallow ranges westwards into Europe and Africa, and is also found in Central Asia.

825. **Hirundo hyperythra.**   *The Ceylon Swallow.*

Hirundo hyperythra, *Layard, Blyth, J. A. S. B.* xviii, p. 814 (1849); *id. Cat.* p. 198; *Legge, Birds Ceyl.* p. 592 ; *Hume Cat.* no. 85 quint. ; *Sharpe, Cat. B. M.* x, p. 167 ; *Oates in Hume's N. & E.* 2nd ed. ii. p. 201.

Wæhælaniya, Cing.

Coloration. Rump and the shorter upper tail-coverts deep chestnut ; with this exception, the whole upper plumage, wings, and tail glossy steel-blue ; sides of the head and whole lower plumage deep chestnut with narrow shaft-streaks ; terminal half of under tail-coverts black.

Iris sepia-brown ; bill deep brown, in some blackish, base of lower mandible reddish ; legs and feet vinous-brown (*Legge*).

Length about 6·5 ; tail 3·3 ; wing 4·7 ; tarsus ·55 ; bill from gape ·6 ; bifurcation of tail 1·4.

Distribution. Confined to Ceylon. A similar but very much larger Swallow (*H. badia*) occurs at Malacca.

Habits, &c. Breeds from March to June, constructing a cup-shaped nest of mud under bridges or in outhouses. The eggs are pure white, and measure about ·85 by ·56.

Family MOTACILLIDÆ.

The intrinsic muscles of the syrinx fixed to the ends of the bronchial semi-rings; the edges of both mandibles perfectly smooth except for the presence of a small notch in the upper near the tip; the hinder part of the tarsus longitudinally bilaminated, the laminæ entire and smooth; wing with nine primaries, the first and second nearly equal; bill long and slender; the longest secondaries reaching nearly or quite to the tip of the wing; a complete autumn and a partial spring moult; young not very dissimilar to the adult; tail of twelve feathers; rictal bristles present and fairly well developed; sexes alike or nearly so; tarsus slightly scutellated.

The *Motacillidæ* comprise the Wagtails and Pipits, birds of wide distribution, and, in nearly all cases, migratory.

The Indian species of this family resemble each other very closely in structure, and there are few characters by which to divide them into genera. I shall content myself with making use of four genera, two of them being of large extent, and two restricted to a single species each.

The Wagtails and Pipits are chiefly found on the ground; a few species have the habit of flying up into trees when disturbed. They feed entirely on insects, and their deportment is very graceful. They have no great power of song.

Key to the Genera.

A. Upper plumage neither streaked nor mottled, but plain.
    a'. Middle pair of tail-feathers as long as the others, or longer .... MOTACILLA, p. 285.
    b'. Middle pair of tail-feathers abruptly shorter than the next, and of a markedly different colour .... LIMONIDROMUS, p. 300.
B. Upper plumage streaked or mottled.
    c'. Tips of tail-feathers rounded and of normal shape .... ANTHUS, p. 301.
    d'. Tips of tail-feathers sharply pointed .... OREOCORYS, p. 313.

Genus MOTACILLA, Linn., 1766.

The genus *Motacilla* contains the typical Wagtails, which are found over the whole of the Old World.

In *Motacilla* the upper plumage is quite plain, being characterized by an utter absence of all streaks, spots, or mottlings. The

sexes are alike or nearly so, and the difference between the summer and winter plumage in most of the species is very striking.
The Pied Wagtails are constantly undergoing a change of colour,
and it is hardly possible to find two birds at the same date in
the same state of plumage. This causes them to be very difficult
to identify by any set description. The Yellow Wagtails do not
undergo such complete changes as the Pied, but their similarity
to each other is so great in winter and immature plumage, that
their recognition is still more difficult.

The Wagtails have the tail and wing of nearly equal length, and
they have the habit of vibrating the former repeatedly. They frequent open land, fields, and the banks of rivers and ponds, some
of the species of Yellow Wagtails being only found on marshy land.
They construct their nests on or near the ground or in holes of
walls and banks, and their eggs are much spotted with brown.

Key to the Species.

A. Hind claw much curved and shorter than
   hind toe (fig. 78).
   a'. Plumage black, white, and grey.
     a''. Ear-coverts and sides of neck always
          white-washed.
       a'''. No black streak through eye.
         a⁴. Greater wing-coverts merely mar-
             gined with white ; back never
             black .... *M. alba*, p. 287.
         b⁴. Outer webs of greater wing-coverts
             entirely white; back black or dusky
             on the mantle................ *M. leucopsis*, p. 288.
       b'''. A broad black streak through the eye. *M. ocularis*, p. 289.
     b''. Ear-coverts and sides of neck always
          black.
       c'''. Forehead entirely white.
         c⁴ Back always grey .... *M. personata*, p. 290.
         d⁴. Back black or with traces of black
             or dusky .... *M. hodgsoni*, p. 291.
       d'''. Forehead with black of the crown
             produced to base of bill........... *M. maderaspatensis*, [p. 291.
   b'. Plumage largely yellow and green ...... *M. melanope*, p. 293.
B. Hind claw little curved and much longer than
   hind toe (fig. 81, p. 294).
   c'. Tarsus unmistakably shorter than one inch.
     c''. Crown dark slaty grey : supercilium
          absent or obsolete; cheeks blackish .. *M. borealis*, p. 294.
     d''. Crown dark slaty blue ; supercilium
          very broad and distinct ; cheeks black-
          ish and ear-coverts streaked with white. *M. flava*, p. 295.
     e''. Crown pale grey ; supercilium broad
          and distinct ; cheeks white .... *M. beema*, p. 296.
     f''. Crown black ; supercilium absent or
          obsolete; cheeks and ear-coverts deep
          black .... *M. feldeggi*, p. 297.

d'. Tarsus considerably longer than one inch.
  g". Back always ashy grey .... *M. citreola*, p. 298.
  h". Back either black or with some black or
    dusky feathers .... *M. citreoloides*, p. 299.

826. Motacilla alba. *The White Wagtail.*

Motacilla alba, *Linn. Syst. Nat.* i, p. 331 (1766); *Hume, Cat.* no. 591
  ter; *Scully, S. F.* viii, p. 314; *Oates, B. B.* i, p. 156; *Sharpe,
  Cat. B. M.* x, p. 464.
Motacilla dukhunensis, *Sykes, P. Z. S.* 1832, p. 91; *Blyth, Cat.*
  p. 137; *Horsf. & M. Cat.* i, p. 349; *Brooks, S. F.* ii, p. 457, vii,
  p. 137; *Hume, Cat.* no. 591 bis; *Barnes, Birds Bom.* p. 236.
Dholin. Hind.

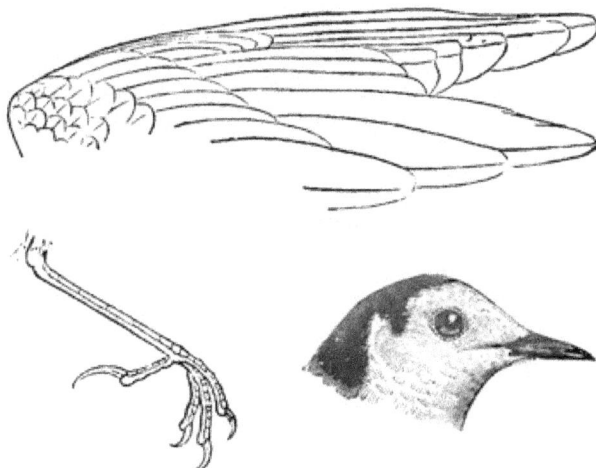

Figs. 77, 78, 79.—Wing, foot, and head of *M. alba*.

Coloration. In normal full summer plumage, the forehead, anterior portion of crown, sides of the head and of the neck are pure white; remainder of crown, nape, and hind neck, chin, throat, fore neck, and breast deep black; upper plumage, scapulars, and lesser wing-coverts grey; upper tail-coverts more or less black, margined exteriorly with white; wing-coverts and tertiaries blackish, broadly margined with white; primaries and secondaries black, narrowly margined with whitish; the four middle pairs of tail-feathers black, the others nearly entirely white; lower plumage from the breast downwards pure white.

In normal winter plumage the chin, throat, and fore neck become white, and the black on the breast is reduced to a narrow crescentic patch, sometimes extending narrowly up the sides of the fore neck.

The nestling is uniform greenish ashy above, and the lower

plumage is yellowish grey with indications of a pale pectoral crescent.

After the first autumn moult, the young bird resembles the adult in normal winter plumage, but has the posterior part of the crown, the nape, and the hind neck grey like the back, and the white parts of the head tinged with primrose-yellow.

The full summer plumage is probably assumed the first spring, except in the case of the females, which appear to be much longer than the males in acquiring the black on the crown, nape, and hind neck, these parts being at first a pale brown, and afterwards for some time black mottled with brown.

Bill black, bluish below; iris brown; legs and claws dark brown or nearly black.

Length nearly 8; tail 3·6; wing 3·5; tarsus ·85; bill from gape ·75.

Distribution. A winter visitor to the whole Empire, as far south in the peninsula of India as Belgaum, and in Burma as Moulmein. A few specimens of this species are occasionally killed in summer, and I have seen a July specimen from Sámbhar. Indian birds of this species appear to winter in Northern Asia. The range of this Wagtail extends to Europe and Northern Africa.

M. baicalensis is a race of *M. alba* with the wing-coverts almost entirely white. It inhabits Eastern Siberia, and in the British Museum there is a specimen said to have been killed in "India." No reliance can, however, be placed on this locality.

M. persica is another race of *M. alba*, with the wing-coverts almost entirely white, and the black of the hind neck almost in contact with the black of the breast. It inhabits Persia.

827. Motacilla leucopsis. *The White-faced Wagtail.*

Motacilla leucopsis, *Gould*, P. Z. S. 1837, p. 78; *Brooks, S. F.* vii, p. 139; *Hume, S. F.* vii, p. 519; *id. Cat.* no. 590; *Scully, S. F.* viii, p. 313; *Oates, B. B.* i. p. 154; *Sharpe, Cat. B. M.* x, p. 482; *Barnes, Birds Bom.* p. 235.

Motacilla luzoniensis (*Scop.*) *apud Blyth, Cat.* p. 137; *Horsf. & M. Cat.* i, p. 348; *Jerd. B. I.* ii, p. 218; *Anders. Yunnan Exped., Aves,* p. 609.

Motacilla felix, *Swinh.* P. Z. S. 1870, p. 121.

Dhobin, Hind.; *Tangzhengleu,* Lepch.

Coloration. In normal full summer plumage the whole upper plumage, scapulars, and lesser wing-coverts are deep black except the forehead and the anterior portion of the crown, which, with the sides of the head and neck, cheeks, chin, and upper throat, are pure white; lower throat, fore neck, and upper breast deep black; remainder of lower plumage white; median and greater wing-coverts entirely white except a small portion of the inner webs; quills black, a large portion of the inner webs white and the outer webs margined with white; the four middle pairs of tail-feathers black, the others white.

In normal winter plumage the whole back, scapulars, rump, and shorter tail-coverts become grey; the lesser wing-coverts grey mixed with black; the black on the lower throat and fore neck gives place to white, and the black on the upper breast is reduced to a crescentic patch.

The young in first plumage are like the adults in winter, but have the grey of the upper parts much paler and the crown, nape, and hind neck grey like the back: the head and lower plumage are suffused with yellow and the black patch on the breast is very small and ill-defined.

The summer plumage of the adult is probably assumed in the first spring, except in the case of the female, which appears to have the black of the crown and nape mixed with grey at first, probably during the whole of the first summer.

Bill black, bluish below; iris brown; legs and claws dark brown or nearly black.

Length nearly 8; tail 3·6; wing 3·5; tarsus ·85; bill from gape ·75.

Distribution. A winter visitor to the eastern portion of the Empire from Assam down to Central Tenasserim. The western limit of this species appears to be Nepal on the Himalayas and Mirzapur in the plains. It is also found in the Andamans. This Wagtail is found throughout Eastern Asia, breeding in Eastern Siberia and China.

828. **Motacilla ocularis.** *The Streak-eyed Wagtail.*

Motacilla ocularis, *Swinh. Ibis,* 1860, p. 55; *id. P. Z. S.* 1870, p. 129; 1871, p. 364; *Hume & Dav. S. F.* vi, p. 518; *Hume, Cat. no.* 591 quat.; *Scully, S. F.* viii, p. 315; *Oates, B. B.* i, p. 155; *Sharpe, Cat. B. M.* x, p. 471, pl. iv, figs. 5, 6.

Coloration. In normal full summer plumage the forehead, anterior part of crown, a broad supercilium, cheeks, ear-coverts, and sides of the neck white; remainder of crown and nape, a streak from the lores through the eye and over the ear-coverts, chin, throat, and upper breast black; remainder of lower plumage white shaded with grey on the flanks; upper plumage grey, turning to black on the upper tail-coverts; lesser wing-coverts grey; median coverts dark brown broadly tipped with white; greater coverts with the outer webs and a considerable portion of the inner white; quills dark brown edged with white, the later secondaries very broadly so; the two outer pairs of tail-feathers nearly entirely white, the others black.

In normal winter plumage the chin and throat are white, and the black on the breast is reduced to a narrow crescent extending laterally along the sides of the throat.

Young birds resemble the adults in winter plumage, but the whole forehead and crown are grey like the back and the white parts, especially of the head, are suffused with yellow. The eye-streak is less distinct and brown.

Bill black, plumbeous at the base; iris brown; legs and claws black.

Length about 8; tail 3·8; wing 3·7; tarsus ·95: bill from gape ·8; the dimensions of this species vary much.

Distribution. A winter visitor to the greater portion of Pegu and Tenasserim, where this species is abundant. It has been obtained at Dibrughar in Assam, in Manipur, and in Nepal, and may therefore be expected to occur throughout the whole eastern half of the Empire. This Wagtail has a very wide range, extending to China and Eastern Siberia and being occasionally observed in North America.

M. lugens, from China and N.E. Asia, resembles *M. ocularis,* but has the back black in summer and the wings with an immense amount of white on them. The note of this species is a prolonged " Pooh."

829. Motacilla personata. *The Masked Wagtail.*

Motacilla personata, *Gould, Birds As.* iv, pl. 63 (1861); *Hume & Henders. Lah. to Yark.* p. 224; *Scully, S. F.* iv, p. 150; *Hume, Cat.* no. 591; *Biddulph, Ibis,* 1881, p. 68; *Scully, Ibis,* 1881, p. 451; *Sharpe, Cat. B. M.* x, p. 479, pl. v. figs. 3, 4; *Barnes, Birds Bom.* p. 266; *Oates in Hume's N. & E.* 2nd ed. ii, p. 201.
Motacilla dukhunensis, *Sykes, apud Jerd. B. I.* ii, p. 218.
Motacilla cashmeriensis, *Brooks, P. A. S. B.* 1871, p. 210; *id. J. A. S. B.* xli, pt. ii, p. 82; xliii, pt. ii, p. 250; *id. S. F.* ii, p. 456.

The Black-faced Wagtail, Jerd.; *Dhobin,* Hind.

Coloration. In normal summer plumage, the forehead, anterior portion of crown, the upper part of the lores, round the eye, and a broad supercilium are white; remainder of the head, nape, hind neck, the neck all round, and the lower plumage from the chin to the upper breast black; rest of lower plumage white; upper plumage grey turning to black on the upper tail-coverts; lesser wing-coverts grey; median and greater coverts entirely white in the closed wing; quills dark brown, the inner webs largely white and the outer margined with white; the later secondaries with the outer webs almost entirely white; the two outer pairs of tail-feathers white, the others black.

In winter plumage, the chin and throat become white, but a black moustachial streak remains dividing the white of the throat from the white round the eye; the ear-coverts are always black; the feathers of the lower throat have their bases white.

The young have the whole upper plumage dull grey, and throat and breast dull brownish; the wings and tail are from the first very similar to the same parts in the adult. The black plumage of the head and neck is assumed very slowly and probably not completely till the second spring.

Iris blackish brown; legs, feet, and bill black (*Butler*).

Length about 8; tail 3·6; wing 3·6; tarsus ·9; bill from gape ·65.

Distribution. A winter visitor to the whole of India proper down to

Belgaum on the south and to Calcutta on the east. This species is a constant resident in Gilgit and probably other parts of Kashmir. It extends west to Afghanistan and Persia and north to Turkestan and Central Asia.

Habits, &c. Breeds in Afghanistan in May and June, constructing its nest near water under large stones and in similar localities. The eggs have not, however, been described.

830. Motacilla hodgsoni. *Hodgson's Pied Wagtail.*

Motacilla alboides, *Hodgs. As. Res.* xix, 191, part. (1836).
Motacilla hodgsoni, *G. R. Gray, Blyth, Ibis,* 1865, p. 49; *Blanf. J. A. S. B.* xli, pt. ii, p. 59; *Brooks, S. F.* iii, pp. 246, 278; *Godw.-Aust. J. A. S. B.* xlv, pt. ii, p. 81; *Hume, Cat.* no. 589 bis; *Scully, S. F.* viii, p. 312; *Sharpe, Cat. B. M.* x, p. 486, pl. v, figs. 1, 2; *Oates in Hume's N. & E.* 2nd ed. ii, p. 202.
Motacilla maderaspatensis, *Gmel. apud Anders. Yunnan Exped. Aves,* p. 610.

Coloration. In normal summer plumage resembles *M. personata* in every respect except that the whole back, rump, scapulars, and lesser wing-coverts are black.

In normal winter plumage this species is hardly distinguishable from *M. personata* in winter plumage. The only apparent difference, and it is very slight, is that in *M. hodgsoni* the grey of the upper parts is darker and frequently mottled with traces of black or brown on the shoulders and lesser wing-coverts.

Bill and legs black; iris brown (*Cockburn*).

Of the same size as *M. personata.*

Distribution. A winter visitor to the lower ranges of the Himalayas from Kashmir to Assam, extending through the Khási hills, Cachar, and Manipur to the Salween district of Tenasserim. This species summers in the higher parts of the Himalayas and in Turkestan. A few birds seem to visit the plains of India, where this species has been procured at Etáwah for instance, but its range in the plains is distinctly eastern, just as that of *M. personata* is western.

Habits, &c. Breeds in Kashmir about May, constructing its nest in holes under stones, among shingle and pebbles, and in heaps of driftwood and rubbish. The eggs are greyish white speckled with brown and grey, and measure about ·78 by ·62.

A specimen of a Wagtail from Toungngoo in the British Museum cannot be referred to any known species. It resembles *M. leucopsis* in having the black of the upper breast divided off from the black of the hind neck by a band of white running down the side of the neck, but the ear-coverts are black.

831. Motacilla maderaspatensis. *The Large Pied Wagtail.*

Motacilla maderaspatensis, *Gmel. Syst. Nat.* i, p. 961 (1788); *Hume, Cat.* no. 589; *Legge, Birds Ceyl.* p. 607; *Sharpe, Cat. B. M.* x, p. 490; *Barnes, Birds Bom.* p. 234; *Oates, in Hume's N. & E.* 2nd ed. ii, p. 202.

Motacilla maderaspatana, *Briss. apud Blyth, Cat.* p. 137 ; *Horsf. & M. Cat.* i, p. 347 ; *Jerd. B. I.* ii, p. 217 ; *Hume, N. & E.* p. 377.

The Pied Wagtail, Jerd.; *Mamula, Bhuin Mamula, Khanjan*, Hind. ; *Sakala sarela-gadu*, Tel.

Coloration. Male. A broad supercilium from the nostril to the end of the ear-coverts, the whole head, neck, upper plumage, and lesser and median wing-coverts black ; greater wing-coverts almost entirely white ; quills black, edged with white, the edging on some of the secondaries extending to half the outer web ; the middle four pairs of tail-feathers black, narrowly margined with white, the other two pairs white, with a small portion of the inner web black ; breast and lower plumage white, the sides of the breast and body infuscated.

Fig. 80.—Head of *M. maderaspatensis.*

Female. Appears to resemble the male, but the upper plumage is more or less tinged with grey.

I cannot discover that there is any seasonal change of plumage in this species.

The young bird has the same pattern of colour as the adult, but the black is everywhere replaced by greyish brown, the supercilium in front of the eye is not indicated, and the white parts of the plumage are tinged with fulvous. In the first spring the change to adult plumage first takes place by the assumption of some black feathers on the head, and the full plumage is entirely assumed at the succeeding autumn moult. Some males retain a few white feathers on the chin and throat to a late period.

Iris dark brown ; legs, feet, and bill black (*Butler*).

Length about 9 ; tail 4·3 ; wing 3·9 ; tarsus 1 ; bill from gape ·8 ; the female is considerably smaller than the male, of which the above are average dimensions.

Distribution. A permanent resident throughout India, from Kashmir and Sind on the west to Sikhim and Western Bengal on the east, and from the lower ranges of the Himalayas down to Ceylon.

Habits, &c. Breeds throughout the plains of India, and also in the Himalayas up to 3000 feet, as well as in the hills of Southern India, up to 7000 or 8000 feet, constructing a nest of grass and various other materials in holes of banks, under stones, amongst the timbers of bridges, on roofs, and in other similar localities. The nest is merely a pad, or sometimes cup-shaped. The eggs, four in number, are dull white or pale greenish marked with brown of various shades, and measure about ·9 by ·66.

832. Motacilla melanope. *The Gray Wagtail.*

Motacilla melanope, *Pall. Reis. Russ. Reichs.* iii, p. 696, (1776);
Legge, Birds Ceyl. p. 610; *Sharpe, Cat. B. M.* x. p. 497; *Oates in Hume's N. & E.* 2nd ed. ii, p. 207.
Motacilla boarula, *Linn. Mant.* p. 527 (1771); *Blyth, Cat.* p. 137.
Motacilla sulphurea, *Bechst. Naturg. Deutschl.* iii, p. 459 (1807).
Calobates sulphurea (*Bechst.*), *Horsf. & M. Cat.* i, p. 349; *Jerd. B. I.* ii, p. 220; *Hume & Henders. Lah. to Yark.* p. 224.
Calobates boarula (*Penn.*), *apud Hume, N. & E.* p. 381.
Calobates melanope (*Pall.*), *Hume, Cat.* no. 592; *Oates, B. B.* i, p. 159; *Barnes, Birds Bom.* p. 257.

The Gray and Yellow Wagtail, Jerd.; Mudi-tippudu-jitta, Tel.

Coloration. Male. After the autumn moult the whole crown and sides of the head, the hind neck, back, scapulars, and lesser wing-coverts are bluish grey, tinged with green; the rump and upper tail-coverts yellowish green; the three middle pairs of tail-feathers black, margined with greenish; the next two pairs white, with the greater portion of the outer webs black; outermost pair entirely white; median and greater coverts and quills dark brown, edged with yellowish white; a narrow dull white supercilium from the lores to the end of the ear-coverts; chin, throat, and fore neck white; remaining lower plumage bright yellow, rather deeper on the vent and under tail-coverts.

In the spring the lores become dark brown, and the supercilium becomes much broader and clearer; the lower eyelid is clothed with whitish feathers, and the ear-coverts are dark slaty; there is a very broad white moustachial streak extending down the sides of the neck; and the chin, throat, and part of the fore neck are black, with small white edges to the feathers; as the summer passes, these edges become worn but seldom entirely disappear.

Female. After the autumn moult resembles the male; and in the spring merely acquires two rows of dark brown spots, one on each side of the chin, throat, and fore neck, the two sometimes meeting like a gorget on the upper breast; the colour of the lower plumage is less brilliant than that of the male.

The young bird resembles the adult in winter plumage, but the white parts are strongly tinged with buff.

Bill horn-colour, paler at the base of the lower mandible; iris brown; legs flesh-colour.

Length about 7·5; tail 3·7; wing 3·2; tarsus ·75; bill from gape ·7; hind toe and claw ·55.

Distribution. A winter visitor to every portion of the Empire, retiring in summer to those parts of the Himalayas which are above 6000 feet elevation, where a few birds of this species may also be found in winter. In the winter this Wagtail extends down to Malayana. In summer it has a very large range over the greater part of Central and Northern Asia, and it is found in Europe.

Habits, &c. Breeds in Kashmir above 6000 feet, and in Afghanistan in May and June, making a nest of moss and fibres under large stones, or sometimes in a bush. The eggs, usually five in number, are yellowish or brownish white, closely marked with yellowish brown, and measure about ·7 by ·54.

833. Motacilla borealis *. *The Grey-headed Wagtail.*

Motacilla flava borealis, *Sundev. Œfv. K. Vet.-Acad. Förh. Stockh.* 1840, p. 53.
Budytes viridis (*Gmel.*), *apud Blyth, Cat.* p. 138; *Horsf. & M. Cat.* i, p. 350; *Jerd. B. I.* ii, p. 222; *Anders. Yunnan Exped., Aves,* p. 608; *Legge, Birds Ceyl.* p. 617; *Oates, B. B.* i, p. 161.
Budytes cinereocapilla (*Savi*), *apud Hume, Cat.* no. 583; *Brooks, J. A. S. B.* xliii, pt. ii, p. 248; *Barnes, Birds Bom.* p. 238.
Motacilla borealis, *Sundev. Sharpe, Cat. B. M.* x. p. 522, pl. vii, figs. 1–3.

The Indian Field-Wagtail, Jerd.; *Pilkya,* Hind.

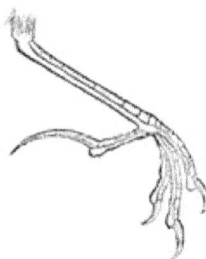

Fig. 81.—Foot of *M. borealis.*

Coloration. Male. In normal winter plumage, on first arrival in India, the forehead, crown, nape, and hind neck are bluish grey, a few of the feathers with greenish tips; the back, scapulars, and rump dull olive-green; upper tail-coverts dark brown edged with olive-green: the four middle pairs of tail-feathers black, narrowly edged with olivaceous; the two outer pairs almost wholly white: coverts and quills dark brown or black, margined with pale fulvous, sometimes with a greenish tinge: lores, cheeks, and ear-coverts dark slaty black, the ear-coverts paling to bluish grey posteriorly: the whole lower plumage yellow, tinged with ochraceous across the breast, and the feathers of that part with dark bases showing through, and giving the breast a mottled appearance; traces of a white interrupted supercilium are frequently visible over the lores and ear-coverts, but these traces are quite absent in most birds.

---

* I agree with Sharpe that Brown's figure of the Green Wagtail is quite unrecognizable, and that consequently Gmelin's name of *viridis,* founded on this figure, must be rejected.

As the winter passes the upper plumage becomes worn and browner in colour, and the black bases of the breast-feathers larger and more distinct.

After the spring moult the forehead, crown, nape, and hind neck are dark slaty grey; the back, scapulars, and rump yellowish green; the upper tail-coverts dark brown, with greenish margins; the wings and tail as in winter, but with the margins of the feathers of the former decidedly yellow; lores, cheeks, round the eye, and the ear-coverts black; traces of a narrow supercilium sometimes present; the whole lower plumage very bright yellow, with concealed black bases to the feathers of the breast; these bases become more conspicuous as the summer passes.

Female. In winter does not differ from the male; in summer it has the upper green parts duller, the crown and nape very faintly tinged with slaty, and almost concolorous with the green back, the lower plumage less brilliant yellow, with the mottlings on the breast more developed, the lores, cheeks, and ear-coverts brown, not black, and the supercilium generally slightly developed and of a pale fulvous colour. As the summer goes on, the head becomes greyer owing to the green tips wearing away.

The young bird on first arrival in India has the entire upper plumage greyish brown, tinged with blue on the rump; the upper tail-coverts black, edged with grey; tail and wings as in the adult, but with the margins of the wing-feathers very pale and almost white; a very broad and nearly white supercilium; lores and ear-coverts greyish brown; lower plumage white, with a row of brown spots on either side of the throat, meeting and forming a gorget across the breast.

During the winter a series of changes are undergone, tending to make the young resemble the adult, and the full plumage appears to be assumed by the first spring.

Iris brown; bill blackish brown, the base of the lower mandible yellowish; legs, feet, and claws dark horn-colour.

Length about 7; tail 3·1; wing 3·2; tarsus ·9; bill from gape ·7.

Distribution. A winter visitor to every portion of the Empire except the higher parts of the Himalayas, where this species is only found on migration. It occurs in winter in the Malay Peninsula. In summer it ranges to Northern Siberia, and it is also found according to season over a considerable portion of Europe and Africa.

The true *M. cinereicapilla*, a closely allied species, is confined to Southern Europe and portions of Africa.

834. Motacilla flava. *The Blue-headed Wagtail.*

Motacilla flava, *Linn. Syst. Nat.* i, p. 331 (1766); *Sharpe, Cat. B. M.* x, p. 516, pl. vi, figs. 3-5.
Budytes flava (*Linn.*), *Hume, Cat.* no. 593 ter (part.); *Oates, B. B.* i, p. 162.
Budytes beema (*Sykes*), *Hume, S. F.* x, p. 227 note.

Coloration. Resembles *M. borealis* very closely in general appearance at all seasons and with regard to both sexes, but may be recognized:—the male by having the crown darker slaty blue, the ear-coverts streaked with white, the chin and a moustachial streak bordering the nearly black cheeks white, and by the presence of a large and pure white supercilium extending from the nostril to the nape:—the female by the darker green of the upper plumage, the presence of white streaks on the ear-coverts, and the large distinct white or pale fulvous supercilium.

The dimensions and the colour of the soft parts are the same as those of *M. borealis.*

An allied species, *M. tairana,* Swinhoe, from China and the Malay peninsula, is extremely likely to be found in Burma. In this species the crown is green, concolorous with the back, and the supercilium, which is very broad and distinct, is bright yellow. A specimen of a Wagtail in the Hume Collection killed at Howrah, Calcutta, would appear to belong to this species, but unless supported by other specimens it would be premature to pronounce it such.

Distribution. Occurs in winter in the eastern portion of the Empire. I have examined unmistakable specimens of this species procured at various localities ranging from Shillong on the Khási hills to the extreme south of Tenasserim. I have also seen it from the Andamans and Nicobars. Hume (S. F., xi, p. 232) gives this species from Cachar and Dibrugarh in Assam. This Wagtail in winter is found in China and the countries to the south and in summer in Siberia. A precisely similar bird is found in Europe and Africa, but it is probable that the two colonies meet on common ground in Northern Asia in summer and take two different routes in the autumn when proceeding to their winter-quarters.

835. **Motacilla beema.** *The Indian Blue-headed Wagtail.*

Motacilla beema, *Sykes, P. Z. S.* 1832, p. 90; *Sharpe, Cat. B. M. x,* p. 521, pl. vi, fig. 6.
Budytes dubius vel anthoides, *Hodgs. in Gray's Zool. Misc.* p. 83 (1844).
Budytes flava (*Linn.*), *Brooks, J. A. S. B.* xliii, pt. ii, p. 248; *Hume, Cat.* no. 593 ter (pt.); *Barnes, Birds Bom.* p. 239.
Budytes dubius, *Hodgs., Brooks, S. F.* vii, p. 130 (1878).

Coloration. Resembles *M. flava* at all seasons and in both sexes, but may be distinguished from that species by the colour of the cheeks, and of the lower half of the ear-coverts, which are white and not dark slaty blue or black, and further by the extreme paleness and purity of the bluish grey of the forehead, crown, and nape.

The dimensions and colour of the soft parts in this species are the same as those of *M. flava.*

Distribution. A winter visitor to the plains of India as far south as Belgaum, ranging from Sind to the longitude of Calcutta. I have examined specimens of this Wagtail from Calcutta itself.

but from no point further east. This species is found on the lower
ranges of the Himalayas in winter and it extends at that season to
Afghanistan. It passes through Kashmir on migration and summers in Central Asia and Southern Siberia, but does not apparently
extend so far north at this season as to meet with *M. flava*.

836. Motacilla feldeggi. *The Black-headed Wagtail.*

Motacilla melanocephala, *Licht. Verz. Doubl.* p. 36 (1823), *nec
Gmelin.*
Motacilla feldeggi, *Michah., Isis,* 1830, p. 814; *Sharpe, Cat. B. M.*
x, p. 527, pl. viii, figs. 1–4.
Budytes melanocephala (*Sykes*), *Blyth, Cat.* p. 138; *Horsf. & M.
Cat. i,* p. 351.
Budytes melanocephala (*Bonap.*), *Brooks, J. A. S. B.* xliii, pt. ii,
p. 218.
Budytes melanocephala (*Licht.*), *Hume, Cat.* no. 593 bis; *Barnes,
Birds Bom.* p. 289.

Coloration. Resembles *M. borealis* in general style of coloration,
but in summer plumage and in both sexes the forehead, crown,
nape, hind neck, lores, cheeks, and ear-coverts are deep black. The
chin and a band below the cheek are usually, but not always,
white. There is occasionally a slight trace of a white supercilium.

In winter plumage the two species are very much alike, but the
crown in *M. feldeggi* is generally concolorous with the back and
not bluish grey, and in the male there is always a certain duski-
ness about the coronal region, and not unfrequently a few black
feathers which suffice to indicate the species.

The young of both species appear to resemble each other closely,
but it appears that in *M. feldeggi* there is seldom or never a well-
marked supercilium as in the young of *M. borealis, M. flava,* and
M. beema.

The dimensions and colours of the soft parts are the same as in
M. borealis.

Owing to Lichtenstein having bestowed on this bird a name
previously given by Gmelin to a very different species it seems
advisable to follow Sharpe and discard the term *melanocephala,* notwithstanding that Gmelin's name has been shown to be applicable
to *Sylvia melanocephala.* The greatest confusion has prevailed
regarding these Yellow Wagtails and the adoption of an unused
name, from which a new start may be taken for any one Wagtail,
is to be hailed with satisfaction.

Distribution. A winter visitor to the plains of India as far south
as Belgaum, and east to Benares. This species passes through
Kashmir on migration and summers in Central Asia. It extends
to Southern Europe and North-east Africa.

M. campestris, Pall. (*Budytes rayi,* auct.), has been recorded
doubtfully from India, but I cannot find any trustworthy instance
of its occurrence within Indian limits and I accordingly exclude it.

This bird much resembles *M. citreola*, but is much yellower above, the entire head seldom becomes uniform yellow as in that species, and there is never any black band on the mantle.

837. Motacilla citreola. *The Yellow-headed Wagtail.*

Motacilla citreola, *Pall. Reis. Russ. Reichs,* iii, p. 696 (1776);
   Sharpe, Cat. B. M. x, p. 503.
Budytes citreola (*Pall.*), *Blyth, Cat.* p. 138; *Horsf. & M. Cat.* i,
   p. 352 (pt.); *Brooks, J. A. S. B.* xli, pt. ii, p. 82; *Scully, S. F.* iv,
   p. 151; *Hume, Cat.* no. 594 bis; *Biddulph, Ibis,* 1881, p. 69;
   Scully, Ibis, 1881, p. 452; *Barnes, Birds Bom.* p. 241.
? Budytes calcarata, *Hodgs., As. Res.* xix, p. 190 (1836); *Hume,
   S. F.* vii, p. 401.

Coloration. In normal winter plumage the sexes are alike. The forehead, a broad supercilium, the sides of the head, and the whole lower plumage are yellow, the ear-coverts more or less streaked with dusky and with a blackish line bordering them below; lores dusky; the sides of the throat and breast with more or less concealed black bases to the feathers; the upper plumage ashy grey, tinged with green on the head; the wings dark brown, the coverts and tertiaries very broadly edged with white, the other quills narrowly; tail black, the two outer pairs of feathers nearly entirely white.

In summer plumage the sexes are also alike except that the female is slightly paler: the entire head and neck become deep yellow, the feathers of the crown with tiny black tips which soon wear off; entire lower plumage the same deep yellow; the yellow on the hind neck is bounded by a broad black band and the upper plumage is tinged with bluish ashy; the other parts remain unchanged.

The young bird is ashy brown above with very narrow paler margins to the feathers; a very broad pale fulvous supercilium, bordered above, on the sides of the crown, by a black band; lores and ear-coverts streaked with dusky and fulvous; the lower plumage a very pale fulvous with a gorget of black spots down the sides of the throat and across the upper breast; wings and tail as in the adult. The young bird probably assumes the adult plumage at the first spring moult.

Bill, legs, and claws black; iris dark brown (*Bingham*).

Length about 7; tail 3·3; wing 3·5; tarsus 1·05; bill from gape ·8; hind claw ·5.

Distribution. A winter visitor to the greater portion of the Empire, descending in India to about the latitude of Belgaum and in the eastern part of the Empire as far as Manipur. This bird summers in Central and Northern Asia, and it also occurs in Europe.

838. **Motacilla citreoloides.** *Hodgson's Yellow-headed Wagtail.*

Budytes citreoloides, *Hodgs., Gould, Birds As.* iv, pl. 64 (1865) ; *Hume & Henders. Lah. to Yark.* p. 224.

Budytes citreola (*Pall.*), *Horsf. & M. Cat.* i, p. 352 (pt.) ; *Jerd. B. I.* ii, p. 225.

Budytes calcaratus (*Hodgs.*), *Brooks, J. A. S. B.* xli, pt. ii, p. 82 ; *Stoliczka, J. A. S. B.* xli, pt. ii, p. 244 ; *Hume, N. & E.* p. 382 ; *Anders. Yunnan Exped., Aves,* p. 609 : *Hume, Cat.* no. 594 ; *Biddulph, Ibis,* 1881, p. 69 ; *Scully, Ibis,* 1881, p. 452 ; *Oates, B. B.* i, p. 163 ; *Barnes, Birds Bom.* p. 240.

Motacilla citreoloides (*Hodgs.*), *Sharpe, Cat. B. M.* x, p. 507 ; *Oates in Hume's N. & E.* 2nd ed. ii, p. 208.

The Yellow-headed Wagtail, Jerd. ; *Pani-ka-pilkya,* Hind.

Coloration. Resembles *M. citreola,* but in summer has the entire back, scapulars, rump, and upper tail-coverts deep black. In the winter months the two species are very close to each other, but *M. citreoloides* has generally a few black or dusky feathers on the upper plumage, by which it may be infallibly recognized. It is pretty certain that the two sexes of this species are alike in plumage, but reliably sexed females are very scarce in collections.

The young do not appear to differ in any respect from those of *M. citreola.*

The dimensions and colours of the soft parts are the same as in *M. citreola.*

This species and the preceding cannot always be separated from each other, but they can always, even in their youngest stage, be separated from the other yellow Wagtails by the greater length of the tarsus.

Both this species and *M. citreola* occur in Nepal and Hodgson procured a large series of the two. It is not clear to me to which species he assigned the names of *calcaratus* and *citreoloides,* but Gould identified the latter name with the present species and figured the bird. It is therefore convenient to discard *calcaratus* altogether as being unnecessary and adopt *citreoloides* of Hodgson, *apud* 'Gould. Birds of Asia.'

Distribution. A winter visitor to the southern slopes of the Himalayas from Kashmir to Assam, extending on the west to Afghanistan. This species also visits the plains of India, and I have examined specimens procured at Sámbhar, Etáwah, and Calcutta, but from no point further south. Stoliczka, however, records it from Cutch, and Dr. Fairbank from Khandála. On the east its range extends from Assam down to Northern Tenasserim and Pegu. A few birds of this species apparently breed in Kashmir, but the majority retire to Central Asia.

Habits, &c. Little is known of the nidification of this Wagtail. Theobald found the nest in Kashmir in May, in a depression in soft earth beneath a rock, with four eggs, which were pale grey marked with greyish brown and greyish neutral tint, and measured about ·95 by ·7.

Genus LIMONIDROMUS, Gould, 1862.

The genus *Limonidromus* contains one species of Wagtail some-what resembling the Pied Wagtails in colour, but the whole upper plumage is suffused with green. The structure of the tail in this genus is peculiar, inasmuch as the middle pair of feathers is very markedly shorter than the others and of a different colour. The sexes are quite alike.

The Forest-Wagtail is found in well-wooded parts of the country and frequently runs about under the shade of trees. On being disturbed it has the habit of perching on a branch. It wags its tail incessantly and does not differ from the other Wagtails in its general habits.

839. Limonidromus indicus. *The Forest-Wagtail.*

Motacilla indica, *Gmel. Syst. Nat.* i, p. 962 (1788).
Nemoricola indica (*Gm.*), *Blyth, Cat.* p. 136; *Horsf. & M. Cat.* i, p. 353; *Jerd. B. I.* ii, p. 226; *Stoliczka, J. A. S. B.* xxxvii, pt. ii, p. 48.
Limonidromus indicus (*Gm.*), *Hume & Dav. S. F.* vi, p. 364; *Legge, Birds Ceyl.* p. 614; *Hume, Cat.* no. 595; *Sharpe, Cat. B. M.* x, p. 532; *Oates, B. B.* i, p. 164; *Barnes, Birds Bom.* p. 241.

The Black-breasted Wagtail, Jerd.; *Uzlallu-jitta*, Tel.

Fig. 82.—Head of *L. indicus.*

Coloration. Plumage above dull olive-green, the tail-coverts dark brown or sometimes black; ear-coverts and lesser wing-coverts also olive-green; a supercilium from the bill over the eye to the nape, the cheeks, chin, throat, and all the lower plumage yellowish white; two black bands across the breast, the upper one entire, the lower one broken in the middle; median and greater wing-coverts black, with broad yellowish-white tips forming two bands across the coverts; quills brown, the second to the seventh primaries with a patch of yellowish white on the outer web near the base; all the primaries and secondaries with an abrupt margin of yellowish white near the tip on the outer web; tertiaries brown, broadly tipped with olive-green; middle pair of tail-feathers similar to the back; the next three pairs dark brown; the next pair brown with a large white tip; and the outer pair all white, except at their base, where they are brown.

Irides nearly black; upper mandible dusky brown, lower mandible fleshy white; legs and feet purplish white; claws horny white (*Armstrong*).

Length about 6·5; tail 2·7; wing 3·1; tarsus ·85; bill from gape ·75; hind toe and claw ·6.

Distribution. A somewhat rare winter visitor to all parts of the Empire except the portion lying west of a line drawn roughly from the Sutlej valley to the head of the Gulf of Cambay. This species extends to Ceylon and is also found in the Andamans. In winter it ranges down the Malay peninsula to the larger islands and eastward to Siam and Southern China. In summer this Wagtail retires to Northern China and Eastern Siberia.

Genus ANTHUS, Bechst., 1807.

The genus *Anthus* contains the Pipits, which may be recognized by their streaked upper plumage and comparatively short tail.

The Pipits are found over nearly the entire world. They resemble each other greatly in their pattern of colour, and consequently they are difficult to describe, and in fact long descriptions of them are useless, their identification depending entirely on a few characters which are easily learnt. Each species is very constant to one type. The young are very much spotted beneath and these spots become reduced in size and number at each successive spring moult and in a few species disappear altogether, the lower plumage in these adults becoming uniform. The difference between summer and winter plumage in the Pipits is very slight, and in my opinion it is useless attempting to treat the two plumages as distinct, although some authors do so. I have, therefore, only described the bird in its fresh autumn plumage. In summer this plumage becomes abraded and faded, and the black marks on the upper plumage more pronounced. Those Pipits which have bright colours about the head and breast assume this colour slowly and permanently and not seasonally.

The Pipits frequent the ground, but a few species occasionally perch on trees and even run along the larger boughs in pursuit of insects. They build their nests on the ground and lay eggs which, like those of the Wagtails, are much spotted with brown. The sexes are invariably alike.

Key to the Species.

a. Hind claw not exceeding hind toe in length.
  a'. Pale tip of inner web of penultimate tail-feather very small, less than a quarter the length of feather; next feather never tipped.
    a''. Streaks on lower plumage large and black, well defined; light parts of tail-feathers white.
      a'''. Upper plumage brown with very large streaks; supercilium fulvous throughout .... *A. trivialis*, p. 302.

840. **Anthus trivialis.**　*The Tree-Pipit.*

Alauda trivialis, *Linn. Syst. Nat.* i, p. 288 (1766).
Alauda plumata, *P. L. S. Müll. Natursyst. Suppl.* p. 137 (1776).
Anthus arboreus, *Bechst. Naturg. Deutschl.* iii, p. 706 (1807); *Horsf.
& M. Cat.* i, p. 354.
Anthus agilis, *Sykes, P. Z. S.* 1832, p. 91; *Brooks, S. F.* iv, p. 278.
Dendronanthus trivialis (*Linn.*), *Blyth, Cat.* p. 135.

Pipastes arboreus (*Bechst.*), *Jerd. B. I.* ii, p. 229 ; *Hume & Henders. Lah. to Yark.* p. 226.
Pipastes plumatus (*Müll.*), *Hume, N. & E.* p. 383.
Anthus trivialis (*Linn.*), *Hume, Cat.* no. 597 ; *Sharpe, Cat. B. M.* x, p. 543; *Barnes, Birds Bom.* p. 242; *Oates in Hume's N. & E.* 2nd ed. ii, p. 208.
Pipastes trivialis (*Linn.*), *Oates, B. B.* i, p. 172.

The European Tree-Pipit, Jerd.

Figs. 83, 84.—Head and foot of *A. trivialis*.

Coloration. Upper plumage sandy brown with large black streaks or centres to the feathers except the rump and upper tail-coverts which are very faintly marked ; coverts and quills of wing dark brown margined with pale fulvous; tail dark brown with narrow pale margins, the outermost feather about half white, the white and brown meeting diagonally ; the penultimate feather with a small white tip ; a pale fulvous supercilium ; sides of the head mixed brown and fulvous ; lower plumage white tinged with fulvous ; a short black moustachial streak ; the whole breast and the sides of the throat with large, well-defined black streaks ; the sides of the body tinged with olivaceous and indistinctly streaked.

The young bird after the autumn moult resembles the adult, but is tinged with bright fulvous, especially on the throat and breast.

Iris dark brown ; legs and feet flesh-colour ; bill dark brown above, pale brown below tipped with dusky.

Length about 6·5 ; tail 2·7 ; wing 3·5 ; tarsus ·85 ; bill from gape ·6 ; hind claw about ·3.

Distribution. A winter visitor to the western portions of the Empire. Judging from the specimens I have examined this species is found as far south as Belgaum and as far east as Manbhoom in Chutia Nágpur. Many years ago Hume identified a Pipit I sent to him from Pegu as this species, but I remember that the skin was a very bad one and I think it not improbable that some mistake was made regarding it. This bird winters in South-western Asia and in Africa, and summers in Europe and Northern Asia. A few birds of this species appear to breed in the Himalayas.

Habits, &c. A nest supposed to belong to this Pipit was a circular, shallow saucer, made of grass, lined with a few hairs and placed on the ground at the foot of a tuft of grass. It was found at Kotgarh in May and contained three eggs, which were greyish white, marked with dull purple and purplish brown, and measured about ·85 by ·63.

Anthus pratensis (Linn.), the Meadow-Pipit, is not unlikely to be found in the north-western parts of the Empire. It bears a close resemblance to *A. trivialis*, but may be recognized at once by its long hind claw.

841. Anthus maculatus. *The Indian Tree-Pipit.*

Anthus maculatus[*], *Hodgs. in Gray's Zool. Misc.* p. 83 (1844);
  A. Anderson, S. F. iii, p. 353; *Hume, Cat.* no. 593; *Sharpe, Cat.*
  B. M. x, p. 547; *Barnes, Birds Bom.* p. 242; *Oates in Hume's*
  N. & E. 2nd ed. ii, p. 203.
Dendronanthus maculatus (*Hodgs.*), *Blyth, Cat.* p. 135.
Anthus agilis, *Sykes, Horsf. & M. Cat.* i, p. 354 (*nec Sykes*).
Pipastes agilis (*Sykes*), *Jerd. B. I.* ii, p. 228; *Hume, N. & E.*
  p. 382.
Pipastes maculatus (*Hodgs.*), *Brooks, S. F.* iii, p. 279; *Anders.*
  Yunnan Exped., Aves, p. 608; *Scully, S. F.* viii, p. 316; *Oates,*
  B. B. i, p. 171.

Liku-jitta, Tel.

Coloration. Resembles *A. trivialis*, but has the whole upper plumage strongly suffused with green, and the streaks and centres to the feathers much narrower and less well-defined; the supercilium is pale fulvous anteriorly and white posteriorly.

In the summer the green tinge is much subdued, and the supercilium becomes very white and distinct.

Bill bluish black, darker above, and yellowish at base of the lower mandible; iris brown; legs and feet flesh-colour.

Generally smaller than *A. trivialis*, the wing being seldom so much as 3·3, and frequently under 3·2; tail 2·5; tarsus ·85.

Distribution. A winter visitor to the plains of the greater portion of the Empire, being found as far west as the Sutlej valley in the Himalayas and Rajputana and Guzerat in the plains. This species appears to breed in considerable numbers on the higher ranges of the Himalayas. To the south this Pipit extends down to the Palni hills, and probably to Cape Comorin. Its winter range extends to China and Cochin China. In summer it is found in Siberia, North China, and Japan.

Habits, &c. Breeds in the higher parts of the Himalayas (7000 to 12,000 feet) from May to July, constructing a nest of moss or grass on the ground under the shelter of a tussock of grass, and laying four eggs, which are thickly marked with dark brown and dingy purple, and appear to measure about ·93 by ·68.

This species and the preceding frequent gardens and localities which are well wooded, feeding on the ground and flying up to a branch when disturbed. They are somewhat social in the winter. The males of both are fine songsters in the breeding-season.

---

[*] Hodgson never described this species himself, and I should reject this name were a prior one available, which there is not.

ANTHUS. 305

842. **Anthus nilgiriensis.** *The Nilgiri Pipit.*

Anthus rufescens*, Jerd. Madr. Journ. L. S. xi, p. 34 (1840).
Anthus montanus*, Jerd., Blyth, J. A. S. B. xvi, p. 435 (1847); id.
  Cat. p. 136; Hume, S. F. vii, p. 461; id. Cat. no. 598; Davison,
  S. F. x, p. 397.
Pipastes montanus (Jerd.), Jerd. B. I. ii, p. 230; Hume, N. & E.
  p. 383; Fairbank, S. F. v, p. 407.
Anthus nilghiriensis, Sharpe, Cat. B. M. x, p. 559 (1885); Oates in
  Hume's N. & E. 2nd ed. ii, p. 211.

The Hill Tree-Pipit, Jerd.

Coloration. Upper plumage fulvous, tinged with olivaceous, each
feather broadly streaked or centred with black, except on the rump,
where the marks are brown and less distinct; wings blackish margined with fulvous; tail black, edged with olivaceous, the two
outer pairs of feathers dull white on the terminal half or third of
their length, the next pair with a dull white tip; a light rufous
supercilium; lores brown; sides of head mixed rufous and brown;
lower plumage tawny fulvous, the sides of the neck, the whole
breast, and the sides of the body with short, narrow, but very
distinct and well-defined black streaks.

Upper mandible dull black, apical half of lower mandible dark
fleshy; iris deep brown; legs and feet fleshy; claws pale brown
(*Davison*).

Length nearly 7; tail 2·6; wing 3·4; tarsus 1; bill from gape
·7; hind claw about ·35.

Distribution. The higher parts of the Nilgiri and Palni hills in
Southern India, where this Pipit is a permanent resident.

Habits, &c. Breeds on the Nilgiris above 6000 feet in May,
making a nest of grass under shelter of a tuft or bush. The eggs
are greenish brown mottled with a darker shade, and measure about
·85 by ·6.

843. **Anthus cockburniæ.** *The Rufous Rock-Pipit.*

Anthus similis, Jerd. Ill. Ind. Orn. pl. 45 (1847), nec Madr. Journ.
  L. S. (1840).
Agrodroma cinnamomea (Rüpp.), apud Jerd. B. I. ii, p. 235.
Agrodroma similis, Jerd., Hume, N. & E. p. 385; Fairbank, S. F.
  iv, p. 260; Hume, Cat. no. 603; Davison, S. F. x, p. 397; Barnes,
  Birds Bom. p. 246.
Anthus sordidus, Rüpp., apud Sharpe, Cat. B. M. x, p. 560; Oates
  in Hume's N. & E. 2nd ed. ii, p. 212.

Coloration. Upper plumage dark brown, the feathers narrowly
margined with fulvous; wings brown, broadly edged with bright
fulvous; tail black margined with fulvous, the outermost feather

---

* Neither of the names given by Jerdon to this species can stand, as the first
had previously been applied by Temminck, and the second by Koch, to other
species of Pipits.

VOL. II.   X

with the outer web and the terminal half of the inner pale rufous;
the penultimate feathers tipped with pale rufous; supercilium and
lower plumage fulvous or sandy buff; a narrow black moustachial
streak; the breast with small, narrow, but very distinct triangular
brown streaks.

Iris wood-brown; upper mandible black; lower mandible fleshy,
with the tip blackish; tarsus reddish fleshy, the feet darker; claws
dark reddish brown; gape yellow (*Davison*).

Length about 8; tail 3·1; wing 3·7; tarsus 1·1; bill from gape
·9; hind claw ·35.

There has been much confusion regarding the name of this Pipit.
It has been identified with two names of Rüppell's, but wrongly
so. Jerdon figured it as *A. similis* in his 'Illustrations,' and in
the accompanying letterpress confounded it with that species. As
there is, so far as I can ascertain, no specific term that applies to
the present species, I have much pleasure in connecting this fine
Pipit with the name of Miss Cockburn, a lady who has for so many
years successfully worked the Nilgiri hills, and whose specimens
enrich the Hume Collection.

Distribution. A permanent resident in the Nilgiri hills, on the
higher portions of which this Pipit appears to be not uncommon,
frequenting grassy land and occasionally perching on bushes when
disturbed. This species appears to extend some distance north, as
the Hume Collection contains a specimen obtained at Ahmednagar
by Dr. Fairbank.

Habits, &c. A nest of this species was found in the Nilgiris by
Miss Cockburn in March. It was placed under a shelving rock,
and was composed of fine grass. The eggs are creamy white densely
speckled with yellowish brown and purplish grey, and measure
about ·85 by ·65.

814. Anthus similis. *The Brown Rock-Pipit.*

Agrodroma similis, *Jerd. Madr. Journ. L. S.* xi, p. 35 (1840).
Anthus similis (*Jerd.*), *Blyth, Cat.* p. 135 (pt.); *Horsf. & M. Cat.* i,
    p. 356.
Agrodroma sordida (*Rüpp.*), *apud Jerd. B. I.* ii, p. 238; *Hume,
    J. A. S. B.* xxxix, pt. ii, p. 119; *Hume, Cat.* no. 604; *Barnes,
    Birds Bom.* p. 246.
Agrodroma jerdoni, *Finsch, Trans. Z. S.* vii, p. 241 (1870); *Hume
    & Henders. Lah. to Yark.* p. 227, pl. xxi; *Hume, N. & E.* p. 386;
    id. *S. F.* i, p. 203; *Butler, S. F.* iii, p. 491.
Corydalla grisescrnfescens, *Hume, Ibis,* 1870, pp. 286, 400.
Anthus jerdoni (*Finsch*), *Sharpe, Cat. B. M.* x, p. 562; *Oates in
    Hume's N. & E.* 2nd ed. ii, p. 212.

Coloration. Upper plumage ashy brown, the feathers narrowly
edged with fulvous, and with dark shaft-streaks; wings brown,
broadly edged with bright fulvous; tail brown or black margined
with fulvous, the outermost feather with the outer web and the
terminal half of the inner pale rufous; the penultimate pair of
feathers tipped with pale rufous; supercilium and lower plumage

fulvous or sandy buff, the breast with a few ill-defined and pale brown streaks.

Iris dark brown; legs and feet yellowish flesh; bill dark brown above, flesh-colour below (*Hume Coll.*).

Length about 8·5; tail 3·6; wing 4·1; tarsus 1·1; bill from gape ·9; hind claw about ·45.

This species very closely resembles *A. sordidus*, Rüpp., from Palestine, Arabia, and Africa, but is much larger and more brightly coloured than that bird.

Agrodroma similis of Jerdon was founded on a single specimen, the locality of which was not mentioned in the original description, but was subsequently, in the 'Illustrations of Indian Ornithology,' stated to be Jálna. In the 'Birds of India' the Jálna bird was referred to *A. sordida*, the name there adopted for the present species.

Distribution. A winter visitor to the plains of the north-west of India, extending to the east as far as the Sikhim terai and Mughal Sarai, and to the south as far as Khandesh, Jálna, and Nágpur. This Pipit retires in summer to the Himalayas, where it breeds from Hazara to Sikhim, up to about 6000 feet elevation. The range of this bird extends to Afghanistan, Baluchistan, and Persia.

Habits, &c. Breeds at Murree and in Afghanistan, below 6000 feet, from May to July, making a rough nest of grass on a hill-side, and laying four eggs, which are brownish or greyish white marked with brown and purple, and measure about ·85 by ·63.

845. Anthus richardi. *Richard's Pipit.*

Anthus richardi, *Vieill. Nouv. Dict. d'Hist. Nat.* xxvi, p. 491 (1818);
  Blyth, Cat. p. 135; *Horsf. & M. Cat.* i, p. 355; *Sharpe, Cat. B. M.* x, p. 564.
Corydalla richardi (*Vieill.*), *Jerd. B. I.* ii, p. 234; *Brooks, S. F.* i,
  p. 358; *Anders. Yunnan Exped., Aves,* p. 606; *Hume, Cat.* no. 589;
  Legge, Birds Ceyl. p. 624; *Oates, B. B.* i, p. 166.

The Large Marsh-Pipit, Jerd.; *Palla puraki, Meta kále,* Tam.

Fig. 85.—Foot of *A. richardi.*

Coloration. Upper plumage fulvous-brown, the feathers centred with blackish, the rump more uniform; wings dark brown mar-

x 2

gined with fulvous; tail dark brown, with pale margins, the outermost feather almost entirely white, the penultimate with an oblique portion of the inner web, about an inch and a half in length, also white; superciliuma and lower plumage pale fulvous, the sides of the throat and fore neck and the whole breast streaked with dark brown; sides of the body darker fulvous, with a few indistinct streaks.

Bill brown, yellowish at the base of lower mandible; mouth yellow; iris brown; legs flesh-colour; the claws darker.

Length about 7·5; tail 3·4; wing 3·7; tarsus 1·2; bill from gape ·85; hind claw about ·8.

Distribution. A winter visitor to the whole of the Eastern part of the Empire from Assam to Tenasserim, extending along the Himalayas as far west as the Sutlej valley, and southwards through Bengal and Chutia Nágpur along the eastern side of India to Ceylon. In winter this species is found in China on the one side and in Europe on the other. It summers in Central and Northern Asia.

846. **Anthus striolatus.** *Blyth's Pipit.*

Cichlops thermophilus, *Hodgs. in Gray's Zool. Misc.* p. 83 (1844, desc. nullá).
Anthus striolatus, *Blyth, J. A. S. B.* xvi. p. 435 (1847); *id. Cat.* p. 136; *Blanf. J. A. S. B.* xli, pt. ii, p. 61; *Sharpe, Cat. B. M.* x, p. 568.
Anthus thermophilus (*Hodgs.*), *Horsf. & M. Cat.* i, p. 356.
Corydalla striolata (*Blyth*), *Jerd. B. I.* ii, p. 233; *Brooks, S. F.* i, p. 350; *Hume, Cat. no.* 601; *Legge, Birds Ceyl.* p. 628; *Oates, B. B.* i, p. 167; *Barnes, Birds Bom.* p. 245.

The Large Titlark, Jerd.

Coloration. Resembles *A. richardi,* but is considerably smaller, the tarsus and feet being conspicuously smaller, and the hind claw hardly longer than the hind toe; the amount of white on the penultimate tail-feathers is much less, varying from half to a whole inch in length.

Length about 7; tail 3; wing 3·5; tarsus 1; bill from gape ·75; hind claw ·5.

Distribution. Occurs in every portion of the Empire from the Himalayas to Ceylon and the extreme south of Tenasserim, wintering in the plains and retiring to the higher parts of the Himalayas for the summer. This species is, however, met with in the plains up to a very late date (June), and a few pairs may breed in suitable localities. Blanford observed this Pipit as high as 15,000 feet in Sikhim in October. Its nest has not yet been discovered.

847. **Anthus rufulus.** *The Indian Pipit.*

Anthus rufulus, *Vieill. Nouv. Dict. d'Hist. Nat.* xxvi, p. 494 (1818); *Blyth, Cat.* p. 135; *Horsf. & M. Cat.* i, p. 356; *Sharpe, Cat. B. M.* x, p. 574; *Oates in Hume's N. & E.* 2nd ed. ii, p. 213.

Anthus cinnamomeus, *Rüpp. Neue Wirb., Vögel*, p. 103 (1835).
Anthus malayensis, *Eyton, P. Z. S.* 1839, p. 104; *Horsf. & M. Cat.* i, p. 357.
Corydalla rufula (*Vieill.*), *Jerd. B. I.* ii, p. 232; *Hume, N. & E.* p. 384; *id. Cat.* no. 600; *Scully, S. F.* viii, p. 317; *Legge, Birds Ceyl.* p. 625; *Oates, B. B.* i, p. 168; *Barnes, Birds Bom.* p. 214.
Corydalla malayensis (*Eyton*), *Hume & Dav. S. F.* vi, p. 366; *Hume, Cat.* no. 600 bis.
Corydalla ubiquitaria (*Hodgs.*), *apud Anders. Yunnan Exped., Aves*, p. 607.

The Indian Titlark, Jerd.; *Rayel, Chachari*, Hind.; *Gurapa-madi jitta*, Tel.; *Meta kāle*, Tam.

Coloration. An exact miniature of *A. richardi*, from which this species differs in nothing but size. It has, however, a proportionally larger bill.

Bill dark brown, yellowish at the base of lower mandible; iris brown; legs flesh-colour; claws brownish.

Length about 6·5; tail 2·4; wing 3; tarsus 1; bill from gape ·75; hind claw ·5.

Distribution. A permanent resident in every portion of the Empire and Ceylon, ascending the Himalayas to about 6000 feet. This species has not yet been found in the Andamans or Nicobars, but probably occurs there. It extends through the Malay peninsula and islands to Lombock and Timor and it is largely distributed in Africa.

Habits, &c. Breeds all over the Empire (up to 6000 feet in the Himalayas) from March to July and perhaps later. The nest is a small structure of grass placed on the ground under shelter of a tuft of grass or clod of earth. The eggs, three in number, are brownish or greenish stone-colour, thickly marked with brown and purplish red, and measure about ·8 by ·6.

848. **Anthus campestris.** *The Tawny Pipit.*

Alauda campestris, *Linn. Syst. Nat.* i, p. 288 (1766).
Anthus campestris (*Linn.*), *Blyth, Cat.* p. 136; *Sharpe, Cat. B. M.* x, p. 569.
Agrodroma campestris (*Linn.*), *Jerd. B. I.* ii, p. 234; *Hume, S. F.* i, p. 202; *id. Cat.* no. 602; *Barnes, Birds Bom.* p. 245.

The Stone-Pipit, Jerd.; *Chillu*, Hind.

Coloration. The fully adult bird is pale sandy brown above, with darker centres or streaks to all the feathers except those of the rump; wings dark brown, margined with bright sandy buff; tail black, the outermost feather with the terminal half obliquely white, the penultimate with an oblique patch of white about an inch in length; lores and sides of the head mixed brown and fulvous; a broad supercilium and the whole lower plumage sandy buff or fulvous entirely unmarked.

The young bird has the upper plumage darker, a row of brown spots down each side of the throat and the whole breast streaked

with rather well-defined brown marks and a few indistinct streaks
on the sides of the body. At each spring moult the streaks become
reduced in number and ultimately disappear.

Iris blackish brown; bill dark horny brown above, pale flesh
below; legs and feet pale yellowish flesh (*Butler*).

Length about 7·5; tail 3; wing 3·5; tarsus 1; bill from gape ·8;
hind claw ·45, very slightly longer than hind toe.

Distribution. A winter visitor to the plains of the north-west
portion of India, ranging to Ahmednagar on the south and to
Manbhoom and Mughal Sarai on the east. At this season this
species ranges westwards to Northern Africa. It summers in
Europe, Central Asia, and Siberia.

849. Anthus cervinus.   *The Red-throated Pipit.*

Motacilla cervina, *Pall. Zoogr. Ross.-Asiat.* i, p. 511 (1811).
Anthus cervinus (*Pall.*), *Hume, S. F.* ii, p. 239; *Hume & Dav. S. F.*
vi, p. 367; *Hume, Cat.* no. 605 bis; *Oates, B. B.* i, p. 169; *Sharpe,*
Cat. B. M. x, p. 585.

Coloration. The fully adult has the whole upper plumage black
with fulvous or pale rufous margins to all the feathers; wings and
tail dark brown, edged with pale fulvous, the outermost tail-feather
diagonally white on the terminal two thirds of its length, the
penultimate with a small white tip; a distinct supercilium, cheeks,
chin, throat, and upper breast vinous or cinnamon-red, the breast
with a few black streaks; sides of the breast densely streaked;
remainder of lower plumage fulvous, suffused with pink, the sides
of the body densely and coarsely streaked with black; lores and
ear-coverts vinous-brown; under wing-coverts and axillaries buff.

The young bird has the upper plumage, wings, and tail like the
adult; the supercilium indistinct and fulvous; lores and ear-coverts
rufous-brown; the whole lower plumage fulvous, the chin, throat, and
cheeks unspotted, a broad black band down each side of the throat,
the whole breast and sides of the body with very broad black streaks;
middle of abdomen, vent, and under tail-coverts unmarked.

At each successive spring moult the young bird acquires more
and more vinous on the head and breast and probably becomes fully
adult in three years.

Iris brown; bill dark brown, the gape and base of lower mandible
yellowish; legs yellowish flesh-colour; claws horn-colour.

Length about 6·5; tail 2·4; wing 3·3; tarsus ·9; bill from
gape ·65; hind claw ·45.

Distribution. A common winter visitor to the eastern part of the
Empire from Assam down to Tenasserim, extending along the
Himalayas to Gilgit and Kashmir. This Pipit is also found in the
Andamans in winter. At this season it ranges west to North-east
Africa and East China and the Malayan islands. It summers in
Northern Europe and Siberia.

850. Anthus rosaceus. *Hodgson's Pipit.*

Anthus rosaceus *vel* rufogularis, *Hodgs. in Gray's Zool. Misc.* p. 83 (1844, descr. nulla).

Anthus cervinus (*Pall.*), *apud Blyth, Cat.* p. 136 ; *Jerd. B. I.* i, p. 237.

Anthus rosaceus, *Hodgs., Horsf. & M. Cat.* i, p. 357.

Anthus rosaceus, *Hodgs., Blanf. J. A. S. B.* xli, pt. ii, p. 61 ; *Brooks, J. A. S. B.* xli, pt. ii, p. 83 ; *Hume, N. & E.* p. 386 ; *id. S. F.* ii, p. 241 ; *Brooks, S. F.* iii, p. 252 ; *Hume, Cat.* no. 605 ; *Scully, S. F.* viii, p. 317 ; *Oates, B. B.* i, p. 170 ; *Sharpe, Cat. B. M.* x, p. 589 ; *Hume, S. F.* xi, p. 236 ; *Oates in Hume's N. & E.* 2nd ed. ii, p. 216.

The Vinous-throated Pipit, Jerd.

Coloration. The fully adult bird has the upper plumage dark brown or blackish, the feathers broadly margined with olivaceous brown ; wings dark brown, margined with olivaceous and the median coverts broadly tipped with the same ; tail dark brown, half the outermost feather diagonally white, the penultimate with a triangular white tip ; lores dusky ; ear-coverts brown streaked with yellowish ; a broad supercilium, chin, throat, fore neck, and middle of the breast vinous-pink ; sides of the breast vinous pink-grey, streaked with black ; remainder of lower plumage pale fulvous, the sides of the body boldly and coarsely streaked with black ; axillaries and under wing-coverts yellow ; during the summer the green tinge on the upper plumage fades away and is almost entirely absent.

The young bird has the axillaries and under wing-coverts yellow as in the adult, but has no trace of vinous-pink on the head and breast. The upper plumage, wings, and tail are as in the adult, and the whole fore neck and breast are thickly and coarsely streaked with black. The amount of streaking diminishes at each successive spring moult and the amount of pinkish grey acquired increases.

Bill dusky, blackish on culmen and fleshy brown at base of lower mandible ; iris dark brown ; feet brownish fleshy ; claws dusky (*Scully*).

Length about 6·5 ; tail 2·8 ; wing 3·6 ; tarsus ·9 ; bill from gape ·75 ; hind claw ·4.

Distribution. Found in winter on the lower slopes of the Himalayas from Kashmir to Assam and in the plains of the Punjab and N.W. Provinces, extending on the east down to Manipur and, according to Blyth, to Arrakan. This species is found in summer on the higher parts of the Himalayas from 12,000 to 15,000 feet, and at this season it appears to be found also in Western China and probably in Turkestan. Westwards it extends to Afghanistan.

Habits, &c. There is little authentic information regarding the nidification of this Pipit. A nest found in Nepal in May was a pad of grass placed on the ground and contained two eggs. An egg from Darjiling is described as being greyish white, marked with earthy brown, and as measuring ·85 by ·6.

851. **Anthus spinoletta.** *The Water-Pipit.*

Alauda spinoletta, *Linn. Syst. Nat.* i, p. 288 (1766).
Anthus aquaticus, *Bechst. Naturg. Deutschl.* iii, p. 745 (1807); *Scully, S. F.* iv, p. 152.
Anthus blakistoni, *Swinh. P. Z. S.* 1863, pp. 90, 273; *id. Ibis,* 1867, p. 389; *Hume, S. F.* v, p. 345; *id. Cat.* no. 605 quat.; *Brooks, S. F.* viii, p. 484; *Biddulph, Ibis,* 1881, p. 70; *Scully, Ibis,* 1881, p. 453; *Hume, S. F.* xi, p. 257; *Barnes, Birds Bom.* p. 244.
Anthus neglectus, *Brooks, Ibis,* 1876, p. 501; *Hume, S. F.* v, p. 345.
Anthus spinoletta (*Linn.*), *Hume, Cat.* no. 605 ter; *Barnes, Birds Bom.* p. 243.
Anthus spipoletta (*Linn.*), *Sharpe, Cat. B. M.* x, p. 592.

Coloration. The adult has the upper plumage ashy brown, with the feathers streaked or centred darker; the wings brown margined with pale fulvous; tail brown, the outermost feather with the terminal half obliquely white, the penultimate tipped white; lores and sides of the head rufous; supercilium and the whole lower plumage uniform fulvous or vinous.

The young bird has the upper plumage, wings, and tail like the adult; the supercilium and lower plumage pale isabelline, with narrow ill-defined pale brown streaks on the fore neck and breast, and streaks of the same kind, but much larger, on the sides of the body.

Iris brown; bill, legs, feet, and claws black (*Bingham*).

Length about 6·5; tail 2·7; wing 3·5; tarsus ·85; bill from gape ·7; hind claw ·4.

This Pipit varies much in size and colour of plumage, and an attempt has been made to establish two species on these variable characters. After examining a very large series of these birds collected in all parts from Europe to China, I am quite unable to discover any grounds for separating the Chinese and Indian bird (*A. blakistoni*) from the European (*A. spinoletta*). It is noteworthy that nearly all the Pipits of this species procured in the North-west Provinces of India have the wing under 3·3 in length, whereas European birds have it 3·6 on the average. In the Punjab, however, the large and small birds are met with together, and birds of intermediate size occur everywhere. Chinese birds have the wing 3·4. The alleged differences of striation on the upper and lower plumages are apparently merely matters of age.

Distribution. A winter visitor to Sind, the Punjab, and the North-west Provinces, the whole of Kashmir and the Himalayas as far east as the Sutlej valley. This Pipit ranges into Europe on the one hand, and into China on the other. It appears to breed in Northern and Central Asia.

852. **Anthus japonicus.** *The Eastern Water-Pipit.*

Anthus pratensis japonicus, *Temm. & Schleg. Faun. Japon., Aves,* p. 59, pl. 24 (1850).
Anthus blakistoni, *Swinh., apud Butler, S. F.* vii, p. 177.

OREOCORYS. 313

Anthus ludovicianus (*Gmel.*), *apud Brooks, S. F.* viii, p. 485; *Hume, S. F.* xi, p. 238.
Anthus japonicus, *Temm. & Schleg., Sharpe, Cat. B. M.* x, p. 598.

Coloration. The adult of this species appears to be of a uniform unspotted vinous colour below, but I have been unable to examine any but young birds.

The young bird as usually found in India is ashy brown above, tinged with green, all the feathers except those of the rump and upper tail-coverts centred darker; the wings blackish margined with fulvous white; the tail dark brown margined paler, the outermost feather with the terminal half obliquely white, the penultimate with the inner web broadly tipped white; sides of the head brown and fulvous; a supercilium and the whole lower plumage pale fulvous, every portion of the latter except the chin, throat, and middle of the abdomen streaked with very wide coarse black streaks, those on the breast being particularly large.

Legs and feet brown; bill black; iris blackish brown (*Butler*).

Length about 6·5; tail 2·6; wing 3·5; tarsus ·85; bill from gape ·65; hind claw ·4.

Distribution. Undoubted specimens of this Pipit are in the British Museum from Karáchi, Mooltan, and Darjiling, killed in the winter months; a specimen from Nepal and another from Umballa in indifferent order appear to be referable to this species. This Pipit winters in China and Japan, and apparently summers in Eastern Siberia and Kamtschatka.

Anthus pennsylvanicus, Lath., of North America is an allied species. It differs in having the lower plumage in all stages of the young very fulvous, and in having the streaks on the breast narrow, well-defined, and detached from one another.

Anthus gustavi, Swinhoe, from China, Northern Asia, and other parts, is very richly streaked with black above on a bright fulvous ground, and the breast and sides of the body are streaked with long well-defined black streaks.

Genus **OREOCORYS**, Sharpe, 1885.

The only Pipit belonging to this genus is remarkable for the pointed shape of the tail-feathers. Otherwise it does not differ from the true Pipits.

The Upland Pipit is confined to the higher parts of the Himalayas, where it is found on grassy slopes.

853. **Oreocorys sylvanus.** *The Upland Pipit.*

Heterura sylvana, *Hodgs., Blyth, J. A. S. B.* xiv, p. 556 (1845); *id. P. Z. S.* 1845, p. 35; *Blyth, Cat.* p. 134; *Jerd. B. I.* ii, p. 239; *Ball, S. F.* iii, p. 207; *Brooks, S. F.* iii, p. 252; *Hume, N. & E.* p. 387; *id. Cat. no.* 606; *Scully, S. F.* viii, p. 318.
Oreocorys sylvanus (*Hodgs.*), *Sharpe, Cat. B. M.* x, p. 622; *Oates in Hume's N. & E.* 2nd ed. ii, p. 217.

Coloration. The whole upper plumage, wings, and tail blackish brown, each feather margined with rich rufous buff, the terminal two thirds of the outermost tail-feather diagonally white, the penultimate feather with about an inch of white at the tip, the next feather with a narrow white tip, the next tipped very faintly at the extreme tip ; an indistinct pale fulvous supercilium ; sides of the head pale fulvous streaked with blackish ; chin and throat white, bounded on either side by a narrow black moustachial streak ; fore neck, breast, sides of the body, vent, and under tail-coverts buff streaked with black, the streaks on the breast more or less triangular, those on the other parts narrow and long ; abdomen unspotted pale fulvous.

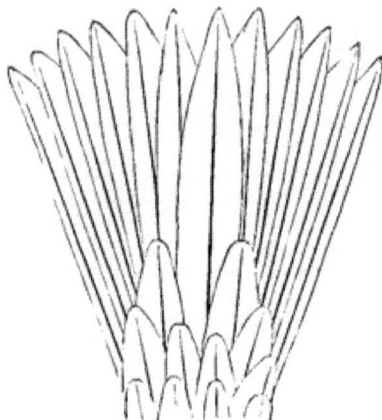

Fig. 86.—Tail of *O. sylvanus.*

In the summer the rufous margins on the upper plumage and the triangular marks on the breast become much worn away.

Upper mandible blackish brown ; lower mandible, legs, and feet fleshy, tip of lower mandible and claws shaded dusky ; iris dark brown (*Hume*).

Length about 7 ; tail 3 ; wing 3·2 ; tarsus ·95 ; bill from gape ·8 ; hind claw ·35.

Distribution. The Himalayas from Murree to Nepal, where this species appears to be a permanent resident. A specimen in the Hume Collection is stated to have been obtained at Etáwah in the plains, but probably there is some error.

Habits, &c. Breeds on the Himalayas from 4000 to 8000 feet about May and June, making a shallow open nest of grass on the ground under a tuft of grass or a stone, and laying four or five eggs, which are whitish marked with red and purple, and measure about ·9 by ·68.

Fig. 87.—*Alauda arvensis.*

Family ALAUDIDÆ.

The intrinsic muscles of the syrinx fixed to the ends of the bronchial semi-rings; the edges of both mandibles perfectly smooth, except for the presence of a notch in the upper mandible: the hinder part of the tarsus transversely scutellated; the front part of the tarsus also scutellated: wing of nine or ten primaries: tail of twelve feathers; one moult a year, in the autumn; plumage of the nestling spotted; sexes usually alike; rictal bristles well-developed; head usually crested; hind claw usually long.

The *Alaudidæ* or Larks form a group of birds which bear a close general resemblance to the Pipits, but which differ from them, as well as from all other Indian Passeres, in having the hinder part of the tarsus transversely scutellated.

The Larks have only one moult a year, but many undergo a
seasonal change of plumage through casting off the margins of the
feathers in spring. The variations of colour produced by this,
combined with the plain coloration of many of the species, render
the study of the Larks rather difficult. Specimens of Larks killed
at the same time of the year should therefore be compared with
each other. Some of the Larks are subject to much variation in
size as well as in colour.

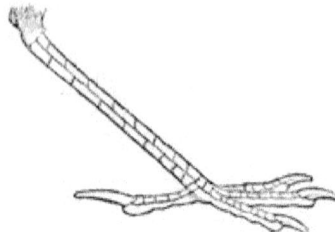

Fig. 88.—Foot of *Alæmon deserticum*, to show scutellations on
hinder part of tarsus.

The Larks for the most part frequent open plains and cultivated
land, but some are found only in arid deserts, and others again
affect the borders of woods. They generally sing whilst soaring in
the air, and their song is always agreeable and in many cases fine.
Many are migratory, others are resident or very locally migratory.
They are generally sociable and occasionally gregarious. They
all breed on the ground, constructing a slight grass nest in a hollow, and their eggs are marked with brown.

I have had the great advantage of studying the Larks with
Mr. Bowdler Sharpe, who was engaged at the time in writing the
Catalogue of these birds. We have in every case arrived at the
same conclusions with regard to each species, and the only point
on which I subsequently found reason to differ from him was in
the suppression of the genus *Spizalauda*.

Key to the Genera.

a. Nine primaries, the first reaching to
about tip of wing.
  a'. A tuft of pointed feathers springing
from each side of crown............. Otocorys, p. 319.
  b'. No tuft of pointed or other feathers
springing from side of crown.
    a''. The longer secondaries or tertiaries
reaching to about tip of wing.... Calandrella, p. 327.
    b''. The longer secondaries or tertiaries
falling short of tip of wing by a considerable interval................ Alaudula, p. 330.

b. Ten primaries, the first minute.
 c. First primary large, considerably
  exceeding the primary-coverts.
  c″. Bill as long as, or longer than, the
   head, and very slender .... ALÆMON, p. 317.
  d″. Bill much shorter than the head,
   and thick.
   a‴. Nostrils not covered by plumelets,
    but clearly visible .... MIRAFRA, p. 332.
   b‴. Nostrils quite concealed by dense
    plumelets .... AMMOMANES, p. 339.
 d′. First primary very small, not exceeding
  the primary-coverts.
  e″. Crest, if any, short, and covering the
   whole crown.
   e‴. Hind claw long and straight.
    a⁴. Wings reaching nearly to tip of
     tail; tertiaries falling short of
     tip of wing by more than
     length of tarsus .... MELANOCORYPHA, p. 322.
    b⁴. Wings falling short of tip of tail
     by a considerable distance;
     tertiaries falling short of tip of
     wing by less than length of
     tarsus .... ALAUDA, p. 324.
   d‴. Hind claw very short and curved. PYRRHULAUDA, p. 341.
  f″. Crest consisting of a few very
   elongate feathers, springing from
   centre of crown .... GALERITA, p. 336.

Genus **ALÆMON**, Keys. & Blas., 1840.

The genus *Alæmon* contains one Indian Lark of large size which is found in the desert tracts of Sind as a permanent resident.

In *Alæmon* the bill is very long and slender, and gently curved on its terminal half, and the nostrils are fully exposed to view.

Fig. 89.—Head of *A. desertorum.*

The first of the ten primaries of the wing is small, but exceeds the primary-coverts. The toes and claws are very short, and the latter are very stout. The sexes are quite alike.

The Desert-Lark is said to run with great speed. The male at the breeding-season rises about fifty feet, and sings a short song of a few notes, after which he descends and perches on a bush, prior to reaching the ground.

854. Alæmon desertorum. *The Desert-Lark.*

Alauda desertorum, Stanley in Salt's Exped. Abyss. App. p. lx (1814).
Certhilauda desertorum (Stanl.), Blyth, Cat. p. 133; *Horsf. & M. Cat.* ii, p. 464; *Jerd. B. I.* ii, p. 438; *Stoliczka, J. A. S. B.* xli, pt. ii, p. 248; *Hume, Cat.* no. 770; *Doig, S. F.* ix, p. 281; *Barnes, Birds Bom.* p. 284.
Alæmon desertorum (Stanl.), Hume, S. F. i, p. 216; *Sharpe, Cat. B. M.* xiii, p. 519; *Oates in Hume's N. & E.* 2nd ed. ii, p. 219.

Coloration. Upper plumage isabelline grey, tinged with ashy on the forehead and upper tail-coverts, which latter have dusky shafts; middle pair of tail-feathers sandy brown, very broadly margined on both webs with bright fulvous; the other tail-feathers black, narrowly margined with fulvous, the outermost feather with a well-defined white margin occupying half the outer web; wing-coverts brown, with broad fulvous margins, some of the greater coverts tipped white; primary-coverts black, tipped with white; the first few primaries black, with white bases; the remaining primaries and secondaries white, with a black spot on the outer web, and a portion of the shaft also black; tertiaries like middle pair of tail-feathers; feathers immediately next the nostrils fulvous; a black streak through the lores, with a white band above and below, the lower band continued under the eye; a black band behind the eye, with a broad pale fulvous supercilium above it; cheeks and ear-coverts bright fulvous, divided by a blackish patch; chin and throat white; fore neck and breast pale fulvous, with large black spots; remainder of lower plumage white, the flanks shaded with brown.

Legs and feet china-white; iris brown; bill horny brown above, darkening at the tip; lower mandible fleshy (*Butler*).

Size extremely variable, the female being apparently much smaller than the male. Length 8·5 to 11; tail 3·3 to 4; wing 4·5 to 5·5; tarsus 1·2 to 1·3; bill from gape about 1·4.

Distribution. A permanent resident in Sind and Cutch, extending west through Afghanistan and Persia to Arabia and North-eastern Africa, and ranging in a modified form through Northern Africa.

Habits, &c. This Lark affects the desert and runs like a Plover. Stoliczka observed it on tracts of mud in the Rann of Cutch. It breeds in May and June, making a small nest of grass on the sand, and laying three eggs which are greyish white, marked with yellowish brown, and measure about 1·02 by ·74.

Genus OTOCORYS, Bonap., 1858.

The genus *Otocorys* contains the Horned Larks, three species of which are found at considerable altitudes in the Himalayas. They occur in flocks, and at the breeding-season the males are said to rise in the air, and sing in the same manner as the Sky-Lark.

The Horned Larks have a few lengthened feathers springing from each side of the crown in both sexes. The bill is of median length, and the nostrils are densely covered by plumes; the wing has nine primaries, of which the first reaches nearly to the tip of the wing; the hind claw is straight and about as long as the hind toe. The sexes are slightly different in colour, and both sexes undergo a considerable variation of plumage according to season, caused by the gradual wearing away of the margins of the feathers in winter and spring.

Key to the Species.

a. Black of the side of the head connected with
   black of breast.......................... *O. penicillata*, p. 319.
b. Black of the side of the head divided from
   black of breast by a white band.
   a'. Wing about 5; bill at front ·55 to ·6 ...... *O. longirostris*, p. 320.
   b'. Wing about 4·5; bill at front ·4 to ·45 .... *O. elwesi*, p. 321.

855. Otocorys penicillata. *Gould's Horned Lark.*

Alauda penicillata, *Gould, P. Z. S.* 1837, p. 126.
Otocorys penicillata (*Gould*), *Blanf. E. Pers.* ii, p. 240; *Biddulph, Ibis,* 1881, p. 89; *Scully, Ibis,* 1881, p. 580; *Biddulph, Ibis,* 1882, p. 285; *Sharpe, Cat. B. M.* xiii, p. 530.

Coloration. After the autumn moult the feathers at the base of the upper mandible and the lores are deep black; the forehead, a broad supercilium, the posterior part of the ear-coverts, the chin, and upper throat are dull yellow; anterior part of crown black, each feather broadly margined with vinous; coronal horns black, with pale tips; upper plumage ashy brown, the crown, mantle, and upper tail-coverts with the major portion of the feathers rich vinous, imparting a tinge of this colour to those parts; the rump and upper tail-coverts with blackish shaft-streaks; middle tail-feathers dark brown, margined with paler brown; the other feathers black, narrowly margined paler, the outermost feather more broadly margined with pure white; wing-coverts brown, margined with ashy and washed with vinous; primaries and secondaries dark brown, narrowly margined with whitish; tertiaries like the middle pair of tail-feathers; cheeks, anterior part of ear-coverts, sides of throat, fore neck, upper breast, and sides of the neck deep black, each feather margined with yellowish; remainder of lower plumage white, washed with vinous on the breast and sides of the body.

Soon after the autumn moult the yellow parts of the head become pure white, and the black parts lose all their fringes and become deep black ; the crown, mantle, and upper tail-coverts become deep vinous, and the whole of the upper plumage and wing-coverts become more or less tinged with this colour ; the coronal horns become very distinct and deep black.

Fig. 90.– Head of *O. penicillata*.

The female has no black band on the front part of the crown, the whole crown being streaked with black ; the black on the other parts of the head is much duller, and there never is any trace of vinous or pink on the plumage, except perhaps immediately after the autumn moult ; the black patch on the fore neck and breast is somewhat smaller, and the streaks on the upper plumage are more pronounced.

Iris deep reddish brown ; bill blackish above, bluish grey below ; tarsi and upper surface of toes (in May) black in the male, dusky in the female ; claws the same ; soles of feet whitish (*Blanford*).

Length about 8 ; tail 3·3 ; wing 4·6 ; tarsus ·9 ; bill from gape ·75 ; the female is considerably smaller.

Distribution. A winter visitor to Gilgit, where this species is found at a height of about 5000 feet. Biddulph procured this Lark on the Shandur plateau, where it no doubt breeds. It ranges north into Turkestan, and west into Asia Minor.

856. Otocorys longirostris. *The Long-billed Horned Lark.*

Otocorys longirostris, *Gould, Moore, P. Z. S.* 1855, p. 215, pl. 3 ; *Horsf. & M. Cat.* ii, p. 470 ; *Jerd. B. I.* ii, p. 431 ; *Hume, Cat.* no. 764 ; *Scully, Ibis,* 1881, p. 581 ; *Biddulph, Ibis,* 1882, p. 285 ; *Sharpe, Cat. B. M.* xiii, p. 536.

Coloration. Very similar to *O. penicillata*, but differing in having the white of the throat passing under the cheeks and ear-coverts and merging into the pale colour of the side of the neck, thus isolating the black on the fore neck and upper breast ; the present species is also much larger and the feathers of the upper surface have, in the worn summer plumage, very much larger and more conspicuous dark shaft-streaks. The pale parts of the head are at all seasons white, never yellow.

The colours of the soft parts of this Lark do not appear to have been recorded.

Length about 8·5; tail 3·7; wing 5; tarsus ·95; bill from gape ·8; bill at front ·55 to ·6.

Fig. 91.— Head of *O. longirostris*.

Distribution. The higher parts of the Himalayas from Kashmir and Ladák down to Kumaun, extending into the adjoining parts of Tibet. This species appears to be found only at very high altitudes.

857. Otocorys elwesi. *Elwes's Horned Lark.*

Otocorys penicillata (*Gould*), *Horsf. & M. Cat.* ii, p. 469; *Jerd. B. I.* ii, p. 429; *Hume, S. F.* vii, p. 422; *id. Cat.* no. 763.
Otocorys elwesi, *Blanf. J. A. S. B.* xli, pt. ii, p. 62; *Sharpe, Cat. B. M.* xiii, p. 534; *Oates in Hume's N. & E.* 2nd ed. ii, p. 220.
Otocorys longirostris, *Gould, apud Hume & Henders. Lah. to Yark.* p. 267.

The Horned Lark, Jerd.

Coloration. Of the same coloration as *O. longirostris.* The present species differs in its much smaller size; its conspicuously smaller bill; in the large amount and intensity of the vinous or pink tinge, which suffuses the whole of the upper plumage and wings; and in the small extent and paler tone of the streaks on the back and rump.

The nestlings of this and other species of *Otocorys* have the whole upper plumage fulvous, each feather with a subterminal black bar and a white tip: the wing-feathers are all broadly margined with fulvous; lower plumage pale yellowish white, spotted with brown on the throat, fore neck, and breast. The adult plumage is assumed at the first autumn moult.

Bill black above, pale near the base below; legs black; soles of feet yellowish (*Blanford*).

Length less than 8; tail 3·4; wing 4·3 to 4·8; tarsus ·9; bill from gape ·7.

This species is easily distinguished from *O. longirostris* by the characters pointed out above, the best of which are the smaller bill and general smaller size.

Distribution. A permanent resident in the highest parts of the Himalayas from Sikhim to Ladák, extending into Central Asia from Turkestan to Mongolia and China. Blanford met with this species in Sikhim at nearly 18,000 feet elevation.

Habits, &c. Some eggs, presumably of this Lark, found in Native Sikhim are described as being greyish white freckled and mottled all over with pale olive-brown and purplish grey and as measuring about ·89 by ·64.

Genus **MELANOCORYPHA**, Boie, 1828.

The genus *Melanocorypha* contains the Calandra Larks, which are birds of heavy build and large size. One of the Indian species is found only on the highest parts of the Himalayas, but the other is found in the cultivated parts of the plains.

The Alpine species has a much longer bill than the lowland species and has hardly any trace of the black pectoral patches which characterize the Calandras. It has also a much longer and a straighter claw. It is, however, hardly advisable to place the two in separate genera as there are many points of resemblance between them.

In *Melanocorypha* the bill is thick and gently curved and the nostrils are covered by plumelets; the wing has ten primaries, the first of which is very minute, and the wing is very long, reaching, when folded, nearly to the tip of the tail; the hind claw is very long and straight. The sexes are alike or nearly so.

Key to the Species.

a. Wing about 6 and largely white .... *M. maxima*, p. 322.
b. Wing about 4·5 and without any white .... *M. bimaculata*, p. 323.

858. **Melanocorypha maxima.** *The Long-billed Calandra Lark.*

Melanocorypha maxima, *Gould, Birds As.* iv, pl. 72 (1867); *Hume, S. F.* i, p. 492; *id. Cat.* no. 761 quat.; *Sharpe, Cat. B. M.* xiii, p. 554.

Coloration. Forehead, crown and nape, lower rump, and upper tail-feathers rufous brown, each feather mesially darker and edged with pale fulvous; back, scapulars, and upper rump dark brown, with broad lateral fulvous margins and narrow tips; the hind neck and mantle suffused with ashy; middle pair of tail-feathers brown margined with tawny, the others black increasingly tipped with white, the penultimate feather with half the outer web white, and a broad tip to the inner web, the outermost feather with the terminal two-thirds of both webs white; wing-coverts and tertiaries brown, edged with fulvous; primaries and secondaries dark brown, margined paler and with the outer web of the first primary almost entirely white; all the quills except the first few primaries broadly tipped with white; ear-coverts rufous; lores, supercilium, and cheeks white, mottled with rufous; chin and throat whitish; breast pale ashy with some ill-defined brown spots, and the sides of the breast with an obsolete dark patch; remainder of lower plumage white, tinged with fulvous on the flanks.

In the spring and summer months the plumage is much darker owing to the wearing away of the margins of the feathers.

The young bird is nearly black above with yellowish or fulvous margins to all the feathers; the breast is also very dark or blackish with yellowish margins and the remainder of the lower plumage is pale yellow.

The colours of the soft parts of this Lark do not appear to have been recorded.

Length about 9·5; tail 3·5; wing 6; tarsus 1·1; bill from gape 1·2; hind claw up to ·8.

Distribution. The higher parts of Sikhim, extending into the adjoining parts of Tibet and the mountain region of Western China.

859. **Melanocorypha bimaculata.** *The Eastern Calandra Lark.*

Alauda bimaculata, *Ménétr. Cat. Rais. Cauc.* p. 37 (1832).
Melanocorypha bimaculata (*Ménétr.*), *Blanf. S. F.* v, p. 246; *Hume, S. F.* vii, p. 421; *id. Cat.* no. 761 ter; *Biddulph, Ibis,* 1881. p. 89; *Scully, Ibis,* 1881, p. 580; *Barnes, Birds Bom.* p. 279; *Sharpe, Cat. B. M.* xiii, p. 555.
Melanocorypha torquata, *Blyth, J. A. S. B.* xvi, p. 476 (1847); *Hume & Henders. Lah. to Yark.* p. 265, pl. xxvii; *Scully, S. F.* iv, p. 173.

Fig. 92.—Head of *M. bimaculata.*

Coloration. The upper plumage dark brown, each feather laterally margined with fulvous, the character of the upper plumage being streaky; tail dark brown, margined with fulvous, the inner web of all the feathers, except the middle pair, with a terminal white spot;

Y 2

wing-coverts and quills dark brown, margined with fulvous; no
white whatever about the wing; a dark line running through the
lores and behind the eye; a broad pale fulvous supercilium;
cheeks and ear-coverts rufous, streaked with brown; a patch of
white under the black band through the lores and under
the eye; chin, throat, and a lateral band behind the ear-
coverts white; a broad black band across the upper breast interrupted in the middle; remainder of breast fulvous streaked with
brown; other parts of lower plumage white, the flanks tinged
with fulvous.

The female has the black pectoral band reduced in size.

Legs and feet fleshy or yellowish fleshy, more or less dusky at
joints; claws dusky; iris brown, in some light brown; bill horny
brown or blackish horny on upper mandible, lower mandible
greenish horny changing to yellow at base and gape.

Length about 7·5; tail 2·3; wing 4·5; tarsus 1; bill from gape
·8; the female is smaller than the male.

Distribution. A winter visitor to Sind, Rajputana, Baháwalpur,
the Punjab, the N.W. Provinces and Oudh, passing through
Kashmir on migration and summering in Central Asia. This Lark
extends on the west to Afghanistan and Persia and ranges as far
as North-eastern Africa.

Genus **ALAUDA**, Linn., 1766.

The genus *Alauda* is now restricted to the Sky-Larks, two species
or races of which occur in India. These Larks vary very much in
size and plumage and I quite agree with Sharpe in not recognizing
more than these two species as occurring in India.

In *Alauda* the bill is slender and feeble and the nostrils are
covered by plumelets; there are ten primaries, the first of which is
very minute, and the wing is somewhat short, not reaching to far
beyond the middle of the tail, and the tertiaries are lengthened; the
hind claw is very long and nearly straight. The sexes are alike.

The Sky-Larks frequent cultivation chiefly, and are noted for
the excellence of their song, which is given forth at a great height
from the ground. They are abundant in most parts of India, but
become less frequent to the east in Burma.

Key to the Species.

a. Of larger size; wing generally over 4 .... *A. arvensis*, p. 324.
b. Of smaller size; wing seldom exceeding 3·5 .... *A. gulgula*, p. 326.

860. **Alauda arvensis.** *The Sky-Lark.*

Alauda arvensis, *Linn. Syst. Nat.* i, p. 287 (1766): *Blyth, Cat.*
p. 131; *Horsf. & M. Cat.* ii, p. 465; *Hume, S. F.* i, p. 39; *id.
N. & E.* p. 485; *Sharpe, Cat. B. M.* xiii, p. 567; *Oates in Hume's
N. & E.* 2nd ed. ii, p. 220.

Alauda leiopus vel orientalis, *Hodgs. in Gray's Zool. Misc.* p. 84
(1844, descr. nulla).
Alauda dulcivox, *Hodgs. in Gray's Zool. Misc.* p. 84 (1844, descr.
nulla); *Brooks, S. F.* i, p. 484; *id. J. A. S. B.* xliii, pt. ii, p. 253;
Hume, Cat. no. 766; *Scully, S. F.* viii, p. 338; *id. Ibis*, 1881,
p. 582; *Biddulph, Ibis*, 1881, p. 89.
Alauda triborhyncha, *Hodgs. apud Horsf. & M. Cat.* ii, p. 467 (part.);
Jerd. B. I. ii, p. 433; *Hume & Henders. Lah. to Yark.* p. 268,
pl. xxviii.
Alauda guttata, *Brooks, J. A. S. B.* xli, pt. ii, p. 85 (1872); *id. S. F.*
i, p. 485; *Biddulph, Ibis*, 1881, p. 90; *Scully, Ibis*, 1881, p. 583.

The Himalayan Sky-Lark, Jerd.

Coloration. Upper plumage and wing-coverts dark brown, each
feather broadly edged with fulvous; quills brown edged with fulvous,
and with a tinge of rufous near the base of some of the quills; tail
brown, edged with fulvous, the penultimate feather with the
outer web almost entirely white, the outermost all white, except
the base of the inner web; a pale supercilium from the nostrils to
the end of the ear-coverts; ear-coverts streaked brown and
rufous; lower plumage pale fulvous, the cheeks and throat slightly,
the breast boldly, streaked with black; the sides of the body less
distinctly streaked with brown; remainder of lower plumage very
pale fulvous and at times almost white.

The young have the feathers of the crown much rounded and
tipped with white, and most of the feathers of the upper plumage
are very rufous and also tipped with white; the wing-coverts are
much more broadly margined with fulvous or rufous.

Bill dusky above, lower mandible greyish horny, faintly yellowish
at the extreme tip; iris dark brown; legs and feet brownish fleshy
(*Scully*).

Size very variable; length about 7; tail 2·5 to 2·9; wing 3·7 to
4·4; tarsus ·95; bill from gape ·65; these represent the extreme
measurements of numerous Indian birds.

The Sky-Lark is as variable in size as the Raven, and the shades
of brown, fulvous, and rufous of which its plumage is composed also
vary exceedingly throughout its great range.

Distribution. The whole extent of the Himalayas from Hazára
and Kashmir to Assam, where the Sky-Lark appears to be a constant resident, moving about to different levels according to season.
In the winter many birds appear to visit the plains of the Punjab
and North-west Provinces, and a Lark killed by Dr. Anderson near
Bhámo in Upper Burma appears referable to this species.

The Sky-Lark is spread over Europe and the greater portion of
Asia as far as China.

Habits, &c. Breeds in the Himalayas in May and June, constructing a nest of fine grass on the ground and laying three to
five eggs, which are marked with yellowish and purplish brown
and measure about ·95 by ·67.

861. **Alauda gulgula.** *The Indian Sky-Lark.*

Alauda gulgula, *Frankl. P. Z. S.* 1831, p. 119; *Blyth, Cat.* p. 132; *Jerd. B. I.* ii, p. 434; *Hume & Henders. Lah. to Yark.* p. 269, pl. xxix; *Hume, S. F.* i, p. 40; *Brooks, S. F.* i, p. 485; *Hume, N. & E.* p. 486; *Legge, Birds Ceyl.* p. 630; *Hume, Cat.* no. 767; *Scully, S. F.* viii, p. 288; *Oates, B. B.* i, p. 373; *Barnes, Birds Bom.* p. 282; *Sharpe, Cat. B. M.* xiii, p. 575; *Oates in Hume's N. & E.* 2nd ed. ii, p. 221.
Alauda triborhyncha, *Hodgs. in Gray's Zool. Misc.* p. 84 (1844, descr. nullâ); *Horsf. & M. Cat.* ii, p. 467 (part.).
Alauda malabarica, *Scop. apud Horsf. & M. Cat.* ii, p. 467.
Alauda australis, *Brooks, S. F.* i, p. 486 (1873); *Hume, Cat.* no. 768.
Alauda peguensis, *Oates, S. F.* iii, p. 343 (1875).

Baruta-pitta, Niala pichiké, Tel.; *Manam-badi,* Tam.; *Bharut,* Hind.; *Bee-lone,* Burm.; *Gomarita,* Cing.

Coloration. So similar to *A. arvensis* as to require no separate description. Differs in being constantly smaller, the wing seldom exceeding 3·5.

There are as many races of this Lark as there are of *A. arvensis*, but they are equally unworthy of recognition, as they are based on points of size and colour which are by no means constant or even definable.

Sharpe is of opinion that *A. gulgula* differs from *A. arvensis* in having paler under wing-coverts, and by the almost entire absence of flank-stripes. These points may be of service in discriminating between the two birds, but the only character which is of real use seems to me to be that of size.

Mouth yellowish; upper mandible dark horn; lower mandible pinkish fleshy, dusky at the tip; iris brown; eyelids plumbeous; legs fleshy brown; claws pale horn-colour.

Length about 6·5; tail 2·3; wing 3·4; tarsus ·9; bill from gape ·75.

Distribution. Every portion of the Empire and Ceylon, except Tenasserim, south of Moulmein, and the middle ranges of the Himalayas, where this Lark is absent or comparatively rare. A specimen of a Lark procured by Brooks at Almorah, however, appears to be referable to this species. It has not been recorded from the Andamans or Nicobars.

Habits, &c. Breeds throughout India apparently from April to June and in Burma from December to April, constructing a nest similar to that of *A. arvensis*, and laying similar eggs, which are, however, somewhat smaller, and measure about ·8 by ·61. The habits of this species closely resemble those of the European Sky-lark, the song is similar and is uttered in the same manner as the bird soars. Both species associate in flocks in the winter.

Genus CALANDRELLA, Kaup, 1829.

The genus *Calandrella* contains the Short-toed Larks, of which four fairly distinct and recognizable species occur in India.

In *Calandrella* the bill is somewhat deep and short, and the nostrils are concealed from view by plumelets; there are nine primaries only in the wing, and the first is long and reaches to the tip; the tertiaries are lengthened and reach to the end of the primaries or nearly so; the hind claw is rather longer than the hind toe and gently curved.

The Short-toed Larks are fond of dry sandy ground, and their habits are similar to those of the Sky-Larks. The song, however, is weak and monotonous.

Key to the Species.

a. Pale part of inner web of outermost tail-
   feather of large extent, more than an
   inch in length.
   a'. Light part of outermost tail-feather
       pale buff.
       a''. General colour of lower plumage
            whitish; wing generally under 3·6. *C. brachydactyla*, p. 327.
       b''. General colour of lower plumage
            fulvous; wing generally over 3·9. . *C. dukhunensis*, p. 328.
   b'. Light part of outermost tail-feather
       white .... *C. tibetana*, p. 329.
b. Pale part of inner web of outermost tail-
   feather of very small extent, much less
   than an inch in length .... *C. acutirostris*, p. 329.

862. Calandrella brachydactyla. *The Short-toed Lark.*

Alauda brachydactyla, *Leisler, Wetterau Gesellsch. Ann.* iii, pp. 357–350 (1812).
Calandrella brachydactyla (*Leisl.*), *Hume, S. F.* i, p. 213; *Adam, S. F.* i, p. 389; *Butler, S. F.* iii, p. 500; *Hume, S. F.* vii, p. 66; *id. Cat.* no. 764 (pt.); *Biddulph, Ibis,* 1881, p. 88 (pt.); *Scully, Ibis,* 1881, p. 579; *Biddulph, Ibis,* 1882, p. 285 (pt.); *Barnes, Birds Bom.* p. 279(pt.); *Sharpe, Cat. B. M.* xiii, p. 580.

Fig. 95.—Head of *C. brachydactyla.*

Coloration. Upper plumage sandy buff, each feather broadly centred or streaked with dark brown or black, the upper tail-

coverts more sandy than the other parts; wing-coverts and quill-
dark brown broadly edged with fulvous; middle pair of tail-feathers
dark brown broadly edged with rufous; the next three pairs dark
brown, very narrowly margined with fulvous; the penultimate dark
brown, with the terminal half of the outer web very pale buff; the
outermost with the outer web nearly entirely pale buff, and
the inner web with the inner half brown and the outer half whitish;
lores and a supercilium pale buff; ear-coverts hair-brown; sides
of the neck pale brown; lower plumage dull white, the breast
washed with brown; the breast is frequently streaked with darker
brown, but sometimes perfectly plain; there is almost always a
dusky or blackish patch on each side of the breast.

Iris brown; legs and feet brownish flesh-colour; upper mandible
dark horny brown; lower mandible pale fleshy (*Butler*).

Length about 6·5; tail 2·4; wing 3·5 to 3·8; tarsus ·8; bill from
gape ·55.

Distribution. A winter visitor to the north-west portion of the
Empire, west of a line drawn roughly from Bombay to Kumaun.
This species ranges into Turkestan and extends westwards to
Europe and Northern Africa.

863. Calandrella dukhunensis. *The Rufous Short-toed Lark.*

Emberiza baghaira, *Frankl. P. Z. S.* 1831, p. 119 (descr. nullâ).
Alauda dukhunensis, *Sykes, P. Z. S.* 1832, p. 93.
Calandrella brachydactyla (*Temm.*), *Blyth, Cat.* p. 132; *Jerd. B. I.*
ii, p. 426; *Ball, S. F.* ii, p. 423; *Fairbank, S. F.* iv, p. 261; *Ball,
S. F.* vii, p. 223; *Hume, Cat.* no. 761 (pt.); *Scully, S. F.* viii,
p. 337; *Butler, S. F.* ix, p. 418; *Reid, S. F.* x, p. 58; *Davidson,
S. F.* x, p. 314; *Barnes, Birds Bom.* p. 279 (pt.).
Coryphidea calandrella (*Bonelli*), *Horsf. & M. Cat.* ii, p. 472.
Alauda (Calandrella) brachydactyla, *Temm. Blanf. J. A. S. B.* xli,
pt. ii, p. 62.
Calandrella dukhunensis, *Sharpe, Cat. B. M.* xiii, p. 584.

Baghaira, Bagheyri, Baghoda, Hind.

Coloration. Resembles *C. brachydactyla,* but has the whole lower
plumage fulvous, darker on the breast and sides of the body; the
upper plumage also a rich fulvous; has a longer wing.

Iris brown; legs and feet brownish flesh-colour, dusky at the
joints; bill dark horny brown above, pale flesh below (*Butler*).

Length about 6·5; tail 2·4; wing 3·8 to 4·1; tarsus ·8; bill
from gape ·6.

This appears to me to be an easily recognizable form of Short-
toed Lark with quite a distinct area of distribution from *C. brachydactyla.* I have examined specimens killed in India from August
to April. The deep fulvous lower plumage combined with the
longer wing suffice to separate this species from *C. brachydactyla*
when the plumage is in good order.

Distribution. The whole of India east of a line drawn roughly
from Bombay to Kumaun, and as far south as Belgaum, extending

into Assam, and more rarely into Pegu. This species also extends
into Tibet and probably breeds there or in Central Asia. It is
very probable that some of these Larks may also remain in the
plains of India or in the Himalayas to breed, but a great majority
are winter visitors.

Habits, &c. This Short-toed Lark frequents open ground, cultivated or waste, and is generally found in small flocks, which, about
the end of March, associate together, often forming assemblages
of many thousand birds. At this season these Larks are fat and
are killed in great numbers for food; they are commonly known
by Europeans throughout India as Ortolans.

864. Calandrella tibetana. *Brooks's Short-toed Lark.*

Calandrella brachydactyla (*Temm.*), *Hume & Henders. Lah. to
Yark.* p. 264.
Calandrella tibetana, *Brooks, S. F.* xiii, p. 488 (1879); *Hume, S. F.*
ix, p. 97; *Sharpe, Cat. B. M.* xiii, p. 585.

Coloration. Resembles *C. brachydactyla*, but is much greyer or
less fulvous above; has a smaller bill; and has the pale part of the
outermost tail-feather pure white, not pale buff; the supercilium
and cheeks nearly white.
Length about 6; tail 2·5; wing 3·7; tarsus ·8; bill from
gape ·6.
Distribution. The Himalayas from Gilgit and Kashmir to Sikhim,
extending into the neighbouring parts of Tibet. This Lark appears
to descend to the plains of Upper India in the winter, the British
Museum containing specimens collected at Cawnpore and in the
Rohtak District of the Punjab.

865. Calandrella acutirostris. *Hume's Short-toed Lark.*

Calandrella acutirostris. *Hume, Lah. to Yark.* p. 265 (1873); *Sharpe,
Cat. B. M.* xiii, p. 585.
Calandrella brachydactyla (*Leisl.*), *Scully, S. F.* iv, p. 172;
Biddulph, Ibis, 1881, p. 88 (pt.), 1882, p. 285 (pt.).

Coloration. Resembles *C. brachydactyla*, but has the upper
plumage decidedly tinged with ashy; the extent of light colour on
the inner web of the outermost tail-feather of very small extent,
much less than an inch in length; and a much more slender bill.

Bill dusky along ridge of upper mandible, at tip, and along half
of under surface of lower mandible; rest of bill yellowish horny;
iris brown; legs and feet brownish fleshy; claws dusky brown
(*Scully*).
Length about 6; tail 2·5 to 2·8; wing 3·6 to 3·9; tarsus ·85;
bill from gape ·6.
Distribution. Visits the plains of Upper India in the winter. I
have examined specimens obtained at Delhi, Etáwah, and Mughal
Sarai. Summers in Afghanistan, Gilgit, and Turkestan.

Genus ALAUDULA, Blyth, 1856.

The genus *Alaudula* contains the Sand-Larks, which differ structurally from the Short-toed Larks only in having a more slender bill and shorter tertiaries.

The Sand-Larks frequent the sandy banks of rivers, running about near the edge of the water. They have a poor song.

Key to the Species.

a. Wing under 3·5.
   a'. Bill at front ·5 .... *A. raytal*, p. 330.
   b'. Bill at front ·4 .... *A. adamsi*, p. 331.
b. Wing about 4 .... *A. persica*, p. 331.

866. Alaudula raytal. *The Ganges Sand-Lark.*

Alauda raytal, *Buch. Hamilton, Blyth, J. A. S. B.* xiii, p. 962 (1844).
Calandrella raytal (*Blyth*), *Blyth, Cat.* p. 132.
Alaudala raytal (*Blyth*), *Horsf. & M. Cat.* ii, p. 471; *Jerd. B. I.* ii, p. 428.
Alaudula raytal, *Blyth, Ibis*, 1867, p. 46; *Hume, N. & E.* p. 481; *id. S. F.* iii, p. 159; *Anders. Yunnan Exped., Aves*, p. 606; *Cripps, S. F.* vii, p. 295; *Hume, Cat. no.* 762; *Oates, B. B.* i, p. 374; *Barnes, Birds Bom.* p. 280; *Sharpe, Cat. B. M.* xiii, p. 591; *Oates in Hume's N. & E.* 2nd ed. ii, p. 225.

The Indian Sand-Lark, Jerd.; *Retal,* Hind.

Fig. 94.—Head of *A. raytal*.

Coloration. Upper plumage greyish brown, with dark brown shaft-streaks; lores, supercilium, and under the eye white; ear-coverts streaked with grey and brown; lower plumage white, with a few distinct and well-defined brown streaks on the breast, most numerous on the sides; wings dark brown, edged with greyish brown; middle tail-feathers brown, broadly edged with fulvous white, the others dark brown, the penultimate feather with the outer web almost entirely white, the outermost with the outer web entirely white, the outer half of the inner web white and the inner half black.

Legs fleshy yellow; claws pale horn; bill horn-colour, with a greenish tinge, the tip dusky, the gape yellowish; iris brown; eyelids plumbeous; mouth flesh-colour.

Length about 5·5; tail 2; wing 3·3; tarsus ·75; bill from gape ·65; bill from tip to forehead ·5.

Distribution. Found on the sand-banks of all the large rivers of the North-west Provinces, the Nepal Terai, Oudh, Behar, and Bengal. This Lark is also found along the banks of the Brahmaputra, and it occurs on the Irrawaddy river, on which it has been procured near Bhámo and Thayetmyo. Barnes records this Lark from Neemuch in Rajputana, and Hume from the sand-banks of the Nerbudda river, but I have seen no specimens from these localities.

Habits, &c. Breeds from March to May or June, making a small cup-shaped nest of grass or leaflets in a hollow on a sand-bank, under shelter of a shrub or stone, and laying two or three eggs, which are greyish white, speckled with yellowish brown, and measure about ·75 by ·55.

867. **Alaudula adamsi.** *The Indus Sand-Lark.*

Alauda adamsi, *Hume, Ibis,* 1871, p. 405.
Alaudula adamsi (*Hume*), *Hume, S. F.* i, p. 213; *id. N. & E.* p. 482; *id. Cat.* no. 762 ter; *Barnes, Birds Bom.* p. 280; *Sharpe, Cat. B. M.* xiii, p. 592; *Oates in Hume's N. & E.* 2nd ed. ii, p. 226.

Coloration. Resembles *A. raytal,* but has a very much smaller bill, and, generally, a shorter wing and tail [*].

Bill fleshy, dark brown on culmen and tip, with a slight shade of horny blue on lower mandible; legs and feet brownish flesh; iris brown (*Butler*).

Length about 5·5; tail 1·8; wing 3·2; tarsus ·75; bill from gape ·55; bill from tip to forehead ·4.

Distribution. Sind, and along all the large rivers of the Punjab, as far east as the Jumna.

Habits, &c. Breeds in March, April, and May: the mode of nidification of this species does not appear to differ in any important particular from that of *A. raytal.*

868. **Alaudula persica.** *Sharpe's Sand-Lark.*

Alauda pispoletta, *Pall., Hume, Ibis,* 1870, p. 531.
Alaudula pispoletta (*Pall.*), *Hume, S. F.* vii, p. 528; *id. Cat.* no. 762 bis.
Alaudula persica, *Sharpe, Cat. B. M.* xiii, p. 590 (1890).

Coloration. Upper plumage sandy fulvous, each feather with a narrow dark brown shaft-streak; wings dark brown margined with sandy fulvous; middle pair of tail-feathers brown broadly margined with fulvous; the next three pairs dark brown with narrow margins; the penultimate dark brown on the inner web, white on the outer; the outermost feather white, with the inner half of the inner

---

[*] *A. adamsi* is said to be larger than *A. raytal,* but a series of measurements of both taken by me has convinced me that the contrary is the case.

web black; lores, supercilium, and sides of the head pale buff; lower plumage fulvous white, rather darker on the breast and flanks; breast distinctly streaked with brown; flanks indistinctly streaked; the cheeks frequently streaked with brown.

The colours of the soft parts of this Lark do not appear to have been recorded.

Length about 6; tail 2·6; wing 4; tarsus ·8; bill from gape ·5; bill from forehead to tip ·4.

Sharpe has quite properly bestowed a name on this hitherto unnoticed and well-defined species, but in doing so he has selected a Persian example as his type, a bird collected by Blanford at Niriz in Persia. The Larks of this type found in India, although agreeing with this Persian bird in general size, structure, and coloration, differ from it immensely in the size of the bill. The former have the culmen measuring ·4 and the latter ·6. Having seen only one Persian bird I do not wish to insist too much upon this difference in size of bill, otherwise I should be disposed to keep the Indian birds distinct.

Distribution. The Punjab, extending west through Afghanistan to Persia. This species is probably only a winter visitor to the plains of the Punjab.

Genus **MIRAFRA**, Horsf., 1821.

The genus *Mirafra* contains the Bush-Larks, which are found in well-wooded districts. They frequently perch on bushes and low trees and they take short flights in the air. Their song is pleasant but weak.

In *Mirafra* the bill is thick and short and the nostrils are quite exposed to view; there are ten primaries in the wing, the first of which is about a half or a third the length of the second; the hind claw is much longer than the hind toe and gently curved.

Key to the Species.

a. Inner web of outer tail-feather largely white............................... *M. cantillans*, p. 333.
b. Inner web of outer tail-feather dark brown.
  a'. Rufous on inner and outer webs of primaries not reaching to shaft.
    a". General tone of upper plumage ashy brown .... *M. assamica*, p. 334.
    b". General tone of upper plumage rufous.
      a'''. Wing 3·2 to 3·5 .... *M. affinis*, p. 335.
      b'''. Wing 2·7 to 3 .... *M. microptera*, p. 336.
  b'. Rufous on inner and outer webs of primaries reaching to shaft and confluent. *M. erythroptera*, p. 334.

869. **Mirafra cantillans.** *The Singing Bush-Lark.*

Mirafra cantillans, *Jerd., Blyth, J. A. S. B.* xiii, p. 960 (1844); *id. Cat.* p. 134; *Horsf. & M. Cat.* ii, p. 476; *Jerd. B. I.* ii, p. 420; *Hume, N. & E.* p. 476; *id. Cat.* no. 757; *Barnes, Birds Bom.* p. 275; *Sharpe, Cat. B. M.* xiii, p. 605; *Oates in Hume's N. & E.* 2nd ed. ii, p. 227.

Aghnn, Aghin, Hind.; *Burutta pitta, Aghin pitta,* Tel.

Coloration. After the autumn moult the whole upper plumage is dark brown, each feather with rufous lateral margins and a whitish terminal band; wing-coverts and tertiaries brown margined with rufous; primary-coverts, primaries, and secondaries with nearly the entire outer web deep rufous or chestnut; middle pair of tail-feathers brown broadly margined with rufous, the next three pairs

Fig. 95.—Head of *M. cantillans.*

almost entirely brown, the penultimate brown on the inner web, white on the outer, the outermost white with a blackish band on the inner margin of the inner web; lores and supercilium very pale fulvous; sides of the head mottled with fulvous and brown; chin and throat white; remainder of lower plumage fulvous, the sides of the neck and the whole breast streaked with triangular brown marks.

Shortly after the autumn moult the whitish terminal bands or fringes of the feathers of the upper plumage become worn away.

Iris brown; legs, feet, and lower mandible fleshy; upper mandible horny brown (*Butler*).

Length nearly 6; tail 2·1; wing 3·2; tarsus ·75; bill from gape ·55.

Distribution. Locally distributed over a considerable portion of the Indian Peninsula. This species is found in the Punjab, Rajputana, the North-West Provinces, and Western Bengal, extending south to about the latitude of Madras. It appears to ascend the slopes of the Himalayas in suitable spots, as Stoliczka records it from the Sutlej valley.

Habits, &c. Breeds from March to August, probably having two broods in the year, and laying four eggs, which are dull white thickly marked with various shades of brown, and measure about ·81 by ·62. This species is less frequently found amongst bushes than other members of the genus, and it is commonly met with on grass-land about cultivated tracts. It is often kept caged for the sake of its song.

870. **Mirafra assamica.** *The Bengal Bush-Lark.*

Mirafra assamica, *McClell. P. Z. S.* 1839, p. 162; *Horsf. & M. Cat.*
ii, p. 476; *Jerd. B. I.* ii, p. 416; *Hume, N. & E.* p. 473; *Ball, S.*
F. vii, p. 223; *Cripps, S. F.* vii, p. 294; *Hume, Cat.* no. 754;
Anders. Yunnan Exped., Aves, p. 606; *Oates, B. B.* i, p. 375;
Sharpe, Cat. B. M. xiii, p. 600; *Hume, S. F.* xi, p. 287; *Oates in*
Hume's N. & E. 2nd ed. ii, p. 229.
Mirafra assamensis, *McClell., Blyth, Cat.* p. 134.
Mirafra immaculata, *Hume, S. F.* i, p. 12 (1873).

Aggia, Hind.; *Bhiriri* at Bhágalpur.

Coloration. The whole upper plumage dark ashy brown with
blackish streaks on all parts except the rump; wing-coverts blackish
margined with pale ashy; primary-coverts externally rufous; quills
dark brown, most of them externally margined with chestnut and
all of them with a large portion of the inner web chestnut; tail
brown margined with ashy rufous, the penultimate and outer feathers
with the greater part of the outer web pale rufous; lores and an
indistinct supercilium fulvous; sides of the head fulvous barred
with brown; chin and throat pale fulvous white; remainder of
lower plumage darker fulvous, the breast coarsely streaked with
triangular marks of brown; under wing-coverts and axillaries
ferruginous.

Bill dusky above, fleshy white below; legs fleshy white; iris
yellowish brown (*Hume*).

Length rather more than 6; tail 2; wing 3·3; tarsus 1; bill
from gape ·75.

Distribution. The north-eastern part of the Indian Peninsula,
north and east of a line drawn roughly from Garhwál to Cuttack,
extending through Bengal into Assam and thence south on the
one hand to the neighbourhood of Bhámo and on the other to
Arrakan.

Habits, &c. Breeds from March to July, laying four or five eggs,
which are greyish white speckled with brown of different shades,
and measure about ·82 by ·61.

871. **Mirafra erythroptera.** *The Red-winged Bush-Lark.*

Mirafra erythroptera, *Jerd. Madr. Journ. L. S.* xiii, pt. 2, p. 136
(1844); *Blyth, Cat.* p. 133; *Jerd. Ill. Ind. Orn.* pl. 38; *Horsf.*
& M. Cat. ii, p. 474; *Jerd. B. I.* ii, p. 418; *Hume, N. & E.* p. 475;
Ball, S. F. vii, p. 223; *Hume, Cat.* no. 756; *Barnes, Birds Bom.*
p. 274; *Sharpe, Cat. B. M.* xiii, p. 612; *Oates in Hume's N. & E.*
2nd ed. ii, p. 231.

Jungli aggia, Hind.; *Chinna eeli-jitta,* Tel.

Coloration. The whole upper plumage fulvous-brown streaked with
dark brown or black; middle pair of tail-feathers pale brown margined with fulvous; the others blackish, the outermost feather with

the outer web entirely pale rufous, and the inner web tipped with the same colour; wing-coverts brown edged with fulvous; quills brown with a large portion of both webs chestnut; lores and a supercilium pale fulvous; cheeks and ear-coverts fulvous speckled with brown; chin and throat whitish; remainder of lower plumage pale fulvous, the breast spotted with triangular marks of brown or black; under wing-coverts pale rufous.

Bill horny brown above, fleshy below; legs flesh-colour; iris hazel (*Butler*).

Length about 5·5; tail 2·1; wing 3; tarsus ·85; bill from gape ·65.

Distribution. The whole of India from the foot of the Himalayas to about the latitude of Nellore and east to the longitude of Calcutta. This Lark appears to be rare in Sind and the western parts of Rajputana and the Punjab.

Habits, &c. Breeds from March to September. The eggs, four or five in number, are speckled with various shades of red and brown, and measure about ·76 by ·59.

872. **Mirafra affinis.** *The Madras Bush-Lark.*

Mirafra affinis, *Jerd. Madr. Journ. L. S.* xiii, pt. 2, p. 136 (1844); *id. Ill. Ind. Orn.* text to pl. 38; *Blyth, Cat.* p. 133; *Horsf. & M. Cat.* ii, p. 475: *Jerd. B. I.* ii, p. 417; *Hume, N. & E.* p. 474; *Ball, S. F.* vii, p. 223; *Hume, Cat.* no. 755; *Legge, Birds Ceyl.* p. 634: *Davison, S. F.* x, p. 404; *Sharpe, Cat. B. M.* xiii, p. 614; *Oates in Hume's N. & E.* 2nd ed. ii, p. 233.

Eeli-jitta, Tel.; *Chirchira,* Hind.; *Leepee,* in Central India; *Gomarita,* Cing.

Coloration. Upper plumage rufous-brown with very broad median dark brown streaks to all the feathers; tail dark brown narrowly margined with rufous, the outer web of the outermost feather being very broadly margined with this colour; wing-coverts and quills dark brown, margined with rufous, most of the quills with a large band of rufous on the inner web; lores and supercilium pale fulvous; ear-coverts rufous mottled with brown; chin and throat very pale fulvous; remainder of lower plumage deeper fulvous, the breast streaked with large triangular patches of dark brown; under wing-coverts and axillaries rufous.

Legs, feet, claws, lower mandible, and edges of upper mandible fleshy white; rest of upper mandible horny brown; iris sienna-brown (*Davison*).

Length about 6; tail 1·8 to 2·1; wing 3·2 to 3·5; tarsus 1; bill from gape ·7.

Distribution. Ceylon and Southern India, extending north to Midnapore in Bengal on the eastern side of the Peninsula and to the Nilgiris on the western.

Habits, &c. Little is known about the nidification of this Lark. It appears to lay in May and June and its nest and eggs are not likely to differ in any respect from those of the other species, the habits of which are well known.

873. **Mirafra microptera.** *The Burmese Bush-Lark.*

Mirafra microptera, *Hume, S. F.* i, p. 483 (1873); *id. N. & E.* p. 475; *id. S. F.* iii, p. 159; *id. Cat.* no. 755 bis; *Sharpe, Cat. B. M.* xiii, p. 615; *Oates in Hume's N. & E.* 2nd ed. ii, p. 233.
Mirafra affinis, *Jerd. apud Oates, B. B.* i, p. 376.

Coloration. Resembles *M. affinis,* but is considerably smaller.

Iris hazel; lower mandible and margin of upper very pale pinkish fleshy; remainder of upper mandible dark horny; legs light fleshy; claws pinkish.

Length 5·5; tail 1·55; wing 2·7 to 3; tarsus ·85; bill from gape ·55.

Distribution. The northern part of Pegu about the district of Thayetmyo. Specimens of a Bush-Lark from Saigon appear referable to this species.

Habits, &c. A nest of this bird that I found in Pegu in July contained two eggs and one young bird, and was placed on the ground in a hoof-mark. It was partially domed and constructed of grass and fibres. One egg measured ·83 by ·6 and was thickly marked with brown.

Genus **GALERITA**, Boie, 1828.

The genus *Galerita* contains the Crested Lark, and two other Larks which have hitherto been placed in the genus *Spizalauda.* These two genera appear to me to be identical and I therefore unite them.

In *Galerita* the bill is about half the length of the head and pretty strong; the head is furnished with a few long feathers forming a conspicuous crest; the nostrils are completely covered by plumelets; there are ten primaries in the wing, the first of which is very small; the hind claw is about the same length as the hind toe and very slightly curved; and the sexes are quite alike.

The Crested Larks resemble the Sky-Larks closely in their habits.

Key to the Species.

a. Bill at front about ·7; general colour above earthy brown.................... *G. cristata,* p. 337.
b. Bill at front about ·5; general colour above rufous.
  a'. Wing about 3·4; pectoral streaks few and narrow .... *G. deva,* p. 338.
  b'. Wing about 3·8; pectoral streaks numerous and broad .... *G. malabarica,* p. 339.

GALERITA.

574. Galerita cristata. *The Crested Lark.*

Alauda cristata, *Linn. Syst. Nat.* i, p. 288 (1766).
Alauda chendoola, *Frankl. P. Z. S.* 1831, p. 119.
Galerida chendoola (*Frankl.*), *Blyth, Cat.* p. 133.
Galerida cristata (*Linn.*), *Horsf. & M. Cat.* ii, p. 465; *Jerd. B. I.* ii,
   p. 436; *Hume, N. & E.* p. 488; *id. S. F.* i, p. 214; *Butler, S. F.*
   vii, p. 185; *Hume, Cat.* no. 760; *Barnes, Birds Bom.* p. 283;
   Sharpe, Cat. B. M. xiii, p. 626; *Oates in Hume's N. & E.* 2nd ed.
   ii, p. 233.
Galerida magna, *Hume, Ibis,* 1871, p. 407; *id. & Henders. Lah. to
   Yark.* p. 270, pl. 30; *Scully, S. F.* iv, p. 175.

The Large Crested Lark, Jerd.; *Chendul,* Hind.; *Chendul, Jutu-pitta,*
Tel.

Fig. 96. Head of *G. cristata.*

Coloration. Upper plumage earthy brown, with blackish streaks
or centres to most of the feathers; tail-feathers brown, with sandy
margins and tips, the penultimate feather with the greater portion
of the outer web pale rufous, the outermost all pale rufous except
the inner portion of the inner web, which is brown; wing-coverts
and quills brown with sandy margins, the quills with a large patch
of rufous on the inner web; lores brown; supercilium pale fulvous; ear-coverts pale fulvous white, mottled with brown; entire
lower plumage pale fulvous with some brown spots on the cheeks
and numerous brown streaks on the breast; the sides of the body
obsoletely streaked; under wing-coverts and axillaries rufous.

Bill yellowish; feet pale brown; iris dark brown (*Jerdon*).

Length about 7·5; tail about 2·7; wing 3·5 to 4·3; tarsus 1·05;
bill from gape about ·9.

The Crested Lark varies as much as the Common Sky-Lark both
in size and colour, and it is as difficult in the case of the one as of
the other to subdivide it into two or more races.

Distribution. The north-western portion of India, extending east
as far as the 85th degree of east longitude, and south as far generally as the 23rd degree of north latitude, but occasionally further
south in favourable localities, this species having been recorded from
Raipur in the Central Provinces. Many Larks of this species are
resident and breed in India, but the majority appear to migrate in

VOL. II.

spring to Central Asia. This Lark, in a more or less variable form, has an immense range, being found in Europe and Northern Africa and the greater part of Asia, as far east as China.

Habits, &c. Breeds in India from March to June, constructing a small nest of grass on the ground under shelter of a stone or clod of earth. The nest is usually lined with cotton, hair, fibres, and feathers. The eggs, usually three in number, are dull white marked with brown and purple, and measure about ·87 by ·65.

875. **Galerita deva.** *Sykes's Crested Lark.*

Alauda deva, *Sykes, P. Z. S.* 1832, p. 92.
Mirafra hayii, *Jerd. Madr. Journ. L. S.* xiii, pt. 2, p. 136 (1844);
  Blyth, Cat. p. 133.
Spizalauda deva (*Sykes*), *Horsf. & M. Cat.* ii, p. 477; *Jerd. B. I.*
  ii, p. 432; *Hume, Cat.* no. 765; *Davison, S. F.* x, p. 704; *Barnes,*
  Birds Bom. p. 281; *Sharpe, Cat. B. M.* xiii, p. 621.
Alauda simillima (*Hume*), *J. A. S. B.* xxxix, pt. ii, p. 120 (1870).
Spizalauda simillima (*Hume*), *Hume, N. & E.* p. 484.
Alauda (Spizalauda) deva, *Sykes, Blanf. S. F.* iv, p. 240.
Galerita deva (*Sykes*), *Oates in Hume's N. & E.* 2nd ed. ii, p. 236.

The Small Crested Lark, Jerd.; *Chinna chandul,* Tel.

Coloration. Upper plumage rufous, with dark brown streaks or centres to all the feathers; the upper tail-coverts more uniformly rufous; tail dark brown edged with rufous, the penultimate feather with the outer web entirely rufous, the outermost feather all rufous except a small portion of the inner web; wing-coverts and quills brown, edged with rufous, all the quills with a large amount of pale rufous on the inner web; a broad and well-defined pale rufous supercilium; sides of the head pale rufous mottled with brown: the entire lower plumage rufous, with a few spots on the cheeks and a few narrow black streaks on the breast: under wing-coverts and axillaries rufous.

Bill horny brown above, pale flesh-colour below: legs and feet yellowish brown: iris dark brown (*Butler*).

Length about 6; tail 2·1; wing 3·4; tarsus ·8; bill from gape ·6.

Distribution. A considerable portion of the peninsula of India, where this species is a permanent resident. It occurs in Cutch, Rajputana, the eastern portion of the Punjab, the North-west Provinces and Central India, extending south to about the latitude of Bangalore and Madras. Its eastern limits are not known with any precision, but it does not seem to be found east of the 80th degree of longitude. To the west it is found everywhere as far as the sea-coast.

Habits, &c. Breeds from June to August, making its nest at the foot of a tuft of grass or bush, and laying three eggs, which are speckled with reddish brown and purplish, and measure about ·77 by ·6.

876. **Galerita malabarica.** *The Malabar Crested Lark.*

Alauda malabarica, *Scop. Del. Flor. et Faun. Insubr.* ii, p. 94 (1786);
   Jerd. B. I. ii, p. 436.
Spizalauda malabarica (*Scop.*), *Hume, N. & E.* p. 483; *id. Cat. no.* 765
   bis; *Davison, S. F.* x, p. 405; *Barnes, Birds Bom.* p. 282.
Alauda (Spizalauda) malabarica, *Scop. Blanf. S. F.* iv, p. 241.
Galerita malabarica (*Scop.*), *Sharpe, Cat. B. M.* xiii, p. 633; *Oates
   in Hume's N. & E.* 2nd ed. ii, p. 237.

The Crested Malabar Lark, Jerd.

Coloration. Resembles *G. deva* very closely, but is considerably
larger, has the streaks on the breast very broad and coarse and the
light pattern of the tail much deeper rufous.

Iris dark brown; legs and feet livid flesh; bill horny brown
above, whitish flesh below (*Butler*).

Length nearly 7; tail 2·3; wing 3·8; tarsus ·85; bill from
gape ·7.

Distribution. A permanent resident in the western part of the
peninsula of India from Guzerat to Travancore, occurring up to
the summit of the hill-ranges of those parts.

Habits, &c. Breeds from March to September, having two broods.
The mode of nidification of this species appears to be quite the
same as that of *G. deva*, but the eggs are larger and more distinctly
marked and measure about ·87 by ·65.

Genus **AMMOMANES**, Cabanis, 1850.

The genus *Ammomanes* contains those Finch-Larks which are
characterized by a general rufous tone of plumage and by the
sexes being alike in colour.

In *Ammomanes* the bill is thick and slightly curved and resembles
that of *Calandrella* very closely; the nostrils are covered by
plumelets; the wing has ten primaries, the first being small but
exceeding the primary-coverts considerably, and the second quill
is much shorter than the third; the hind claw is short and curved.

The Larks of this genus are found in open plains and arid spots.
They rise singing in the air for a short distance and descend with
a sudden drop.

Key to the Species.

a. Tail deep rufous broadly tipped black. . . .   *A. phœnicura,* p. 339.
b. Tail brown throughout, merely tinged with
   rufous . . . .   *A. phœnicuroides,* p. 340.

877. **Ammomanes phœnicura.** *The Rufous-tailed Finch-Lark.*

Mirafra phœnicura, *Frankl. P. Z. S.* 1831, p. 119; *Blyth, Cat.* p. 131.
Ammomanes phœnicura (*Frankl.*), *Horsf. & M. Cat.* ii, p. 477;
   Jerd. B. I. ii, p. 421; *Hume, N. & E.* p. 477; *id. Cat. no.* 758;

Bull, S. F. vii, p. 223; *Barnes, Birds Bom.* p. 276; *Sharpe, Cat. B. M.* xiii, p. 642; *Oates in Hume's N. & E.* 2nd ed. ii, p. 240.

Aygiya, Retal, Hind.; *Ambali-jori-gadu, Dowa-pitta,* Tel.

Coloration. Upper plumage dark brown, with slightly darker shafts and obsolete pale margins to all the feathers, those of the head with blackish streaks; upper tail-coverts deep rufous; tail deep rufous with a broad black tip; wing-coverts and quills brown margined with sandy brown, the quills with a large amount of rufous on the inner web; a very indistinct supercilium pale rufous; sides of the head rufous streaked with brown; entire lower plumage rufous, the chin, throat, and breast streaked with brown.

Bill horny brown above, fleshy at the base beneath; legs fleshy; iris brown (*Jerdon*).

Length about 6·5; tail 2·4; wing 4·1; tarsus ·9; bill from gape ·65.

Distribution. A permanent resident over a considerable portion of the peninsula of India. The western limit of this species appears to be a line drawn from the head of the Rann of Cutch to Delhi and thence produced to the Ganges; the northern boundary would appear to be the Ganges itself as far as Dinapore, and thence this Lark is spread over the entire country, in suitable localities, down to Coimbatore.

Habits, &c. Breeds from February to April, making its nest of grass on the ground and laying three or four eggs, which are speckled with yellowish and reddish brown and measure about ·85 by ·62.

878. Ammomanes phœnicuroides. *The Desert Finch-Lark.*

Mirafra phœnicuroides, *Blyth, J. A. S. B.* xxii, p. 583 (1853).
Ammomanes phœnicuroides (*Blyth*), *Horsf. & M. Cat.* ii, p. 478; *Sharpe, Cat. B. M.* xiii, p. 617; *Oates in Hume's N. & E.* 2nd ed. ii, p. 242.
Ammomanes lusitanica (*Gmel.*), *Jerd. B. I.* ii, p. 422; *Hume, N. & E.* p. 478; *id. S. F.* i, p. 211.
Ammomanes deserti (*Licht.*), *Hume, Cat.* no. 759; *Barnes, Birds Bom.* p. 276.

The Pale-rufous Finch-Lark, Jerd.

Coloration. Upper plumage greyish brown, tinged with rufous on the upper tail-coverts and slightly streaked with blackish on the crown; tail brown margined with pale rufous, the outer web of the outer tail-feather entirely of this colour, the bases of all the feathers tinged reddish; wing-coverts and quills brown margined with pale fulvous, the inner web of all the quills largely pale rufous; lores brown; a ring round the eye and a line above and below the lores fulvous white; ear-coverts greyish brown; chin and throat pale fulvous white, with a few brown spots on the lower throat; remainder of lower plumage fulvous grey, with a few brown streaks on the breast; under wing-coverts and axillaries rufous.

Bill dusky above, yellowish beneath; feet pale yellow-brown (*Jerdon*).

Length about 7; tail 2·6; wing 4·1; tarsus ·9; bill from gape ·65.

Distribution. A permanent resident throughout Sind and the northern part of the Punjab, ranging west to the Persian Gulf.

Habits, &c. Breeds in April, May, and June, making its nest on the ground and surrounding it with a circle of small pieces of stone. The eggs, three or four in number, resemble those of *A. phœnicura* and measure about ·83 by ·6.

Genus PYRRHULAUDA, Smith, 1839.

The genus *Pyrrhulauda* contains those Finch-Larks in which the sexes are very different, the males being black beneath and the females rufous or white.

In *Pyrrhulauda* the bill is very short and deep, with the culmen well rounded; the nostrils are densely covered with plumelets; the wing has ten primaries, the first very small and not exceeding the primary-coverts, the second reaching to the tip of the wing; the hind claw is very short and curved.

The Larks of this genus affect open country and they take short flights, ascending and descending suddenly. They occasionally perch on houses and trees.

Key to the Species.

a. Lower surface blackish.
  a'. Forehead and crown ashy brown,
    margined with pale grey .... *P. grisea* ♂, p. 341.
  b'. Forehead white, crown blackish .... *P. melanauchen* ♂, p. 343.
b. Lower plumage pale rufous or whitish.
  c'. General aspect of upper plumage
    brown; wing 2·9............... *P. grisea* ♀, p. 341.
  d'. General aspect of upper plumage
    sandy; wing 3·1 or 3·2 .... *P. melanauchen* ♀, p. 343.

879. Pyrrhulauda grisea. *The Ashy-crowned Finch-Lark.*

Alauda grisea, *Scop. Del. Flor. et Faun. Insubr.* ii, p. 95 (1786).
Pyrrhulauda grisea (*Scop.*), *Blyth, Cat.* p. 134; *Horsf. & M. Cat.* ii, p. 479; *Jerd. B. I.* ii, p. 424; *Hume, N. & E.* p. 479; *Ball, S. F.* vii, p. 223; *id. Cat.* no. 760; *Legge, Birds Ceyl.* p. 637; *Barnes, Birds Bom.* p. 277; *Sharpe, Cat. B. M.* xiii, p. 652; *Oates in Hume's N. & E.* 2nd ed. ii, p. 243.

342 ALAUDIDÆ.

The Black-bellied Finch-Lark, Jerd.; *Diyora, Duri, Dalhuk chun, Jothanli*, Hind.; *Chat-bharai, Dhula chata*, Beng.; *Poti-pichike, Piyala pichike*, Tel.

Coloration. Male. Upper plumage ashy brown, each feather margined with pale grey and the forehead and crown more broadly margined than the other parts; middle tail-feathers light brown, the others dark brown, the outermost feather with the outer web and the terminal half of inner whitish; wings dark brown margined with pale grey: lores, front part of cheeks, a supercilium, chin, throat, sides of neck, breast, abdomen, and under tail-coverts dark chocolate-brown; posterior part of cheeks, ear-coverts, and sides of the breast white; sides of body mixed ashy and blackish; under wing-coverts and axillaries chocolate-brown.

Fig. 97.—Head of *P. grisea.*

Female. Darker brown above, with narrower and darker grey margins and with a tinge of rufous throughout; tail as in the male: wings of much the same colour as the upper plumage; lores, a supercilium, and round the eye rufous; ear-coverts mixed rufous and brown; lower plumage pale rufous, with obscure, ill-defined brown striations chiefly on the breast.

The young bird resembles the female, but has the margins of the feathers of the upper plumage very distinct and broad and of a pale rufous colour.

Iris dark brown; legs and feet brownish flesh; bill bluish flesh, horny brown on the culmen (*Butler*).

Length about 5·5; tail 1·8; wing 3; tarsus ·55; bill from gape ·5.

Distribution. The plains of India from Sind to the longitude of Calcutta and from the foot of the Himalayas to Cape Comorin, extending to Ceylon. This species is not recorded from the northern and western portions of the Punjab, but with this exception is found throughout the above area in suitable localities. It is everywhere a permanent resident.

Habits, &c. Breeds from January to August, having two broods in the year. The nest is a small pad of grass, fibres, and feathers placed on the ground. The eggs, two in number, are speckled with brown and grey and measure about ·73 by ·55.

880. **Pyrrhulauda melanauchen.** *The Black-crowned Finch-Lark.*

Coraphites melanauchen, *Cabanis*, Mus. Hein. i. p. 121 (1850).
Pyrrhulauda affinis, *Blyth*, Ibis, 1867, p. 185; *Hume*, S. F. i, p. 212.
Pyrrhulauda melanauchen (*Cab.*), *Hume*, S. F. vii, p. 64; *id. Cat.*
no. 750 bis; *Barnes, Birds Bom.* p. 277; *Sharpe, Cat. B. M. xiii.*
p. 655; *Oates in Hume's N. & E.* 2nd ed. ii, p. 248.

Coloration. *Male.* Resembles male of *P. grisea*, but differs in
having the forehead broadly white, the whole crown dark chocolate-
brown or blackish, and the white of the ear-coverts produced
narrowly round the hind neck to form a collar; the black sides of
the neck are also produced as a collar over the mantle, immediately behind the white collar.

Female. Resembles the female of *P. grisea*, but is much paler and
more sandy; the lower plumage is less rufous and almost pure
white on the abdomen, and the streaks are fewer in number.

Bill pale whity brown, bluish on lower mandible; legs and feet
pale whity brown; iris brown (*Hume*).

Length about 5·5; tail 2·2; wing 3·3; tarsus ·65; bill from
gape ·5.

Distribution. A permanent resident in Sind and the western half
of Rajputana. This species has also been obtained at Muttra, just
within the limits of the Punjab. It extends westwards to Arabia
and North-eastern Africa.

Habits, &c. Breeds apparently throughout the year in Sind,
having three broods. The nest and eggs appear to be very similar to
those of *P. grisea*, and the eggs measure about ·75 by ·54.

Family NECTARINIIDÆ.

The intrinsic muscles of the syrinx fixed to the ends of the
bronchial semi-rings; both mandibles finely and evenly serrated on
the terminal third of their edges; tongue tubular; bill long and
cylindrical; the nestling resembling the adult female; one moult in
the year; wing of ten primaries, the first small; rectrices twelve;
tarsus scutellated; rictal bristles short.

The *Nectariniidæ*, or *Sun-birds*, constitute a family of birds which
are found only in the Old World and chiefly within the tropics. The
Sun-birds are of small and delicate make and the majority are clothed
in resplendent plumage. They are found solitary or in pairs; they
are entirely arboreal in their habits and they feed on minute insects
and on the nectar of flowers. This latter they secure with their
tongues when clinging to flower-stems, as they are unable to poise
themselves in the air, after the manner of Humming-birds, except

on rare occasions, and only then for a very brief interval. The males
have a short but pretty song in the breeding-season. The Sun-birds
build elaborate pensile nests at the end of branches or attach them
to the underside of a broad leaf, such as that of a plantain (*Musa*).
They usually lay two eggs, which are always, so far as is known,
spotted.

After examining all the known species of Sun-birds I find that
without exception they are characterized by having both mandibles
of the bill serrated on the terminal third of their length. This
character suffices to separate them from all the other Passeres

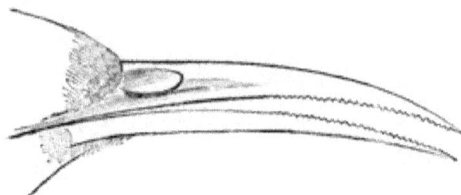

Fig. 98.—Bill of *Anthothreptes malaccensis* (enlarged), to show serrated edges
of mandibles.

except the *Dicæidæ*, and from these they may be distinguished by the
shape of their bill, which is long, fine, and cylindrical, whereas in
the *Dicæidæ* it is short and triangular. Under these circumstances
the key to the families of Passeres (vol. i, pp. 8, 9) is susceptible of
being considerably improved and simplified by deleting section *a'*,
Tongue non-tubular and *b'*, Tongue tubular. The *Nectariniidæ*
may then be entered under section *b'* together with the *Dicæidæ*,
thus :—

   b'. Both mandibles finely and evenly serrated on the
     terminal third of their edges.
    c''. Bill long, fine, and cylindrical; primaries invari-
      ably ten .... **Nectariniidæ.**
    d''. Bill short and triangular; primaries either nine
      or ten .... **Dicæidæ.**

I find also that, for reasons explained in their proper place, the
genus *Chalcoparia* cannot be placed among the *Nectariniidæ*. The
position of this genus is undoubtedly among the *Liotrichinæ* in the
family *Crateropodidæ*, probably near *Myzornis*.

It has frequently been asserted that the males of many species
of Indian Sun-birds have a distinct summer and winter plumage.
After examining the very large series of Sun-birds in the British
Museum, I am convinced that this is never the case. Full-
plumaged males of all the common species, and it is of these that
the assertion has been made, shot in every month of the year,
or at such frequent intervals as to practically amount to the same
thing, are in the National Collection, and prove that the adult males
never change their colours. Young males are to be found throughout their first year in immature plumage, and these have probably

given rise to the belief that a seasonal change takes place in the adult.

The young birds of both sexes resemble the adult female up to the first autumn moult. The males then commence to assume the colours of the adult and the change is effected very slowly and probably extends over a whole year.

The Indian Sun-birds may be conveniently divided into two subfamilies.

Sexes different ; plumage of male in part
    metallic ; bill slender ; nest pensile.   *Nectariniinæ*, p. 345.
Sexes alike ; plumage non-metallic ; bill
    large ; nest cup-shaped, attached by
    a portion of the rim to a broad
    leaf......................   *Arachnotherinæ*, p. 368.

Subfamily NECTARINIINÆ.

The Sun-birds of this subfamily are characterized by a slender body and an attenuated bill ; by the sexes being different, and by the males having bright metallic colours in their plumage, the females being of a dull green or yellow colour.

Key to the Genera.

a. Covering-membrane of nostril feathered.   CHALCOSTETHA, p. 345.
b. Covering-membrane of nostril bare.
  a'. Lower mandible of bill distinctly
    curved downwards.
    a''. Males with middle tail-feathers
      elongated ; rump yellow ; females
      with lower plumage green......   ÆTHOPYGA, p. 346.
    b''. Both sexes with short, rounded
      tails ; females yellow beneath ..   ARACHNECHTHRA. p. 357.
  b'. Lower mandible of bill straight or
    nearly so ....   ANTHOTHREPTES, p. 365.

Genus **CHALCOSTETHA**, Cabanis, 1850.

The genus *Chalcostetha* contains a single species of Sun-bird, which may be recognized by the covering-membrane of the nostril being completely plumed and by the tail being of considerable length and well graduated. The bill is slender and the lower mandible is nearly straight.

881. **Chalcostetha pectoralis**. *Maklot's Sun-bird.*

Nectarinia pectoralis, *Temm. Pl. Col.* pl. 138, fig. 3 (1823).
Nectarinia insignis, *Jard. Naturalist's Libr., Sun-birds,* p. 274
    (1843).
Nectarinia insignis, *Gould, P. Z. S.* 1865, p. 663.
Chalcostetha insignis (*Jard.*), *Walden, Ibis,* 1870, p. 44 ; *Hume, S. F.*

iii, p. 319 ; *Shelley, Mon. Nect.* pp. xxv, 87, pl. 30 ; *Hume & Dav. S. F.* vi, p. 183 ; *Hume, Cat.* no. 231 ter ; *Gadow, Cat. B. M.* ix, p. 12.

Chalcostetha insperata, *Hume, S. F.* iii, p. 320 (1875).
Chalcostetha pectoralis (*Temm.*), *Oates, B. B.* i, p. 317.

Coloration. Male. Forehead, crown, and nape metallic green : lores, sides of the head, and back black ; scapulars, lesser and median wing-coverts, lower back, rump, and upper tail-coverts metallic green suffused with lilac ; greater coverts and quills dark brown margined with purple ; tail deep blue margined with metallic purple ; chin and throat metallic copper-colour, surrounded by lilac-purple, which colour also covers the whole breast ; pectoral tufts bright yellow ; abdomen, sides of the body, and under tail-coverts dull brownish black.

Female. Forehead, crown, and nape brown, edged with grey ; upper plumage dull olive-green ; quills brown, edged with the colour of the back ; tail black, all but the median pair of feathers tipped white ; feathers round the eye, sides of the head, chin, and throat pale grey ; breast, abdomen, and sides of the body yellow ; vent and under tail-coverts pale yellowish white.

The young resemble the female, and the young male moults into adult plumage at the first autumn.

Legs and feet black or bluish black ; bill black ; iris dark brown (*Hume & Davison*).

Length about 5·5 ; tail 2·1 ; wing 2·3 ; tarsus ·55 ; bill from gape ·8.

Distribution. The extreme southern point of Tenasserim and Patoe Island, extending down the Malay peninsula to the islands.

Genus **ÆTHOPYGA**, Cabanis, 1850.

The genus *Æthopyga* contains those Sun-birds, the males of which have the middle pair of tail-feathers produced beyond the next pair and the lower back or rump yellow. The females are not so easy to diagnose. They resemble each other very closely and also the females of the next genus *Arachnechthra*, but they may be distinguished from the latter by the general green tone of the lower plumage.

In this genus the bill is slender and well curved downwards and the covering-membrane of the nostril is bare.

Key to the Species.

a. Chin and throat crimson.
  a'. Middle tail-feathers exceeding next pair
    by more than length of tarsus.
    a''. Crown and tail metallic green; tail
      2·7 .... *Æ. seheriæ* ♂, p. 348.
    b''. Crown and tail metallic violet; tail
      2·4 .... *Æ. andersoni* ♂, p. 349.

    b'. Middle tail-feathers exceeding next pair
        by less than half tarsus.
        c". No metallic patch behind the ear-
            coverts.
            a'''. Moustachial streak entirely violet.   *Æ. cara* ♂, p. 349.
            b'''. Moustachial streak bordered with
                black interiorly ....   *Æ. nicobarica* ♂, p. 350.
        d". A metallic patch behind the ear-
            coverts ....   *Æ. rigorsi* ♂, p. 350.
  b. Chin and throat of a dark colour, not
    crimson.
    c'. Middle tail-feathers red ....   *Æ. ignicauda* ♂, p. 351.
    d'. Middle tail-feathers green or violet.
        e". Back crimson.
            c'''. Dark portion of crown abruptly
               defined posteriorly and not ex-
               tending to nape.
               a'. Breast yellow ....   *Æ. gouldiæ* ♂, p. 352.
               b'. Breast crimson ....   *Æ. dabryi* ♂, p. 353.
            d'''. Dark portion of crown extending
              to hind neck or mantle.
               c'. Crown, upper tail-coverts, and
                 tail steel-blue.
                 a'. Breast black ....   *Æ. saturata* ♂, p. 354.
                 b'. Breast yellow, streaked with   [p. 354.
                    crimson..................   *Æ. sanguinipectus* ♂,
               d'. Crown, upper tail-coverts, and
                 tail green ....   *Æ. nepalensis* ♂, p. 355.
        f". Back olive-yellow.................   *Æ. horsfieldi* ♂, p. 356.
  c. Chin and throat green, like remainder of
    lower plumage.
    e'. A yellow band across the rump.
        g". Upper plumage light green.
            e'''. Bill from gape to tip ·8 ....   *Æ. saturata* ♀, p. 354.
            f'''. Bill from gape to tip ·65 ....   *Æ. gouldiæ* ♀, p. 352.
        h". Upper plumage dull green tinged
            with ashy.   [p. 354.
            g'''. Bill from gape to tip ·8 ....   *Æ. sanguinipectus* ♀,
            h'''. Bill from gape to tip ·7 ....   *Æ. dabryi* ♀, p. 353.
    f'. No yellow band across rump.
        i". Distance from tip of outermost feather
            of tail to tip of tail equal to tarsus
            or more.
            i'''. Pale tips to tail-feathers obsolete.   *Æ. ignicauda* ♀, p. 351.
            k'''. Tips of tail-feathers large, white,   { *Æ. nepalensis* ♀, p. 355.
               and well-defined ....   { *Æ. scheriæ* ♀, p. 348.
        k". Distance from tip of outermost feather
            of tail to tip of tail less than hind
            toe.
            l'''. Tail about 1·7 ; lower plumage
               distinctly ashy green ....   *Æ. rigorsi* ♀, p. 350.
            m'''. Tail about 1·5 ; lower plumage
               distinctly pure green * ....   *Æ. cara* ♀, p. 349.

---

* I have not been able to examine females of *Æ. nicobarica* and *Æ. andersoni*, and the only female of *Æ. horsfieldi* accessible to me has no bill. I consequently omit these three from the Key.

882. Æthopyga seheriæ. *The Himalayan Yellow-backed Sun-bird.*

Nectarinia seheriæ, *Tickell, J. A. S. B.* ii, p. 577 (1833).
Cinnyris miles, *Hodgs. Ind. Rec.* ii, p. 273 (1837).
Certhia goalpariensis, *Royle, Ill. Him. Bot.* p. lxxvii, pl. 7 (1839).
Nectarinia goalpariensis (*Royle*), *Blyth, Cat.* p. 223.
Æthopyga miles (*Hodgs.*), *Horsf. & M. Cat.* ii, p. 732 : *Jerd. B. I.* i,
    p. 362 ; *Ball, S. F.* ii, p. 396 ; *Hume, S. F.* v, p. 122.
Æthopyga goalpariensis (*Lath.*), *Hume, N. & E.* p. 146.
Æthopyga seheriæ (*Tick.*), *Shelley, Mon. Nect.* pp. xxi, xxiii, 67,
    pl. 22 ; *Hume, Cat.* no. 225 ; *Gadow, Cat. B. M.* ix, p. 18 ; *Hume,
    S. F.* xi, p. 80 ; *Oates in Hume's N. & E.* 2nd ed. ii, p. 249.

The Himalayan Red Honey-sucker, Jerd.

Coloration. Forehead and greater part of the crown metallic
green; hinder part of crown and nape brownish green; back,
scapulars, lesser and median wing-coverts, sides of head and neck,
chin, throat, and breast crimson; rump bright yellow; upper tail-
coverts and middle pair of tail-feathers metallic green; the other
tail-feathers brown, suffused with violet and edged with metallic
green; greater coverts of wing and quills dark brown, margined
with olive-yellow; a long narrow moustachial streak metallic
violet; abdomen, flanks, and under tail-coverts slaty greenish
yellow; under wing-coverts and axillaries pale yellowish white.

Female. General colour green, the centres of the feathers of the
crown dark ashy, the back, upper tail-coverts, and the margins of
the wing-feathers suffused with yellow and with a russet tinge;
under wing-coverts, axillaries, and sides of the body clear pale
yellow; chin, throat, and sides of head suffused with ashy; middle
tail-feathers greenish, the laterals blackish tipped broadly with
white.

Legs and feet dark brown; upper mandible dark brown;
lower mandible dark horny brownish yellow; iris dark brown
(*Hume*).

Male: length about 6; tail 2·7; wing 2·2; tarsus ·55; bill from
gape ·8. Female: length about 5; tail 1·8; wing 2. In the male
the middle tail-feathers project one inch beyond the tips of the
next pair; in the female the middle tail-feathers are ·75 longer
than the outermost feathers.

Distribution. The Himalayas from Garhwâl to Dibrugarh in
Assam up to 7000 feet in summer; Cachar; Sylhet; the Khási
hills; Manipur. This species is also found in the plains, having
been recorded from Seheria in Borabhoom by Tickell and an *Æthopyga* was seen in Singbhoom by Ball. It is commonly found along
the base of the Himalayas at all seasons, and it is probably resident
in all parts of its range, except the higher portions of the Himalayas.

Habits, &c. Breeds from April to August, constructing a pear-
shaped nest suspended from the end of a twig not far from the
ground. The materials of which the nest is made are grass and

rootlets externally and fine stems of flowering grasses internally. The eggs, two or three in number, are white speckled with greyish purple, and measure about ·59 by ·46.

883. Æthopyga andersoni, n. sp. *Anderson's Yellow-backed Sun-bird.*

Æthopyga miles (*Hodgs.*), *apud Anderson, Yunnan Exped., Aves,* p. 661.
Æthopyga cara, *Hume, apud Salvadori, Ann. Mus.Civ. Gen.* (2) iv, p. 583.

Coloration. Male. Differs from *Æ. scheriæ* in having the forehead, crown, and the visible portions of the closed tail metallic lilac, not green, and the tail measuring 2·4 inches.
Female. Unknown.
Distribution. There are three specimens of this species in the British Museum—two obtained by Dr. Anderson at Sawaddy east of Bhámo in January, and one obtained by my own collector at Bhámo in November. All three are adult males and agree with each other in the particulars pointed out above.

884. Æthopyga cara. *The Tenasserim Yellow-backed Sun-bird.*

Æthopyga miles (*Hodgs.*), *apud Wald. P. Z. S.* 1866, p. 541; *Beavan, Ibis,* 1869, p. 419.
Æthopyga cara, *Hume, S. F.* ii, p. 473, note (1874); *Hume & Davison, S. F.* vi, p. 179; *Shelley, Mon. Nect.* pp. xxi, xxiii, 63, pl. 21; *Hume, Cat.* no. 225 ter; *Bingham, S. F.* ix, p. 170; *Oates, B. B.* i, p. 316; *Gadow, Cat. B. M.* ix, p. 49; *Salvadori, Ann. Mus. Civ. Gen.* (2) v, p. 500, vii, p. 68.

Coloration. Male. Differs from the males of *Æ. scheriæ* and *Æ. andersoni* in having the crown tinged with violet and the nape crimson, not brown; the middle tail-feathers only ·2 longer than the next pair, whereas in *Æ. scheriæ* these feathers are ·7 longer than the adjoining ones: the exposed portions of the closed tail are metallic violet, not metallic green.
Female. Differs from the female of *Æ. scheriæ* in wanting the pale yellow on the flanks, in having the tail 1·5 long, and in having the outermost tail-feathers only ·2 inch shorter than the middle ones.
Legs and feet dark chocolate-brown; upper mandible black; lower mandible pale reddish brown; iris dark brown; mouth pale salmon-colour.
Male: length about 5; tail 2·1; wing 2·2; tarsus ·5; bill from gape ·75. Female: length about 4·3; tail 1·5; wing 2.
Distribution. Pegu east of the Irrawaddy river, from the sea up to a few miles north of the town of Pegu; Tenasserim from Toungngoo down to Tenasserim town and the Thoungyeen valley. The limits

of this species to the north and west are unknown, and it must remain doubtful for the present whether this Sun-bird extends to Arrakan or not.

885. **Æthopyga nicobarica.** *The Nicobar Yellow-backed Sun-bird.*

Æthopyga nicobarica, *Hume, S. F.* i, p. 412 (1873); *Shelley, Mon. Nect.* pp. xx, 61, pl. 20; *Hume, Cat. no.* 225 bis; *Gadow, Cat. B. M.* ix, p. 22.

Coloration. Male. Differs from *Æ. seheriæ* in having the metallic portion of the crown of very small extent and suffused with violet; the nape crimson; the moustachial stripe lined with black interiorly; the exposed parts of the tail violet, not green.

Female. Differs from *Æ. seheriæ* in wanting the yellow on the flanks; in having the tail 1·3 inches in length, the lateral feathers being only ·2 shorter than the middle pair; and in having a much broader and a pale-coloured bill.

The male agrees with that of *Æ. cara* in having short middle tail-feathers, but differs from it in having a smaller cap and the moustachial stripe lined with black.

Both Hume and Shelley are in error, I think, in asserting that the female of this species has the throat red; three specimens in the Hume Collection, sexed as females it is true, but having the throat red, are in my opinion young males, and the females when obtained will probably prove to be green birds without a trace of red, as is the case with all the other species of this genus.

This species differs from *Æ. siparaja*, which inhabits the Malay peninsula, Sumatra, and Borneo, in having a much longer bill. In *Æ. siparaja* the bill measured at front is ·6.

Male: legs, feet, and upper mandible dark brown; lower mandible pale brown; iris brown. Female: upper mandible horny brown; lower mandible, legs, and feet yellow; iris brown (*Hume*).

Male: length about 5; tail 2; wing 2·1; tarsus ·5; bill at front ·75.

Distribution. The Nicobar Islands.

886. **Æthopyga vigorsi.** *Vigors's Yellow-backed Sun-bird.*

Cinnyris vigorsii, *Sykes, P. Z. S.* 1832, p. 98.
Æthopyga vigorsii (*Sykes*), *Horsf. & M. Cat.* ii, p. 735; *Jerd. B. I.* i, p. 363; *Wald. Ibis,* 1870, p. 35; *Fairbank, S. F.* iv, p. 255; *Hume, S. F.* v, p. 122; *Cat. no.* 226; *Shelley, Mon. Nect.* pp. xxi, xxiii, 71, pl. 23; *Gadow, Cat. B. M.* ix, p. 18; *Barnes, Birds Bom.* p. 135; *Oates in Hume's N. & E.* 2nd ed. ii, p. 250.

The Violet-eared Red Honey-sucker, Jerd.

Coloration. Male. Forehead and central portion of crown metallic green; hinder part of crown and nape dull blackish; sides of the head and neck, hind neck, back, scapulars, and lesser wing-coverts deep red, the concealed black bases of the feathers showing up in places ; rump bright yellow, some of the feathers occasionally tipped crimson ; upper tail-coverts metallic green ; tail black suffused with violet, the middle pair of feathers and the outer margins of the others metallic green ; median wing-coverts black margined with crimson ; greater coverts, primary-coverts, winglet, and quills brown ; a long moustachial streak and a patch behind the ear-coverts metallic violet ; chin, throat, and breast deep red finely streaked with yellow ; the red of the breast bounded by a black band which extends more or less down the middle of the abdomen ; remainder of lower plumage ashy grey ; under wing-coverts and axillaries white.

Female. General colour dull green ; the feathers of the forehead brown margined with green ; the lower plumage suffused with ashy ; the under tail-coverts broadly margined with ashy yellow ; under wing-coverts and axillaries pale yellowish.

The young resemble the adult female and the young male assumes the adult plumage at the first spring by a moult.

Iris red-brown, crimson (*Fairbank*) ; legs and bill dark brown or blackish.

Length nearly 6 ; tail 2·3 ; wing 2·5 ; tarsus ·65 ; bill from gape ·95 ; the female has the tail 1·7 and the wing 2·2.

Distribution. The British Museum series of this Sun-bird contains birds from Western Khandesh, Matheran, Khandála, Mahableshwar, and the Malabar coast. This latter locality is very vague. Hume gives the range of this bird as extending from the valley of the Tapti river to some distance south of Mahableshwar along the line of ghats. Jerdon observed this species in the Bastar country south-east of Nagpore, but this locality requires confirmation.

Habits, &c. The accounts of the nidification of this bird are very incomplete. It breeds in June and in September, and the nest appears to resemble that of *Æ. scheriæ.* An egg is described as being white, very thickly freckled with yellowish brown, and measuring ·63 by ·48.

887. Æthopyga ignicauda. *The Fire-tailed Yellow-backed Sun-bird.*

Cinnyris ignicaudus, *Hodgs. Ind. Rev.* ii, p. 273 (1837).
Nectarinia ignicauda (*Hodgs.*), *Blyth, Cat.* p. 225.
Æthopyga ignicauda (*Hodgs.*), *Horsf. & M. Cat.* ii, p. 734; *Jerd. B. I.* i, p. 365; *Wald. Ibis,* 1870, p. 36; *Blanford, J. A. S. B.* xli, pt. ii, p. 44; *Godw.-Aust. J. A. S. B.* xliii, pt. ii, p. 159; *Shelley, Mon. Nect.* pp. xx, 45, pl. 15; *Hume, Cat.* no. 228; *Gadow, Cat. B. M.* ix, p. 25.

The Fire-tailed Red Honey-sucker, Jerd.

Coloration. Male. Forehead and crown metallic blue ; sides of the crown, nape, hind neck, sides of the neck, back, scapulars, upper tail-coverts, and middle pair of tail-feathers crimson; remainder of tail brown, margined with crimson or rufous exteriorly; rump yellow; wings brown, each feather edged with olive-yellow; scapulars greenish; chin and throat purple, changing to steel-blue at the sides; lores and ear-coverts dull black; breast yellow suffused in the middle with crimson; remainder of lower plumage dull greenish yellow.

Female. General colour green, the crown-feathers with concealed dark centres, the rump and upper tail-coverts margined with greenish yellow; wings dark brown margined with yellowish; middle tail-feathers yellowish brown, the others blackish on the inner web and green on the outer, and obsoletely tipped pale; throat and breast tinged with ashy; middle of abdomen rather bright yellow.

Bill and legs black; iris dark brown (*Hume Coll.*)

Male: length about 8; tail 5; wing 2·3; tarsus ·65; bill from gape ·85; the middle pair of tail-feathers are 2·7 longer than the next pair. Female: length about 5; tail 1·6; wing 2·2; the outermost tail-feathers fall short of tip of tail by ·5.

Distribution. Nepal; Sikhim; Bhutan; the Khási hills; the Nága hills; Manipur. Hume records this species from Kumaun and Garhwál and also from Sylhet and Cachar; but I have seen no specimens from these localities. Blanford observed this bird in Sikhim at 11,000 feet.

888. **Æthopyga gouldiæ.** *Mrs. Gould's Yellow-backed Sun-bird.*

Cinnyris gouldiæ, *Vigors, P. Z. S.* 1831, p. 44; *Gould, Cent.* pl. 56.
Nectarinia gouldiæ (*Vigors*), *Blyth, Cat.* p. 223.
Æthopyga gouldiæ (*Vigors*), *Horsf. & M. Cat.* ii, p. 733; *Jerd. B. I.* i, p. 364; *Stoliczka, J. A. S. B.* xxxvii, pt. ii, p. 23; *Wald. Ibis,* 1870, p. 35; *Godw.-Aust. J. A. S. B.* xliii, pt. ii, p. 156; *Shelley, Mon. Nect.* pp. xx, xxii, 41, pl. 14; *Hume, Cat.* no. 227; *Oates, B. B.* i, p. 315; *Gadow, Cat. B. M.* ix, p. 27; *Hume, S. F.* xi, p. 81.

The Purple-tailed Red Honey-sucker, Jerd.

Coloration. Male. Forehead, crown, chin, throat, and the posterior part of the ear-coverts coppery red or burnished purple according to the light; lores blackish; a line of feathers over the lores, cheeks, sides of the head and neck, lesser wing-coverts, back, and scapulars crimson; rump yellow; upper tail-coverts rich purple or violet; basal three quarters of the median tail-feathers bright purple, terminal quarter brown; the other rectrices brown, tinged with purple on the outer web and tipped with whitish; greater coverts and quills dark brown, edged with yellowish brown; lower plumage bright yellow; the breast more or less streaked with crimson; the sides of the breast crimson, with a patch of bright purple below the ear-coverts; under wing-coverts and axillaries whitish.

Female. Upper plumage light green, the feathers of the crown with brown centres; rump sulphur-yellow; tail brown edged with

olive-green, the lateral feathers broadly tipped with whitish ; quills
brown edged with green ; the whole lower plumage yellowish green.

Bill blackish brown ; iris reddish chocolate-brown ; tarsus deep
brown ; toes a little paler (*Hume*).

Male: length about 6; tail 3; wing 2·1 ; tarsus ·55 ; bill from
gape ·7. Female: tail 1·4 ; wing 2 ; bill from gape ·65.

Distribution. The Himalayas from the Sutlej valley to Sikhim,
up to 10,000 feet ; the Khásí hills ; the Nága hills ; Manipur ;
Arrakan. This last locality I give on the authority of Blyth. Hume
gives this species from Hill Tipperah and Chittagong, but there
are no specimens in his collection from these localities.

889. **Æthopyga dabryi.** *Dabry's Yellow-backed Sun-bird.*

Nectarinia dabryii, *J. Verr. Rev. et Mag. Zool.* 1867, p. 173, pl. 15.
Æthopyga debrii (*J. Verr.*), *Wald. Ibis,* 1870, p. 35.
Æthopyga dabryi (*J. Verr.*), *Shelley, Mon. Nect.* pp. xx, xxi, 39, pl. 13;
  Anders. Yunnan Exped., Aves, p. 662 : *Hume & Dav. S. F.* vi,
  p. 180 : *Hume, Cat.* no. 227 bis ; *Gadow, Cat. B. M.* ix, p. 28 ;
  Oates, B. B. i, p. 314 ; *Salvadori, Ann. Mus. Civ. Gen.* (2) vii,
  p. 395.

Coloration. Male. Forehead, crown, chin, throat, and upper part
of ear-coverts metallic purple or lilac according to the light ; the
nape, sides of the crown, feathers round the eye, sides of the neck,
back, scapulars, and lesser and median wing-coverts deep crimson ;
rump bright yellow ; upper tail-coverts and broad margins to basal
two thirds of the middle tail-feathers metallic purple ; remainder
of the tail black, the outer three pairs of feathers tipped with dull
white ; greater coverts and quills brown, edged with yellowish green ;
breast scarlet, with a patch of metallic purple on either side ; abdomen, vent, sides of the body, and under tail-coverts yellow, tinged
with dusky ; under wing-coverts and axillaries pale yellow.

Female. Upper plumage olive-green ; the feathers of the crown
with dark brown centres ; rump pale yellow ; tail brown, edged
narrowly with olive-green and the three outer pairs of feathers
tipped with dull whitish ; quills brown, edged with dull greenish
yellow ; the whole lower plumage dull pale green.

Legs and feet dark horny brown : bill dusky black ; irides deep
brown (*Davison*). Iris, bill, and legs brown (*Wardlaw Ramsay*).

Male: length 5·7 ; tail 2·6 ; wing 2·2 ; tarsus ·55 : bill from
gape ·7. Female: length 3·5 ; tail 1·3 ; wing 1·75.

Distribution. The higher portions of Muleyit mountain in Tenasserim ; the Karen hills east of Toungngoo ; Karennee ; the hills
east of Bhámo, extending into South-western China. Hume is
certain that he observed this species in Manipur, but he failed to
obtain a specimen. This Sun-bird appears to be a hill-species and
to be found only above 4000 feet.

890. **Æthopyga saturata.** *The Black-breasted Yellow-backed Sun-bird.*

Cinnyris saturatus, *Hodgs. Ind. Rev.* ii, p. 273 (1837).
Nectarinia saturata (*Hodgs.*), *Blyth, Cat.* p. 224.
Æthopyga saturata (*Hodgs.*), *Horsf. & M. Cat.* ii, p. 735; *Jerd. B. I.*
   i. p. 357: *Wald. Ibis*, 1870, p. 96: *Godw.-Aust. J. A. S. B.* xxxix,
   pt. ii, p. 98: *Hume, N. & E.* p. 147; *Godw.-Aust. J. A. S. B.* xlv.
   pt. ii, p. 70: *Shelley, Mon. Nect.* pp. xx, xxi, 35, pl. 11; *Hume, Cat.*
   no. 231; *Gadow, Cat. B. M.* ix, p. 15: *Hume, S. F.* xi. p. 82: *Oates*
   in Hume's N. & E. 2nd ed. ii, p. 250.
The Black-breasted Honey-sucker, Jerd.

Coloration. Male. Forehead, crown, nape, and hind neck metallic
steel-blue; back dark red; lesser wing-coverts, scapulars, and lower
back dull deep black; the black of the lower back succeeded by a
yellow band; rump and upper tail-coverts metallic steel-blue;
tail dull black except the basal two-thirds of the middle pair of
feathers, which are metallic steel-blue; median, greater, primary-
coverts, and winglet dark brown edged with black; quills brown,
edged narrowly with olivaceous; lores and sides of the head black;
sides of neck dull red; chin, throat, breast, and upper abdomen
deep black; a broad moustachial streak metallic steel-blue; remainder of lower plumage pale greenish; under wing-coverts and
axillaries pale yellowish white.
Female. General colour light green: a broad band of light
yellow across the rump: the wings and tail margined with olive-
green; the under wing-coverts and axillaries clear pale yellow; the
three or four outer pairs of tail-feathers blackish, broadly tipped
with dull white.
   Bill black; legs brown; iris brown (*Jerdon*).
   Length of male nearly 6; tail 3; wing 2·15; tarsus ·55; bill
from gape ·8; the female has the tail 1·6 and the wing 1·9.
Distribution. The Himalayas from Garhwál to the extreme east
of Assam; the Khási hills; Manipur. This species appears to be
found up to 5000 feet.
Habits, &c. The nest of this bird appears, from Hodgson's description, to resemble that of *Æ. scheriæ.* The breeding-season
commences in April. The eggs are described as being white marked
with brown and measuring about ·6 by ·43.

891. **Æthopyga sanguinipectus.** *Walden's Yellow-backed Sun-*
bird.

Æthopyga sanguinipectus, *Wald. A. M. N. H.* (4) xv, p. 400 (1875);
   Hume, S. F. iii, p. 402; *Wald. in Blyth's Birds Burm.* p. 142;
   Hume & Dav. S. F. vi, p. 182; *Hume, Cat.* no. 231 bis; *Shelley,*
   Mon. Nect. pp. xx, xxi, 37, pl. 12: *Oates, B. B.* i, p. 313; *Gadow,*
   Cat. B. M. ix, p. 27; *Salvadori, Ann. Mus. Civ. Gen.* (2) v,
   p. 590, vii, p. 395.
Æthopyga waldeni, *Hume, S. F.* v, p. 51 (1877).

Coloration. Male. Forehead, crown, nape, and hind neck purplish steel-blue; sides of the head dull black; sides of the neck, back, and the shorter scapulars red; lesser wing-coverts, the longer scapulars, and a band on the back below and next to the red deep black; next this black band another yellow one; remainder of the rump, upper tail-coverts, and the basal three quarters of the middle tail-feathers steel-blue; remainder of the tail, the median and greater wing-coverts, and the quills blackish brown; lateral tail-feathers tipped with white; chin black; throat purplish steel-blue; upper breast black, the lateral feathers tipped with red; remainder of the lower plumage dull green, the breast streaked with crimson; under wing-coverts and axillaries yellowish white.

Female. Upper plumage dull green tinged with ashy, and the feathers of the crown with dark centres: rump pale yellow; lower plumage ashy green, becoming paler on the abdomen; tail blackish, all the lateral feathers with pale tips; under wing-coverts and axillaries whitish.

Bill black; iris and legs dark brown (*Wardlaw Ramsay*).

Male: length 5·5 to 6; tail 2·7; wing 2·1; tarsus ·55; bill from gape ·8. Female: length about 4; tail 1·1; wing 1·8.

Distribution. The Karen hills east of Toungngoo; Karennee; Muleyit mountain in Tenasserim. This species is found at elevations above 2000 feet.

892. Æthopyga nepalensis. *The Nepal Yellow-backed Sun-bird.*

Cinnyris nipalensis, *Hodgs. Ind. Rev.* ii, p. 273 (1837).
Nectarinia nipalensis (*Hodgs.*), *Blyth, Cat.* p. 224.
Æthopyga nipalensis (*Hodgs.*), *Horsf. & M. Cat.* ii, p. 735: *Jerd. B. I.* i, p. 365; *Wald. Ibis,* 1870, p. 35; *Hume, N. & E.* p. 147; *Godw.-Aust. J. A. S. B.* xlv, pt. ii, p. 70: *Shelley, Mon. Nect.* pp. xx, xxi, 29, pl. 10: *Hume, Cat.* no. 229: *Gadow, Cat. B. M.* ix, p. 26; *Hume, S. F.* xi, p. 82: *Oates in Hume's N. & E.* 2nd ed. ii, p. 251.

The Maroon-backed Honey-sucker, Jerd.

Coloration. Male. Forehead, crown, nape, hind neck, sides of the crown, chin, and throat metallic green; sides of the head black; sides of the neck and back deep red; scapulars, lower back, and the margins of all the wing-feathers olive-yellow; rump bright yellow; upper tail-coverts and the basal three quarters of the middle pair of tail-feathers metallic green; remainder of tail black, all the feathers with broad pale tips except the two middle pairs; lower part of throat pure yellow; breast and upper abdomen yellow suffused with red and streaked with crimson; remainder of lower plumage dull greenish yellow; under wing-coverts and axillaries white.

Female. Quite undistinguishable from the female of *Æ. seheriæ.*
Bill and legs black: iris brown (*Cockburn*).

2 A 2

Male : length about 6 ; tail 2·7 ; wing 2·1 ; tarsus ·6 ; bill from gape ·9. Female : tail 1·7 ; wing 2.

Distribution. The Himalayas from Nepal to the Daphla hills in Assam ; the Khási hills ; Manipur. This species ranges up to 6000 feet. The limits of this Sun-bird on the west are difficult to define ; they may extend to the extreme west of Nepal, but *Æ. horsfieldi* also occurs in this State.

Habits, &c. Judging from Hodgson's account of the nidification of this Sun-bird, the nest and eggs do not differ in any material respect from those of *Æ. seheriæ.* The eggs, however, appear to be less densely marked and measure about ·68 by ·43.

Jerdon remarks that a nest of this species which he found at Darjiling had a projecting roof over the entrance. No other species of this genus, so far as is known, constructs its nest in this manner.

893. Æthopyga horsfieldi. *Horsfield's Yellow-backed Sun-bird.*

Cinnyris horsfieldi, *Blyth, J. A. S. B.* xi, p. 107 (1842).
Nectarinia horsfieldi (*Blyth*), *Blyth, J. A. S. B.* xii, p. 975 ; *id. Cat.* p. 224.
Æthopyga horsfieldi (*Blyth*), *Jerd. B. I.* i, p. 366 ; *Walden, Ibis,* 1870, p. 36 ; *Shelley, Mon. Nect.* pp. xx, 33, pl. 10 ; *Hume, Cat.* no. 230.
Æthopyga nipalensis (*Hodgs.*), *Gadow, Cat. B. M.* ix, p. 26 (part.).

The Green-backed Honey-sucker, Jerd.

Coloration. The male differs from the male of *Æ. nepalensis* in having the back and sides of the neck olive-yellow instead of deep red, there being merely traces of red along the margin of the metallic green of the hind neck ; in having the yellow of the breast and upper abdomen almost pure, there being hardly a trace of red, and in having the streaks of crimson on the breast few and indistinct ; and lastly in having a much shorter bill, measuring only ·8 from gape to tip.

The females of the two species are undistinguishable from each other in colour. The bill of the present species, however, is probably shorter, judging from the length of bill of the males of the two species. The only female *Æ. horsfieldi* that I have been able to examine is without a bill.

Distribution. Garhwál, Kumaun, and Nepal, but probably only the extreme western portion of the latter State. This species descends to the Dehra Dún, and probably is found up to 6000 feet, as is the case with *Æ. nepalensis.*

Genus ARACHNECHTHRA, Cabanis, 1850.

I retain the genus *Arachnechthra* for those Indian Sun-birds in which the tail in both sexes is short and rounded, the covering-membrane of the nostril bare, and the bill slender and much curved downwards. The type of the genus *Cinnyris*, in which genus these Sun-birds have latterly been placed, is *C. splendida* from Africa, and this species has the upper tail-coverts very ample, reaching quite to the tip of the tail. I cannot therefore consider the Indian species, which I retain in *Arachnechthra*, at all congeneric with this African bird.

In the genus *Arachnechthra* the sexes are very different in coloration, the males being clothed in metallic colours and the females being greenish above and yellow beneath. The females of the different species resemble each other very closely and are difficult to separate, except in the case of a few.

Key to the Species.

a. Chin and throat dark-coloured and metallic.
  a'. Lower plumage, below the breast, dark-coloured.
    a''. Upper plumage uniformly of one colour.
      a'''. Abdomen snuff-brown .... *A. lotenia* ♂, p. 358.
      b'''. Abdomen violet-black .... *A. asiatica* ♂, p. 359.
    b''. Upper plumage green, black, and blue .... *A. hasselti* ♂, p. 360.
  b'. Lower plumage, below the breast, yellow.
    c''. Back olive-yellow.
      c'''. Forehead and anterior part of crown violet-blue.......... *A. pectoralis* ♂, p. 361.
      d'''. Forehead and anterior part of crown like back.
        a'. Pectoral tufts yellow and flame-colour; bill from gape 8 .... *A. flammaxillaris* ♂, p. 362.
        b'. Pectoral tufts entirely yellow; bill from gape more than 9 .... *A. andamanica* ♂, p. 363.
    d''. Back crimson.
      e'''. Upper tail-coverts metallic red .... *A. minima* ♂, p. 363.
      f'''. Upper tail-coverts metallic purple .... *A. zeylonica* ♂, p. 364.
b. Chin and throat pale-coloured and non-metallic.
  c'. Entire lower plumage yellow.
    e''. Rump and upper tail-coverts of the same colour as the back.

g''. Bill from gape quite 1 inch
　　　or more.................. *A. lotenia* ♀, p. 358.
h'''. Bill from gape well under 1
　　　inch.
　　c^4. Lower plumage rich yellow. *A. pectoralis* ♀, p. 361.
　　d^4. Lower plumage pale yellow.
　　　a^5. Bill at gape ·65........ *A. hasselti* ♀, p. 360.
　　　b^5. Bill at gape ·8.
　　　　a^6. Lateral tail-feathers
　　　　　very narrowly tipped. *A. asiatica* ♀, p. 359.
　　　　b^6. Lateral tail-feathers
　　　　　very broadly tipped.. *A. flammaxillaris* ♀, p. 362.
　　　c^5. Bill at gape ·9 .... *A. andamanica* ♀, p. 363.
f''. Rump and upper tail-coverts
　　　red .... *A. minima* ♀, p. 363.
d'. Chin and throat ashy white; re-
　　maining lower parts bright yel-
　　low .... *A. zeylonica* ♀, p. 364.

894. Arachnechthra lotenia. *Loten's Sun-bird.*

Certhia lotenia, *Linn. Syst. Nat.* i, p. 188 (1766).
Nectarinia lotenia (*L.*), *Blyth, Cat.* p. 224.
Arachnechthra lotenia (*L.*), *Horsf. & M. Cat.* ii, p. 743; *Jerd. B. I.*
　　i, p. 372; *Walt. Ibis*, 1870, p. 23; *Legge, Birds Ceyl.* p. 563;
　　Oates in Hume's N. & E. 2nd ed. ii, p. 251.
Cinnyris lotenius (*L.*), *Shelley, Mon. Nect.* pp. xxviii, xxxvi, 177,
　　pl. 56; *Hume, Cat.* no. 235; *Gadow, Cat. B. M.* ix, p. 60; *Vidal,*
　　S. F. ix, p. 57; *Butler, S. F.* ix, p. 390; *Davison, S. F.* x, p. 362;
　　Barnes, Birds Bom. p. 137.

The Large Purple Honey-sucker, Jerd.

Coloration. Male. The whole upper plumage metallic green
glossed with lilac, the upper tail-coverts metallic blue; lesser and
median wing-coverts lilac; greater coverts and wings brown; tail
blue; sides of the head and neck green glossed with lilac; cheeks,
chin, and upper throat metallic green; breast rich metallic violet
changing to green at the sides; a band of maroon below the breast;
pectoral tufts rich yellow with a small intermixture of crimson;
remainder of lower plumage snuff-brown.

Female. Whole upper plumage, wings, sides of the head, and neck
greenish brown; entire lower plumage very dull yellow; tail
blackish, the lateral feathers broadly tipped with whitish.

Bill, legs, feet, and claws black; iris deep brown (*Davison*).

Length about 5·5; tail 1·6; wing 2·3; tarsus ·6; bill from gape
1·2. The female is considerably smaller, the tail being about 1·3,
the wing 2, and the bill from gape 1.

Distribution. Ceylon and Southern India. On the west this
species extends north as far as Ratnagiri, but on the east its limits
are undetermined. Davison found this bird at 5000 feet in the
Wynaad.

Habits, &c. A nest found by Mr. E. H. Aitken in November
was similar to the nest of *A. zeylonica*, and contained a young bird

and an egg. The latter is described as being dirty brownish white covered with dull brown marks.

895. Arachnechthra asiatica. *The Purple Sun-bird.*

Certhia asiatica, *Lath. Ind. Orn.* i, p. 288 (1790).
Nectarinia asiatica (*Lath.*), *Blyth, Cat.* pp. 224, 328.
Arachnechthra asiatica (*Lath.*), *Horsf. & M. Cat.* ii, p. 743; *Jerd. B. I.* i, p. 370; *Hume, N. & E.* p. 151; *Wald. Ibis,* 1870, p. 20; *Hume & Dav. S. F.* vi, p. 190; *Oates in Hume's N. & E.* 2nd ed. ii, p. 252.
Arachnechthra intermedia, *Hume, Ibis,* 1870, p. 436; *id. N. & E.* p. 154.
Nectarinia (Arachnechthra) brevirostris, *Blanford, Ibis,* 1873, p. 86.
Arachnechthra edeni, *Anderson, Yunnan Exped., Aves,* p. 661, pl. xlix (1878).
Cinnyris asiaticus (*Lath.*), *Shelley, Mon. Nect.* pp. xxviii, xxxvi, 181, pl. 57; *Hume, Cat. no. 234; Legge, Birds Ceyl.* p. 566; *Gadow, Cat. B. M.* ix, p. 56; *Scully, S. F.* viii, p. 259; *Oates, B. B.* i, p. 321; *Barnes, Birds Bom.* p. 137.

The Purple Honey-sucker, Jerd.; *Shakar khora,* Hind.; *Jugi jugi,* Bhagalpur; *Than kudi,* Tam; *Gewal kurulla,* Cing.

Fig. 89.—Head of *A. asiatica.*

Coloration. Male. The whole upper plumage, sides of the head and neck, and the lesser and median wing-coverts metallic violet-blue or greenish; greater coverts and all the quills brown, edged paler; tail bluish black; chin, throat, and fore neck metallic violet; breast like the sides of the neck; a narrow band below the breast coppery brown, of varying extent, sometimes absent; the large pectoral tufts mixed orange-red and bright yellow; abdomen, vent, and under tail-coverts violet-black.

Female. Upper plumage, wings, and sides of the head and neck greenish brown; lower plumage rather bright yellow; tail dark brown or blackish, the laterals narrowly tipped with white.

Young males have generally a broad stripe from the chin to the abdomen dark metallic violet; the remainder of the lower plumage yellow.

Bill black; iris hazel-brown; eyelids plumbeous; legs black; claws dark horn.

Length 4·5; tail 1·5; wing 2·1; tarsus ·6; bill from gape ·8.

Birds from Burma are remarkable for the rich tone of their coloration, the prevailing tint being rich violet. In India, especially in the dry north-western portions, the prevailing tint is rather green. Intermediate birds are also found; and this variation of colour, coupled with a bill which also varies remarkably in length

has caused this bird to be subdivided into several races, none of
which, however, appears worthy to be upheld.

Distribution. The whole peninsula of India from Cape Comorin
to the Himalayas, where this species is found up to 5000 feet, and
from Sind and the Punjab to the extreme east of Assam, thence
extending south through Burma to Central Tenasserim and the
Thoungyeen valley. The furthest point south in Tenasserim
where this bird has been observed on the sea-board is Yay. This
Sun-bird also occurs in Ceylon.

Outside Indian limits, this species is found on the west as far as
Persia, and on the east it extends to Cochin China.

Habits, &c. Breeds almost the whole year round, having two
or more broods. The nest is a pear-shaped structure suspended
from a low branch and composed principally of grass, with which,
however, are combined various other materials. The outside is
invariably ornamented with cobwebs to which are attached pieces
of bark, dead leaves, and excreta of caterpillars. The entrance, at
the side, is overhung by a small porch in most instances. The eggs,
two or three in number, are dull white, marked with various shades
of brown and measure about ·64 by ·46.

896. Arachnechthra hasselti. *Van Hasselt's Sun-bird.*

Certhia brasiliana, *Gmel. Syst. Nat.* i, p. 474 (1788).
Nectarinia hasseltii, *Temm. Pl. Col.* pl. 376, fig. 3 (1825); *Blyth, Cat.* p. 226.
Nectarinia phayrei, *Blyth, J. A. S. B.* xii, p. 1008 (1843).
Leptocoma hasseltii (*Temm.*), *Horsf. & M. Cat.* ii, p. 740; *Godw.-Aust. J. A. S. B.* xliii, p. 156.
Leptocoma brasiliana (*Gm.*), *Wald. P. Z. S.* 1876, p. 543; *Hume & Dav. S. F.* vi, p. 184.
Nectarophila brasiliana (*Gm.*), *Wald. Ibis,* 1870, p. 41.
Cinnyris hasselti (*Temm.*), *Shelley, Mon. Nect.* pp. xxvii, xxxi, 127, pl. 42; *Oates, S. F.* x, p. 197; *id. B. B.* i, p. 318; *Gadow, Cat. B. M.* ix, p. 67.
Cinnyris brasiliana (*Gm.*), *Hume, Cat.* no. 233 bis.
Arachnechthra hasselti (*Temm.*), *Oates in Hume's N. & E.* 2nd ed. ii, p. 258.

Coloration. Male. Forehead and crown shining golden green;
lores, cheeks, ear-coverts, the neck above and at the sides, the upper
back, tertiaries, and all the wing-coverts except those near the edge
of the wing deep black; the wing-coverts near the edge of the wing,
scapulars, lower back, rump, and upper tail-coverts brilliant purplish blue; primaries and secondaries brownish black; under wing-
coverts deep black; throat and fore neck brilliant amethystine-
purple; breast and upper abdomen rich red; lower abdomen, sides
of the body, vent, and under tail-coverts dull greyish black; tail
brilliant purplish black.

Female. Upper plumage olive-green, the feathers of the crown
dark-centred; wings brown, the coverts edged with greenish, the

quills with pale rufous; tail blackish, the margins greenish and the lateral feathers tipped pale; entire lower plumage, under wing-coverts, and axillaries pale yellow.

Bill dark brown; the gape and mouth cinnamon-red; iris dark hazel; eyelids plumbeous; legs black; claws brown.

Length 4; tail 1·2; wing 1·95; tarsus ·45; bill from gape ·65.

Distribution. This species is spread over the whole of Arrakan, Pegu, and Tenasserim, but appears to be nowhere very common, except in the southern portion of the latter division. Godwin-Austen obtained this Sun-bird in Tipperah, whence Hume also records it as well as from Chittagong. It extends down the Malay peninsula to Sumatra, Java, and Borneo. The nest has not yet been found within Indian limits.

Habits, &c. This species probably breeds from March to June. A nest is described as being of the ordinary type, without a portico over the entrance, and composed of glistening red-brown scales taken from the stems of ferns felted together and covered with black moss-roots and cocoon-silk. The eggs, two in number, are brown with a darker ring round the larger end, and measure about ·58 by ·4.

897. Arachnechthra pectoralis. *The Malay Yellow-breasted Sun-bird.*

Nectarinia pectoralis, *Horsf. Tr. Linn. Soc.* xiii, p. 167 (1822); *Blyth. Cat.* p. 225.

Cyrtostomus pectoralis (*Horsf.*), *Horsf. & M. Cat.* ii, p. 739; *Hume, N. & E.* p. 155.

Arachnechthra pectoralis (*Horsf.*), *Wald. Ibis*, 1870, p. 25; *Ball, S. F.* i, p. 64; *Hume, S. F.* ii, p. 196; *Oates in Hume's N. & E.* 2nd ed. ii, p. 259.

Cinnyris pectoralis (*Horsf.*), *Shelley, Mon. Nect.* pp. xxvii, xxxvi, 87, 165, pl. 53; *Gadow, Cat. B. M.* ix, p. 88; *Hume, Cat.* no. 234 bis.

Coloration. Male. Forehead, anterior part of crown, and cheeks violet-blue; lores black; upper plumage, sides of head and neck olive-yellow; wings brown, each feather margined with olive-yellow; tail blackish, the lateral feathers broadly tipped with whitish; chin and throat metallic violet, bordered by a band of metallic blue, which gets broader on the upper breast and is narrowly margined with maroon; lower plumage bright yellow; pectoral tufts deep yellow tinged with red; under wing-coverts very pale yellow.

Female. Upper plumage olive-yellow, as also the sides of the head and neck; entire lower plumage deep yellow; wings brown, each feather edged with olive-yellow; tail blackish, the lateral feathers broadly tipped with white.

Legs, feet, and bill black; iris brown (*Hume*).

Length about 4·5; tail 1·4; wing 2·1; tarsus ·55; bill from gape ·7 to ·9.

There are two distinct races of this bird in the Nicobar Islands.

Those found in Car Nicobar, Bompoka, Trinkut, Camorta, and Kat-
chal have the culmen short as in birds from the Malay peninsula
and islands; those found in Condul have the bill extremely long,
the culmen measuring about ·85. These two races differ, however,
in no other respect and I do not propose to separate them.

Distribution. The Nicobar Islands as above, extending to the
Malay peninsula and all the adjacent islands.

Habits, &c. Very partial to the flowers of the cocoanut-palm.
Breeds in the Nicobars in January and February, constructing a
nest similar to that of *A. asiatica*, but larger and coarser. An egg
measured ·61 by ·45 and was greyish brown speckled with darker
brown, some of the spots being surrounded by a purplish tinge.

898. Arachnechthra flammaxillaris. *The Burmese Yellow-breasted Sun-bird.*

Nectarinia flammaxillaris, *Blyth, J. A. S. B.* xiv, p. 557 (1845); *id.*
    Cat. p. 226.
Cyrtostomus flammaxillaris (*Bl.*), *Horsf. & M. Cat.* ii, p. 739.
Arachnechthra flammaxillaris (*Bl.*), *Wald. Ibis,* 1870, p. 21; *Hume,
    N. & E.* p. 154; *Hume & Dav. S. F.* vi, p. 192; *Oates in Hume's
    N. & E.* 2nd ed. ii, p. 260.
Cinnyris flammaxillaris (*Bl.*), *Shelley, Mon. Nect.* pp. xxvii, xxxv, 161,
    pl. 51; *Hume, Cat.* no. 234 ter: *Gadow, Cat. B. M.* ix, p. 83;
    Oates, B. B. i, p. 320.

Coloration. Male. Forehead, crown, sides of the head, back,
scapulars, rump, and upper tail-coverts olive-yellow; tail black, the
middle feathers narrowly tipped with white, the others progressively
with larger white tips; chin, throat, and breast rich metallic
purple, bordered by rich steel-blue; below the breast a band of
orange-red, and another, broader, below the orange band black;
axillaries flame-red; abdomen, sides of the body, vent, and under
tail-coverts yellow; wings and coverts brown, edged with greenish
brown; under wing-coverts yellowish white; edge of the wing
bright yellow.

Female. Upper plumage, wings, and tail like the male, but the
lower plumage entirely yellow.

Iris light brown; eyelids plumbeous; legs and claws deep
bluish black; mouth light salmon-colour; bill blackish. In the
breeding-season the mouth becomes livid.

Length 4·5; tail 1·4; wing 2·1; tarsus ·55; bill from gape ·8.

Distribution. The greater part of Pegu and the whole of
Tenasserim. Blyth also records this species from Arrakan. It
extends into Siam, Cochin China, and the Malay peninsula.

Habits, &c. Breeds from February to August, constructing a nest
similar to that of *A. asiatica*, with a porch over the entrance, and
laying two eggs which are greenish white, marked with greyish
ash, and measure about ·6 by ·45.

899. **Arachnechthra andamanica.** *The Andaman Sun-bird.*

Arachnechthra frenata (*Müll.*), *apud Ball, J. A. S. B.* xli, pt. 2,
  p. 280; *id. S. F.* i, p. 65.
Arachnechthra andamanica, *Hume, S. F.* i, p. 404, ii, p. 198; *Oates
  in Hume's N. & E.* 2nd ed. ii, p. 262.
Cinnyris andamanicus (*Hume*), *Shelley, Mon. Nect.* pp. xxvii, 157,
  pl. 50; *Hume, Cat.* no. 234 quat.; *Gadow, Cat. B. M.* ix, p. 83.

Coloration. Male. Differs from the male of *A. flammaxillaris* in
having a longer bill, the pectoral tufts pale yellow, unmixed with
red, the maroon and black bands below the breast nearly absent,
the band surrounding the chin, throat, and breast steel-green
instead of blue, and in frequently having a pale supercilium.

Female. Resembles the female of *A. flammaxillaris*, from which
it only differs in having the bill longer.

Bill, legs, and feet black; iris deep brown (*Hume*).

Of the same size as *A. flammaxillaris*, with the exception of the
bill, which measures ·9 to 1 from gape to tip according to sex.

Distribution. The Andaman Islands.

Habits, &c. A nest of this species was found on the 3rd March
with two eggs. The nest appears to have been very similar to
that of *A. asiatica.* One egg measured ·67 by ·48.

900. **Arachnechthra minima.** *The Small Sun-bird.*

Cinnyris minima, *Sykes, P. Z. S.* 1832, p. 99; *Shelley, Mon. Nect.*
  pp. xxvii, xxxiv, 143, pl. 46; *Legge, Birds Ceyl.* p. 572; *Hume, Cat.*
  no. 233; *Gadow, Cat. B. M.* ix, p. 62; *Davison, S. F.* x, p. 362;
  Barnes, Birds Bom. p. 136.
Nectarinia minima (*Sykes*), *Blyth, Cat.* p. 226.
Leptocoma minima (*Sykes*), *Horsf. & M. Cat.* ii, p. 742; *Jerd. B. I.*
  i, p. 369; *Hume, N. & E.* p. 150; *Fairbank, S. F.* iv, p. 256;
  Hume, S. F. iv, p. 392.
Arachnechthra minima (*Sykes*), *Oates in Hume's N. & E.* 2nd ed. ii,
  p. 264.

The Tiny Honey-sucker, Jerd.

Coloration. Male. Forehead and crown metallic green; back
and scapulars deep crimson; rump and upper tail-coverts metallic
red, glossed with lilac; tail black; lesser and median wing-coverts
black, tipped with crimson; remaining coverts and quills dull black;
sides of the head dull black; chin and throat metallic lilac; sides
of neck and the upper breast crimson, followed by a band of
black; remainder of lower plumage yellow; axillaries and under
wing-coverts pale yellowish white.

Female. Upper plumage and sides of the head and neck olive-
green; rump and upper tail-coverts dull red; tail dark brown,
edged with rufous; whole lower plumage yellow; wings brown,
each feather edged with olive-green.

Bill, legs, and feet black; iris dark brown (*Davison*).

364 NECTARINIIDÆ.

Length 3·5 to 4; tail 1·3; wing 1·9; tarsus ·6; bill from gape ·6.

Distribution. The Western Gháts of India from about the latitude of Bombay down to Cape Comorin; Ceylon. This species is found up to 6000 feet.

Habits, &c. Breeds in the Nilgiris in September and October, making a nest of the usual type, and laying two eggs, which are dull white marked with grey and brown, and measure about ·62 by ·42.

901. Arachnechthra zeylonica. *The Purple-rumped Sun-bird.*

Certhia zeylonica, *Linn. Syst. Nat.* i, p. 188 (1766).
Nectarinia zeylonica (*Linn.*), *Blyth, Cat.* p. 226.
Leptocoma zeylonica (*Linn.*), *Horsf. & M. Cat.* ii, p. 740; *Jerd. B. I.* i, p. 368; *Hume, N. & E.* p. 147.
Cinnyris zeylonicus (*Linn.*), *Shelley, Mon. Nect.* pp. xxvii, xxxiii, 137, pl. 45; *Hume, S. F.* v, pp. 270, 398; *Legge, Birds Ceyl.* p. 569; *Hume, Cat. no.* 232; *Gadow, Cat. B. M.* ix, p. 64; *Barnes, Birds Bom.* p. 136.
Arachnechthra zeylonica (*Linn.*), *Oates in Hume's N. & E.* 2nd ed. ii, p. 263.

The Amethyst-rumped Honey-sucker, Jerd.; *Shakar khora,* Hind.; *Man chungee,* Beng.; *Thaa-kudi,* Tam.: *Mal sutika,* Cing.

Coloration. Male. Forehead, crown, and lesser wing-coverts metallic lilac; hind neck, sides of neck, back, scapulars, and median wing-coverts dull crimson; rump and upper tail-coverts metallic purple; tail black, with pale tips to the lateral feathers; greater coverts and quills brown, edged with rufous; sides of the head coppery brown; chin and throat metallic purple; a collar below the throat maroon; breast, abdomen, and under tail-coverts bright yellow; sides of the body, axillaries, under wing-coverts, and the inner margins of quills white.

Female. Upper plumage ashy brown, the longer rump-feathers tipped with rufous, the upper tail-coverts black; tail black, the lateral feathers tipped pale; wings brown, margined with rufous; an indistinct narrow whitish supercilium; lores and a streak behind the eye dark brown; sides of the head ashy; cheeks, chin, and throat pale ashy white; breast, abdomen, and under tail-coverts yellow; sides of body, axillaries, and under wing-coverts white.

Iris dull red; bill and legs black (*Cripps*).

Length about 4·5; tail 1·4; wing 2·3; tarsus ·65; bill from gape ·6.

Distribution. Ceylon; India proper from Cape Comorin to Bombay on the west; thence the northern limits of this species are difficult to trace, but it occurs at Dhulia in Khándesh, at Raipur and Sambalpur in the Central Provinces, and at Lohardugga, and Burdwan in Bengal. A line drawn through these places will probably indicate the ordinary northern limits of this bird. To the east it is said to be common at Furreedpore in Eastern Bengal and

at Dacca, and it has been obtained in the Khási hills. This latter
locality was probably Jerdon's warrant for stating that this Sun-
bird extended to Assam. This species is found up to 2500 feet in
the Nilgiris.

Habits, &c. Breeds in almost every month of the year according
to locality, making a nest of the usual type with a portico over the
entrance. The eggs, two in number, resemble those of *A. asiatica*
and measure about ·65 by ·47.

Genus ANTHOTHREPTES, Swains., 1831.

The genus *Anthothreptes* contains a few Sun-birds which are
closely allied to *Arachnechthra* in structure. In *Anthothreptes*,
however, the bill is deeper, and the lower mandible is straight or
nearly so instead of being well curved downwards. The sexes
are structurally the same, but differ in colour.

The birds of this genus appear to make nests dissimilar to those
of the genus *Arachnechthra*. That of *A. malacensis* is described as
being oval in shape, with a hole on one side near the top, and constructed of cocoanut-fibres &c. This nest, as figured in Shelley's
Monograph, is attached directly to a branch, and has none of the
cord-like connection between the nest and the point of attachment
so usual in the nests of the other Sun-birds.

Key to the Species.

a. Lower plumage streaked ....   *A. hypogrammica.* p. 365.
b. Lower plumage plain.
  a'. Whole upper plumage dark metallic.
    a''. Sides of head greenish yellow....   *A. malacensis* ♂, p. 366.
    b''. Sides of head rufous ....   *A. rhodolaema* ♂, p. 367.
  b'. Whole, or nearly whole, of the upper
    plumage plain green.
    c''. Front of crown metallic dark
      green ....   *A. simplex* ♂, p. 367.
    d''. Front of crown plain green like
      remainder of upper plumage.
      a'''. Bill from gape ·8............   { *A. malacensis* ♀, p. 366.
        { *A. rhodolaema* ♀, p. 367.
      b'''. Bill from gape ·7 ....   *A. simplex* ♀, p. 367.

902. Anthothreptes hypogrammica.   *The Banded Sun-bird.*

Nectarinia hypogrammica, *S. Müll. Verhand. Nat. Gesch., Zool. Aves,*
  p. 63 (1843); *Blyth, Cat.* p. 225.
Anthreptes hypogrammica (*S. Müll.*), *Horsf. & M. Cat.* ii, p. 738;
  Shelley, Mon. Nect. pp. xliii, xliv, 305, pl. 98; *Hume & Dav. S. F.* vi,
  p. 178; *Hume, Cat.* no. 233 quint.; *Oates, B. B.* i, p. 323.
Anthothreptes hypogrammica (*S. Müll.*), *Gadow, Cat. B. M.* ix,
  p. 112.

Coloration. Male. Forehead, crown, nape, sides of the head and
neck, back, scapulars, and wing-coverts yellowish green ; rump,
upper tail-coverts, and a collar on the upper back metallic blue ;
tail blackish brown, the two or three outer pairs of feathers

Fig. 100.—Head of *A. hypogrammica.*

narrowly tipped with white ; quills brown, edged with the colour
of the back ; chin, throat, breast, abdomen, and sides of the body
yellow, streaked with greenish brown : vent, flanks, and under
tail-coverts greenish brown.

Female. The blue collar is absent, and the rump and upper tail-
coverts are of the same colour as the back.

Legs and feet greenish brown or dark plumbeous green : the bill
horny black, and, in the male, the gape dull yellow ; irides dark
brown (*Davison*).

Length 5·5 ; tail 2 : wing 2·6 ; tarsus ·6 : bill from gape ·9.
The female is a little smaller.

Distribution. Tenasserim from Mergui southwards, extending
down the Malay peninsula to Sumatra and Borneo.

903. Anthothreptes malaccensis. *The Brown-throated Sun-bird.*

Certhia malaccensis, *Scop. Del. Flor. et Faun. Insul.* ii, p. 91 (1786).
Nectarinia malaccensis (*Scop.*), *Blyth, Cat.* p. 225.
Anthreptes malaccensis (*Scop.*), *Horsf. & M. Cat.* ii, p. 737 ; *Shelley,
  Mon. Nect.* pp. xliii, xliv, 315, pl. 102 ; *Hume & Dav. S. F.* vi,
  p. 186 ; *Hume, Cat.* no. 233 ter ; *Oates, B. B.* i, p. 324.
Anthothreptes malaccensis (*Scop.*), *Gadow, Cat. B. M.* ix, p. 122
  (part.).

Coloration. Male. Forehead, crown, nape, back, and sides of the
neck metallic lilac, according to the light ; rump and upper tail-
coverts metallic violet-purple ; lores and sides of the head dull
greenish yellow ; a stripe from the gape down the side of the
throat coppery purple ; chin and throat cinnamon-brown ; lower
plumage yellow, tinged with green on the flanks and vent ; under
wing-coverts and axillaries yellowish white ; tail bluish brown,
edged with metallic purple on the outer webs ; lesser and most of
the median wing-coverts brilliant purple ; the longer median
coverts and scapulars olive-brown, tipped with cinnamon ; greater
coverts olive-brown, edged with cinnamon ; quills brown, edged
with olive-green, with a tinge of ferruginous.

Female. The upper plumage and the sides of the head yellowish green; the ear-coverts with pale shaft-stripes; lower plumage yellow, with a tinge of green on the sides; tail brown, tipped very narrowly with whitish and edged on the outer webs with yellowish green; wings and coverts dark brown, edged like the tail.

According to Davison the legs vary a good deal, but are generally more or less green; claws green; bill dark horny brown or nearly black, the gape orange; irides light red to dark brown.

Length 5·2; tail 1·9; wing 2·7; tarsus ·6; bill from gape ·8.

Distribution. Blyth records this species from Arrakan in general, and Hume from Akyab. It does not appear to occur in Pegu, but it is found in Tenasserim from Amherst southwards. It extends down the Malay peninsula to the islands, and also to Siam.

904. Anthothreptes rhodolæma. *The Rufous-throated Sun-bird.*

Anthreptes rhodolæma, *Shelley, Mon. Nect.* pp. xliii, xliv, 313, pl. 101 (1878); *Salvadori, Ann. Mus. Civ. Gen.* (2) vii, p. 435.

Coloration. Male. Resembles the male of *A. malaccensis* in general appearance. Differs in having the forehead, crown, nape, back, and also the sides of the neck metallic green, sometimes shaded with lilac; the lores and sides of the head rufous; all the wing-coverts rufous except those near the anterior edge of the wing; and the breast tinged with olivaceous.

Female. Resembles the female of *A. malaccensis* so closely as to be undistinguishable from it.

Of the same dimensions as *A. malaccensis.*

Distribution. The extreme south of Tenasserim, whence Count Salvadori records this species. It extends down the Malay peninsula and to Sumatra.

905. Anthothreptes simplex. *The Plain-coloured Sun-bird.*

Nectarinia simplex, *S. Müll. Verhand. Nat. Gesch., Zool. Aves,* p. 62 (1843); *Blyth, Cat.* p. 225.
Anthreptes xanthochlora, *Hume, S. F.* iii, p. 320, note (1875).
Anthreptes simplex (*S. Müll.*), *Shelley, Mon. Nect.* pp. xliii, xliv, 309, pl. 100; *Hume & Dav. S. F.* vi, p. 188; *Hume, Cat.* no. 283 quat.; *Oates, B. B.* i, p. 324.
Anthothreptes simplex (*S. Müll.*), *Gadow, Cat. B. M.* ix, p. 114.

Coloration. Male. A large patch on the forehead metallic green; the whole upper plumage and wing-coverts olive-yellow; tail a deeper tint of the same; quills brown, edged with olive-yellow; sides of the head ashy green; cheeks, chin, throat, and fore neck greenish ashy; remainder of lower plumage dull oily yellow; under wing-coverts whitish.

Female. Differs only in wanting the metallic patch on the forehead.

Legs and feet pale dirty green; the bill dark horny brown; irides wood-brown (*Davison*).

Length 4·5; tail 2·1; wing 2·4; tarsus ·55; bill from gape ·7. The female is slightly smaller than the male.

Distribution. The southern part of Tenasserim from Mergui southwards, extending down the Malay peninsula to Sumatra and Borneo.

Subfamily ARACHNOTHERINÆ.

The Sun-birds of this subfamily are characterized by a somewhat massive body, a long and strong bill, and non-metallic plumage. The sexes are either quite alike or very nearly so. The tail of all the members of this subfamily is short and rounded. There is only one genus, containing five Indian species, represented within Indian limits.

Genus **ARACHNOTHERA**, Temm., 1826.

In the genus *Arachnothera* the bill is extremely long, about twice the length of the head or longer, much curved, stout at base, and with the culmen ridged between the nostrils. In four out of the five Indian species of this genus the sexes are alike, and in the fifth they resemble each other very closely. The plumage of all is more or less green.

The Sun-birds of this genus are generally found in dense evergreen forests or in thick plantain-gardens in retired spots. They affect the flowers of plantain-trees (*Musa*) more than those of any other tree and their nests appear to be frequently attached to the leaves of these.

Key to the Species.

a. No yellow on side of head.
  a'. Upper plumage streaked.
    a''. Back and rump distinctly streaked;
      wing 3·7 in males .... *A. magna*, p. 369.
    b''. Back and rump indistinctly streaked;
      wing 3·4 in males .... *A. aurita*, p. 370.
  b'. Upper plumage un-streaked.
    c''. Lower plumage uniform ashy green,
      obsoletely streaked .... *A. modesta*, p. 370.
    d''. Lower plumage yellow; chin and
      throat dull white................. *A. longirostris*, p. 371.
b. Portion of side of head yellow .... *A. chrysogenys*, p. 371.

906. **Arachnothera magna.** *The Larger Streaked Spider-hunter.*

Cinnyris magna, *Hodgs., Ind. Rev.* 1837, p. 272.
Arachnothera magna (*Hodgs.*), *Blyth, Cat.* p. 221; *Horsf. & M. Cat.*
ii, p. 727; *Jerd. B. I.* i, p. 360; *Stoliczka, J. A. S. B.* xxxvii, pt.
ii, p. 23; *Hume, S. F.* iii, p. 85; *Gammie, S. F.* v, p. 385; *Shelley,
Mon. Nect.* pp. xlix, 317, pl. 112; *Hume & Dav. S. F.* xi, p. 173;
Hume, Cat. no. 223; *Oates, B. B.* i, p. 327; *Bingham, S. F.* ix,
p. 169; *Gadow, Cat. B. M.* ix, p. 105; *Hume, S. F.* xi, p. 79;
Oates in Hume's N. & E. 2nd ed. ii, p. 268.

The Large Spider-hunter, Jerd.; *Dom-siriok-pho,* Lepch.; *Vedong-
pichang,* Bhut.

Fig. 101. —Head of *A. magna.*

Coloration. Forehead and crown olive-yellow, each feather with
a large black patch in the centre; lesser and median wing-coverts
the same; remainder of the upper plumage olive-yellow, with
distinct broad black shaft-stripes; greater wing-coverts and tertiaries olive-yellow with black shafts; primaries and secondaries
dark brown, margined with olive-yellow; tail olive-yellow, each
feather with a band of black near the end, followed on all but the
median pair by a lighter patch of pale yellowish; sides of the head
like the back, but paler; the entire under plumage pale yellowish,
each feather with a broad streak of black.

Bill black; iris brown; legs orange-yellow; claws yellow.

Length 7; tail 2; wing 3·7; tarsus ·8; bill from gape 1·8.

Distribution. The Himalayas, from Biláspur in the Sutlej valley
(according to Stoliczka, *l. c.*) to the extreme east of Assam; the
valley of Assam; the Khási hills; Manipur; Arrakan; Tenasserim
as far south as Tavoy and the Thoungyeen valley. This species is
probably spread over the whole of Burma, but I failed to meet with
it in any part of Pegu west of the Sittoung river. It occurs up to
5000 feet.

Habits, &c. Breeds from May to August, constructing an open
cup-shaped nest of vegetable fibres felted together and mingled
with dead leaves, and lined with grass. The nest is attached by
half its rim to a plantain-leaf, to which it is sewn by very numerous
threads. The eggs, usually three in number, are brown speckled
with purple, and measure about ·95 by ·7.

907. Arachnothera aurata. *The Smaller Streaked Spider-hunter.*

Arachnothera aurata, *Blyth, J. A. S. B.* xxiv, p. 478 (1855); *Hume,
S. F.* iii, p. 85; *Blyth & Walden, Birds Burm.* p. 140; *Shelley,
Mon. Nect.* pp. xlix, 351, pl. 112; *Hume & Dav. S. F.* vi, p. 174;
Hume, Cat. no. 223 bis; *Gadow, Cat. B. M.* ix, p. 105; *Oates,
B. B.* i, p. 328; *Salvadori, Ann. Mus. Civ. Gen.* (2) vii, p. 395.

Coloration. Resembles *A. magna* very closely. Differs in being
smaller and in having the striations on both the upper and lower
plumage much narrower and almost or quite absent on the lower
back.

Bill black, the margins of the lower mandible yellow; mouth
yellow; iris brown; eyelids plumbeous; legs orange-yellow; claws
yellow.

Length about 6·5; tail 1·8; wing 3 to 3·4; tarsus ·8; bill from
gape 1·6.

Distribution. Confined to Pegu. This species is found throughout the Pegu hills and it has also been procured both at Thayetmyo
and Toungngoo. It is also recorded from the Karen hills east of
the latter town. The late Captain Beavan is said to have procured
this bird at Kyodan on the Salween river.

908. Arachnothera modesta. *The Grey-breasted Spider-hunter.*

Anthreptes modesta, *Eyton, P. Z. S.* 1839, p. 105.
Arachnothera modesta (*Eyton*), *Hume, S. F.* iii, p. 85; *Hume &
Dav. S. F.* vi, p. 176; *Shelley, Mon. Nect.* pp. xlix, l, 353, pl. 113;
Hume, Cat. no. 224 bis; *Gadow, Cat. B. M.* ix, p. 107; *Oates,
B. B.* i. p. 329.

Coloration. The whole upper plumage and wing-coverts bright
yellowish green, the feathers of the head dark-centred; quills dark
brown, broadly edged with the colour of the back, the tertiaries
almost wholly of this colour; sides of the neck and the upper part
of the ear-coverts olive-green; cheeks, the lower portion of the ear-
coverts, chin, throat, and fore neck ashy green, obscurely streaked
with brown; remainder of the lower plumage ashy green, paler on
the abdomen, and the under tail-coverts tipped yellowish white;
median tail-feathers yellowish green, broadly tipped with black;
the others blackish, the basal two thirds of the outer webs yellowish
green, and each of the feathers with a spot of white near the tip on
the inner web; edge of the wing bright yellow; under wing-
coverts and axillaries pale yellow.

Legs and feet reddish ochre to pale reddish brown; the upper
mandible black, the lower reddish horny to pale reddish brown;
irides brown (*Davison*).

Length 7; tail 2·2; wing 3·5; tarsus ·75; bill from gape 1·5.

Distribution. Tenasserim from the base of Muleyit mountain to
Malawún, extending to Cochin China, the Malay peninsula and
the islands.

909. **Arachnothera longirostris.** *The Little Spider-hunter.*

Certhia longirostra, *Lath. Ind. Orn.* i, p. 299 (1790).
Arachnothera affinis, *Blyth, J. A. S. B.* xv, p. 43 (1846) ; *id. Cat.* p. 222.
Arachnothera pusilla, *Blyth, Cat.* p. 328 (1849) ; *Horsf. & M. Cat.* ii, p. 730 ; *Jerd. B. I.* i, p. 361 ; *Hume, S. F.* iii, p. 85.
Arachnothera longirostra (*Lath.*), *Shelley, Mon. Nect.* pp. xlix, 1, 357, pl. 114 ; *Fairbank, S. F.* v, p. 397 ; *Hume & Dav. S. F.* vi, p. 174 ; *Hume, Cat.* no. 224 ; *Gadow, Cat. B. M.* ix, p. 103 ; *Oates, B. B.* i, p. 390 ; *Barnes, Birds Bom.* p. 135 ; *Hume, S. F.* xi, p. 80.

Coloration. Male. Upper plumage olive-green, the feathers of the forehead and crown centred with dark brown ; lesser wing-coverts like the back ; greater coverts and the wings brown, edged with olive-green ; tail blackish, tipped with dull white and obsoletely margined with olive-green ; lores whitish ; sides of the head ashy brown ; a short moustachial streak dark brown ; chin and throat dull white ; remainder of lower plumage deep yellow ; a tuft of feathers on each side the breast chrome-yellow.

Female. Differs in wanting the pectoral chrome-yellow tufts.

Bill above brown, below plumbeous ; iris dark brown ; legs plumbeous ; claws horn-colour.

Length 6·3 ; tail 1·6 ; wing 2·6 ; tarsus ·65 ; bill from gape 1·5.

Distribution. The Western Ghâts of India from the Palni hills to about the latitude of Belgaum, up to about 5500 feet ; the extreme eastern part of Assam ; Cachar ; Tipperah ; Sylhet ; Manipur ; Chittagong ; Arrakan ; Pegu and the whole of Tenasserim, extending down the Malay peninsula to the islands.

Habits, &c. The nest of this species has not yet been found within Indian limits, but Bernstein, who procured it elsewhere, describes it as being oval and attached to the underside of a large leaf which forms the back wall of the nest.

910. **Arachnothera chrysogenys.** *The Yellow-eared Spider-hunter.*

Nectarinia chrysogenys, *Temm. Pl. Col.* pl. 388, fig. i (1826).
Arachnothera chrysogenys (*Temm.*), *Blyth, Cat.* pp. 222, 327 ; *Horsf. & M. Cat.* ii, p. 729 ; *Hume, S. F.* iii, p. 85 ; *Hume & Dav. S. F.* vi, p. 177 ; *Shelley, Mon. Nect.* pp. xlix, li, 365, pl. 117 ; *Hume, Cat.* no. 224 ter ; *Oates, B. B.* i, p. 331 ; *Gadow, Cat. B. M.* ix, p. 108.

Coloration. Upper plumage dull olive-green, the feathers of the head dark-centred ; coverts and quills dark brown, broadly edged with the colour of the back ; tail olive-green ; feathers on the edge of the upper eyelid and a bunch of feathers springing from near the angle of the gape bright yellow ; ear-coverts and sides of neck like the back ; cheeks, chin, throat, and upper breast dull brownish green, the centres of the feathers darker ; lower breast, abdomen, vent, and under tail-coverts yellow ; sides of the body yellow, tinged with dusky ; under wing-coverts and axillaries pale yellow.

Legs and feet fleshy white; the bill darker horny brown; the edges of both mandibles to within ·6 of tip dirty yellow; gape fleshy white; irides brown (*Davison*).

Length 7; tail 1·7; wing 3·5; tarsus ·75; bill from gape 1·8.

Distribution. Tenasserim south of Mergui, extending down the Malay peninsula to Sumatra, Java, and Borneo.

Arachnothera flaviyastra is a closely allied species inhabiting the Malay peninsula, and is likely to occur in Tenasserim. It may be recognized by its larger size, by the eye being entirely surrounded by yellow, and by its stouter and more flattened bill.

Arachnothera crassirostris (Reichb.) occurs in the Malay peninsula, and is not unlikely to be found in Tenasserim. This species resembles *Arachnothera longirostris* very closely, but may be known by its much broader and rounder bill and by the chin and throat being of the same colour as the breast.

The following species, on being critically examined, proves to be no Sun-bird. I failed to discover this, however, till I was working the *Nectariniidæ*, with which it has always been associated.

My reasons for excluding this bird from the *Nectariniidæ* are threefold:

It has no serrations on the margins of the mandibles, a character found in all the Sun-birds.

It has, according to Wallace (Ibis, 1870, p. 49), a tongue which is "short, triangular, horny at the tip, and entire."

It has habits which resemble those of no other species of Sun-bird.

I know the bird well in life, but prefer to quote what Davison says on this point:—

"In its habits this species differs conspicuously from all its congeners, reminding one very much of the White-eyed Tit (*Zosterops palpebrosus*) or again of *Timalia* (*Cyanoderma*) *erythroptera*. Except perhaps during the breeding-season, it goes about in small parties of from five to ten in amongst the undergrowth, or the skirts of the forest, or in scrub-jungle, hunting amongst the foliage and roots of the trees for insects, on which it chiefly subsists, and keeping up the while an incessant twittering.

"Of other species of Sun-birds a dozen, or even at times fifty, may be seen about a single tree; but in the case of these there is never any concerted action between more than a single pair. They do not go about in flocks, though many individuals may happen to collect in a single place, but the present species, when not breeding, is almost always seen in flocks working together in concert, invariably moving away from one place to another at the same time and hunting, some high and some low, just as a mob of our Titmice on the Himalayas may often be seen doing."

The nestling bird resembles the female, and therefore the proper position of this species appears to be among the *Crateropodidæ* in the subfamily *Liotrichinæ*, probably near *Myzornis* (Vol. i, p. 233).

Genus CHALCOPARIA, Cabanis, 1850.

The single species of this genus has the bill shorter than the head, entire, without any serrations on the margins of the mandibles; the culmen very slightly curved, the lower mandible straight; the rictal bristles weak; the tarsus short and scutellated; the tail of moderate length, slightly rounded, and consisting of twelve feathers; the wing moderate, with ten primaries, the first of which is small.

The sexes are of different colours and the upper plumage of the male is metallic.

Fig. 102. Head of *C. phœnicotis.*

911. Chalcoparia phœnicotis. *The Ruby-Cheek.*

Motacilla singalensis, *Gmel. Syst. Nat.* i, p. 964 (1788).
Nectarinia phœnicotis, *Temm. Pl. Col.* pl. 108, fig. 1 (1824), pl. 388, fig. 2; *Blyth, Cat.* p. 225.
Chalcoparia phœnicotis (Temm.), *Horsf. & M. Cat.* ii, p. 747; *Oates, S. F.* v, p. 147; *Oates in Hume's N. & E.* 2nd ed. ii, p. 269.
Chalcoparia singalensis (*Gmel.*), *Wald. Ibis,* 1870, p. 48; *Hume & Dav. S. F.* vi, p. 189; *Hume, Cat.* no. 233 sex.
Chalcoparia cingalensis (*Gmel.*), *Anders. Yunnan Exped., Aves,* p. 662.
Anthreptes phœnicotis (Temm.), *Shelley, Mon. Nect.* pp. xliii, xlv, 325, pl. 105.
Anthreptes singalensis (*Gmel.*), *Oates, B. B.* i, p. 326.
Anthothreptes phœnicotis (Temm.), *Gadow, Cat. B. M.* ix, p. 121.

Coloration. Male. The whole upper plumage and lesser wing-coverts brilliant metallic emerald-green; lores blackish; cheeks and ear-coverts rich copper-colour, bordered below by a line of rich metallic violet-purple; chin, throat, and breast ferruginous buff; abdomen, sides of the body, vent, and under tail-coverts yellow; tail black, edged externally with metallic green; under wing-coverts pale yellow; greater wing-coverts black, edged with metallic green; wings black, edged more or less with purple.

Female. The lower plumage like that of the male; the upper plumage and the lesser wing-coverts olive-green; ear-coverts and cheeks slate-colour; greater wing-coverts and wings dark brown, edged with yellowish green; tail brown, broadly edged with yellowish green.

The young are like the female.

Bill black; gape orange-yellow; mouth yellow; iris lake-red; legs yellowish green; claws yellowish horny; eyelids greenish.

Length 4·4; tail 1·6; wing 2·1; tarsus ·65; bill from gape ·6.

Distribution. The Sikhim Terai; the Bhutan Doars; the Dibrugarh district of Assam; the Khási and Gáro hills; Sylhet; Cachar;

Dacca; Tipperah; Manipur; the neighbourhood of Bhámo; Chittagong; Arrakan; the southern portion of Pegu; the whole of Tenasserim, thence extending down the Malay peninsula to the large islands.

Habits, &c. Breeds in Southern Pegu from May to August. The nest is suspended from the tip of a branch at any height from the ground and well surrounded by leaves. It is a pear-shaped structure constructed of hair-like fibres and roots and ornamented outside with various substances. The entrance is about midway up the nest and protected by a very ample portico which extends to the base of the nest. The eggs, two in number, are pinkish white marked with brown and purple; they measure about ·64 by ·45.

Family DICÆIDÆ.

The intrinsic muscles of the syrinx fixed to the end of the bronchial semi-rings: both mandibles finely and evenly serrated on the terminal third of their edges; bill short and triangular; primaries nine or ten: the nestling resembling the adult female; one moult in the year: rectrices twelve; rictal bristles short.

The *Dicæidæ* form a very compact and natural family of birds, which may be known at once, and separated from all other *Passeres* except the *Nectariniidæ*, by the peculiar serrations on the edges of both mandibles, as shown in the accompanying cut.

Fig. 103.—Bill of *D. cruentatum* (enlarged), to show serrations on mandibles.

This character holds good in all the species of this family without exception. A lens is generally necessary to observe the serrations, but frequently they may be seen with the naked eye, especially if the open bill is held against a sheet of white paper.

The *Dicæidæ* are all small birds, generally of brilliant plumage. In most species the sexes differ in colour, in some they are alike. The young resemble the adult female. They are all resident, not even migrating locally.

This family forms a connecting-link between the nine-primaried and the ten-primaried Passeres, some of the genera possessing nine of these feathers, and others ten. They all have twelve tail-feathers. The nostrils are covered by a large oval process leaving a lunar aperture; the rictal bristles are short, but the naral bristles are sometimes greatly developed. The tail is always short, and the tarsus is never lengthened.

The Flower-peckers are remarkable for the beauty of their nests, which are frequently pear-shaped, and suspended from a branch. The eggs are invariably white except in *Piprisoma*, in which they are spotted.

Key to the Genera.

a. With nine primaries, the first reaching
to the tip of the wing.
  a'. Bill slender; the lower line of the
    inferior mandible almost straight .. DICÆUM, p. 375.
  b'. Bill thick; lower mandible swollen;
    its lower edge much angulate.
    a''. Tail rounded; nostrils covered by
      long hairs. . . . . ACMONORHYNCHUS, p. 381.
    b''. Tail square; nostrils perfectly
      bare of hairs. . . . . PIPRISOMA, p. 382.
b. Wing with ten primaries, the first one
small.
  c'. First primary about equal to the
    tarsus. . . . . PRIONOCHILUS, p. 384.
  d'. First primary not longer than the
    hind toe. . . . . PACHYGLOSSA, p. 385.

Genus **DICÆUM**, Cuvier, 1817.

The genus *Dicæum* contains eight species of Indian birds, which are characterized by the possession of nine primaries and a slender bill with the lower line of the inferior mandible nearly straight.

In *Dicæum* the males of many of the species are brightly coloured, and in these cases the sexes differ in coloration; in other species they are more dully coloured and the sexes are alike.

They are all without exception of very small size. They frequent trees, generally at a considerable height above the ground, and feed both on insects and small berries. Their nests are beautiful structures made of the finest and most delicate materials, egg-shaped, and suspended from the tip of a branch. They all lay white eggs, so far as is known.

Key to the Species.

a. Upper plumage with some red in it.
  a'. Whole upper plumage crimson .... *D. cruentatum* ♂, p. 376.
  b'. Rump only crimson . . . . *D. cruentatum* ♀, p. 376.
  c'. Back and rump bright orange-red . *D. trigonostigma* ♂, p. 377.
  d'. Rump only pale orange-red . . . . *D. trigonostigma* ♀, p. 377.

b. Upper plumage without any red in it.

 c'. Lower tail-coverts of a different
  colour to the abdomen; lower
  plumage streaked............... *D. chrysorrhœum* ♂ ♀,
                  [p. 8.

 f'. Lower tail-coverts of the same
  colour as the abdomen; lower
  plumage unstreaked.

  a". A patch of red on the breast .... *D. ignipectus* ♂, p. 378.

  b". No red on the breast.

   a'". Bill black or of a dark colour.

    a⁴. Lower plumage of one
     uniform colour.

     a⁵. Rump yellowish green con-
      trasting with the green of
      the back .... *D. ignipectus* ♀, p. 378.

     b⁵. Rump of the same colour
      as the back.

      a⁶. Forehead and lores
       conspicuously whitish.. *D. concolor* ♂ ♀, p. 379.

      b⁶. Forehead and lores of a
       dark colour .... *D. olivaceum* ♂ ♀, p. 380.

    b⁴. Lower plumage not uniform;
     throat and breast whitish,
     abdomen dull yellow ...... *D. virescens* ♂ ♀, p. 380.

   b'". Bill yellow................. *D. erythrorhynchus* ♂ ♀,
                  [p. 381.

912. Dicæum cruentatum. *The Scarlet-backed Flower-pecker.*

Certhia cruentata, *Linn. Syst. Nat.* i, p. 187 (1766).
Certhia coccinea, *Scop. Del. Fl. et Faun. Insub.* ii, p. 91 (1786).
Dicæum cruentatum (L.), *Blyth, Cat.* p. 226; *Hume, N. & E.* p. 155;
 Hume & Dav. S. F. vi, p. 192; *Anders. Yunnan Exped., Aves,*
 p. 663; *Oates, S. F.* vii, p. 46; *Hume, Cat.* no. 236; *Oates, B. B.* i,
 p. 332; *Sharpe, Cat. B. M.* x, p. 15; *Hume, S. F.* xi, p. 83; *Oates*
 in Hume's N. & E. 2nd ed. ii, p. 270.
Dicæum coccineum (*Scop.*), *Horsf. & M. Cat.* ii, p. 747; *Jerd. B. I.* i,
 p. 373.

Fig. 104.— Head of *D. cruentatum.*

Coloration. *Male.* Forehead, crown, nape, back, rump, and
upper tail-coverts rich crimson; lores, sides of the head and neck,
tail, wings, and wing-coverts black; lower plumage pale buff, the
sides of the breast black, and the sides of the body ashy brown;
under wing-coverts and axillaries white.

Female. Head, nape, and back olive-green, the centres of the
feathers of the crown darker, and the nape with a golden yellow
tinge; rump and upper tail-coverts red; tail black; the whole
lower plumage ashy buff, darker on the sides of the neck and body;

upper wing-coverts dark brown, edged with olive-green; tertiaries the same; primaries and secondaries brown, edged exteriorly with greenish white.

The young resemble the female.

Legs and feet black; bill and mouth black; iris dark brown; eyelids plumbeous; in the female the mouth is flesh-coloured.

Length 3·5; tail 1·05; wing 1·9; tarsus ·5; bill from gape ·45.

Distribution. The western and northern limits of this species have not been determined with any great accuracy. It appears to be common at Calcutta, and it has been obtained in the Khási hills, the Bhutan Doars, and the valley of Assam up to Dibrugarh. South and east of these localities it has been found in Sylhet, Cachar, and Manipur. It is common throughout the greater portion of Burma, and extends down to the southernmost point of Tenasserim.

It is diffused through Southern China, Siam, and the Malay peninsula down to Sumatra.

Habits, &c. Breeds from March to May and probably later, constructing a small egg-shaped nest of vegetable down and grass, which is attached to the tip of a branch at a considerable height from the ground as a rule. The eggs, two or three in number, are glossless white, and measure ·56 by ·4.

913. Dicæum trigonostigma. *The Orange-bellied Flower-pecker.*

Certhia trigonostigma, *Scop. Del. Fl. et Faun. Insub.* ii, p. 91 (1786).
Dicæum trigonostigma (*Scop.*), *Blyth, Cat.* p. 226; *Horsf. & M. Cat.* ii, p. 748; *Wald. P. Z. S.* 1866, p. 545; *id. Ibis,* 1876, p. 349, pl. x, f. 2; *Hume & Dav. S. F.* vi, p. 194; *Hume, Cat.* no. 236 bis; *Oates, B. B.* i, p. 336; *Sharpe, Cat. B. M.* x, p. 38; *Oates in Hume's N. & E.* 2nd ed. ii, p. 272.

Coloration. Male. Forehead, crown, nape, sides of the head and neck, scapulars, and wing-coverts dull blue; back and rump flaming orange-yellow, deeper on the back; upper tail-coverts dull blue; chin, throat, cheeks, and breast ashy grey; abdomen, sides of the body, vent, and under tail-coverts flaming orange; under wing-coverts and axillaries whitish; tail black; wings black, edged with dull blue.

Female. Forehead, crown, nape, back, sides of the neck, and scapulars olive-green; rump and upper tail-coverts yellow, tinged with orange at the tips of the feathers; tail blackish; coverts and wings dark brown, narrowly edged with olive-green; sides of the head pale ashy; chin and throat sordid green; breast and sides of the body ashy green; abdomen, vent, and under tail-coverts bright yellow.

The young resemble the female, but have the abdomen dull yellow.

Male: legs and feet horny black; bill black; iris brown.
Female: legs, feet, and claws greenish to dark plumbeous; upper mandible from tip to nostrils and tip of the lower mandible

blackish horny; base of upper mandible reddish brown; lower mandible (except the tip) and gape pale orange-brown to orange-vermilion; iris grey to dark brown (*Hume & Davison*).

Length 3·6; tail 1; wing 1·9; tarsus ·5; bill from gape ·55.

Distribution. Burma. Wardlaw Ramsay obtained this species in the Karen hills east of Toungngoo at 3000 feet elevation, and I procured it near the town of Pegu. Davison observed it in Tenasserim from Moulmein southwards to Bankasun. It extends to Cochin China and the Malay peninsula.

914. Dicæum chrysorrhœum. *The Yellow-vented Flower-pecker.*

Dicæum chrysorrhœum, *Temm. Pl. Col.* pl. 478, f. 1 (1829); *Blyth, Cat.* p. 227; *Horsf. & M. Cat.* ii, p. 751; *Jerd. B. I.* i, p. 374; *Wald. Ibis,* 1872, p. 380; *Godw.-Aust. J. A. S. B.* xliii, pt. ii, p. 156; *Hume & Dav. S. F.* vi, p. 195; *Anders. Yunnan Exped., Aves,* p. 603; *Hume, Cat.* no. 237; *Bingham, S. F.* ix, p. 170; *Oates, S. F.* x, p. 198; *id. B. B.* i, p. 335; *Sharpe, Cat. B. M.* x, p. 44; *Hume, S. F.* xi, p. 84.

Dicæum chrysochlore, *Blyth, J. A. S. B.* xii, p. 1009 (1843).

Coloration. Upper plumage and lesser wing-coverts yellowish green, brighter on the rump and upper tail-coverts; tail blackish; greater wing-coverts dark brown on the inner webs and yellowish green on the outer; quills blackish brown, the secondaries and tertiaries broadly edged with yellowish green, the primaries very narrowly with whitish; sides of the head and neck yellowish green; cheeks, chin, and throat white, with a greenish-brown mandibular streak below the cheeks; lower plumage whitish, streaked with greenish brown; under tail-coverts golden yellow; under wing-coverts and axillaries white.

Iris orange-red; eyelids pinkish; upper mandible and tip of the lower black; remainder of bill pale plumbeous; legs dark plumbeous; claws dark horn; mouth flesh-colour.

Length 4; tail 1·2; wing 2·3; tarsus ·6; bill from gape ·5.

Distribution. Nepal and Sikhim; Naga hills, Tipperah, Manipur; the whole of Burma to the extreme south. The Nepal habitat is somewhat doubtful, for although Hodgson's specimens are said to have come from that country, they may nevertheless have been obtained in Sikhim. This species extends down the Malay peninsula to the islands.

915. Dicæum ignipectus. *The Fire-breasted Flower-pecker.*

Myzanthe ignipectus, *Hodgs. Blyth, J. A. S. B.* xii, p. 983 (1843); *Blyth, Cat.* p. 227; *Horsf. & M. Cat.* ii, p. 751; *Jerd. B. I.* i, p. 377; *Sto'. J. A. S. B.* xxxvii, pt. ii, p. 24; *Hume, N. & E.* p. 159; *Hume & Dav. S. F.* vi, p. 200; *Hume, Cat.* no. 241; *Scully, S. F.* viii, p. 261; *Oates, B. B.* i, p. 337; *Hume, S. F.* xi, p. 85.

Dicæum ignipectus (*Hodgs.*), *Sharpe, Cat. B. M.* x, p. 41; *Oates in Hume's N. & E.* 2nd ed. ii, p. 272.

Sangti-pro-pho, Lepch.

Coloration. *Male.* Upper plumage, sides of head, and neck black with a green and purple gloss and each feather fringed with yellowish brown ; wings and tail black, edged with glossy green ; lower plumage buff, tinged with green on the sides of the body ; a large patch of crimson on the breast, with a black patch below it sometimes produced down the middle of the abdomen. The yellow fringes of the upper plumage get worn off a good deal during the winter.

Female. Above green, rather glossy on the head and tinged with yellow on the rump and upper tail-coverts ; sides of the head ashy green, also the sides of the throat ; lores and lower plumage buff, tinged with green on the sides of the body ; wings and tail black edged with green : under wing-coverts and axillaries white.

The young resemble the female.

Male : bill black ; iris brown or blackish brown ; feet and claws dull or brownish black. *Female* : bill black ; base of lower mandible plumbeous (*Scully*).

Length rather more than 3 ; tail 1 ; wing 1·9 ; tarsus ·5 ; bill from gape ·4.

Distribution. The Himalayas from the Sutlej valley to Assam, up to 7000 feet ; Khási hills ; Manipur ; Karennee and Muleyit mountain in Tenasserim.

Habits, &c. Breeds in the Himalayas from April to July, constructing a pendent nest of very small size attached to the end of a twig of some large tree. In shape the nest is said to be like a purse and the walls to be like thin felt. The eggs are not known.

916. Dicæum concolor. *The Nilgiri Flower-pecker.*

Dicæum concolor, *Jerd. Madr. Journ. L. S.* xi, p. 227 (1840) ; *id. Ill. Ind. Orn.* pl. 39 ; *id. B. I.* i, p. 375 ; *Blyth, Cat.* p. 227 ; *Hume, N. & E.* p. 156 ; *id. Cat.* no. 239 ; *Davison, S. F.* x, p. 363 ; *Sharpe, Cat. B. M.* x, p. 45 ; *Barnes, Birds Bom.* p. 128 ; *Oates in Hume's N. & E.* 2nd ed. ii, p. 272.

Chitlu-jitta, Tel.

Coloration. In freshly-moulted birds the lores, forehead, and round the eye are conspicuously white ; the whole upper plumage dull green, the centres of the crown-feathers darker ; wings and tail dark brown, edged with dull green ; sides of the head and neck pale ashy green ; lower plumage pale yellowish buff. Soon after the autumnal moult the white of the face becomes dull.

Iris dark brown ; legs and feet dusky slaty ; bill lavender-blue, dusky on the culmen (*Butler*).

Length about 3·5 ; tail 1·1 ; wing 2 ; tarsus ·5 ; bill from gape ·5.

Distribution. The western coast of India from Khandála and Mahableshwar to the Palni hills. Blanford is said to have obtained it at Biláspur in the Central Provinces, but most probably some mistake has occurred about this locality.

Habits, &c. Breeds from January to April, making a pendent nest of vegetable down, lichens, &c., attached to the extremity of a twig of some tree. It lays three eggs, which are glossless white and measure ·64 by ·43.

917. **Dicæum olivaceum.** *The Plain-coloured Flower-pecker.*

Myzanthe inornata, *Hodgs. in Gray's Zool. Misc.* p. 82 (1844, *descr. nullâ*); *id. Gray's Cat. Mamm. &c. Nepal Coll. Hodgs.* pp. 60, 151 (1846, *descr. nullâ*).
Dicæum olivaceum, *Wald. A. M. N. H.* (4) xv, p. 401 (1875); *Godw.-Aust. J. A. S. B.* xlv, pt. ii, p. 194; *Hume, S. F.* iv, p. 498; *Hume & Dav. S. F.* vi, p. 195; *Hume, Cat.* no. 237 ter; *Oates, B. B.* i, p. 333; *Hume, S. F.* xi, p. 84.
Dicæum inornatum (*Hodgs.*), *Sharpe, Cat. B. M.* x, p. 45.

Coloration. The whole upper plumage olive-green, the rump rather brighter and the feathers of the head centred darker; tail dark brown, the feathers faintly edged with olive-green; wing-coverts brown, broadly edged with the colour of the back; wings dark brown, edged with olive-green rather brighter than the back; sides of the head and the whole lower plumage dull oily greenish yellow with an ashy tinge.

Legs and feet very dark plumbeous; upper mandible and tip of lower very dark brown; rest of the bill pale plumbeous; iris deep brown (*Hume & Davison*).

Length 3·3; tail 1; wing 1·8; tarsus ·45; bill from gape ·45.

The name *D. inornatum* cannot be used for this species, for Hodgson never published any description of the bird. He, moreover, confounded together the females of *D. inornatum* and *D. ignipectus*, as is shown by his specimens of both species in the British Museum being numbered 393.

Distribution. Occurs in Nepal, Sikhim, the Bhutan Doars, Shillong, the Nága hills, Manipur, the Toungngoo and Karen hills, at Papwon, on the Salween river, Wimpong, and Meetan near Moulmein. This species ranges into the Malay peninsula and to Sumatra.

918. **Dicæum virescens.** *The Andamanese Flower-pecker.*

Dicæum virescens, *Hume, S. F.* i, p. 482 (1873), ii, p. 198; *id. Cat.* no. 237 bis; *Sharpe, Cat. B. M.* x, p. 46.

Coloration. The whole upper plumage olive-green, brightest on the rump and upper tail-coverts, the feathers of the crown centred darker: wings and tail dark brown, edged with olive-green; sides of the head greenish ashy; chin, throat, and breast ashy white; remainder of the lower plumage yellow.

Length about 3·5; tail 1·1; wing 1·8; tarsus ·5; bill from gape ·5.

Distribution. The Andaman Islands.

919. **Dicæum erythrorhynchus.** *Tickell's Flower-pecker.*

Certhia erythrorhynchos, *Lath. Ind. Orn.* i, p. 299 (1790).
Nectarinia minima, *Tickell, J. A. S. B.* ii, p. 577 (1833).
Dicæum tickelliæ, *Blyth, J. A. S. B.* xii, p. 983 (1843).
Dicæum minimum (*Tick.*), *Blyth, Cat.* p. 227; *Horsf. & M. Cat.* ii,
    p. 750; *Jerd. B. I.* i, p. 374; *Hume, N. & E.* p. 155; *Legge, Birds
    Ceyl.* p. 574.
Dicæum erythrorhynchus (*Lath.*), *Hume, Cat.* no. 238; *Oates, B. B.*
    i. p. 334; *Barnes, Birds Bom.* p. 138; *Sharpe, Cat. B. M.* x, p. 48;
    Oates in Hume's N. & E. 2nd ed. ii, p. 274.

Sungti-pro-pho, Lepch.

Coloration. Upper plumage ashy olive, the feathers of the crown
with dark centres; tail dark brown; wings and coverts brown,
edged with ashy olive; sides of the head and lower plumage buffy
white.

Iris brown; legs and feet bluish plumbeous; bill pale livid
fleshy, dusky brown on the culmen towards the tip of the upper
mandible (*Butler*).

Length 3·2; tail 1; wing 1·8; tarsus ·5; bill from gape ·5.

Distribution. Occurs over the greater part of the peninsula of
India, from the Himalayas to Ceylon. Its western limits are
difficult to define for want of information. It is abundant in
South Guzerat, and I have seen specimens procured at Dehra and
at Dharmsála, but none from intermediate localities. It probably
follows the margin of the arid region of Rajputana, keeping well
outside it. To the east it ranges along the foot of the Himalayas
to Dibrugarh, and it has been procured in the Gáro hills. Blyth
records it from Arrakan and Tenasserim, and in the latter divison
Bingham, as quoted by Hume, once obtained a specimen.

Habits, &c. Breeds in March, April, and May, constructing a
small nest about 3 inches long, which is suspended from the
extremity of a twig on a tree. The nest is made of fine vegetable
fibres and down and generally well concealed under some drooping
leaves. The eggs, usually two in number, are glossless white and
measure ·58 by ·41.

Genus **ACMONORHYNCHUS**, n. gen.

I propose this genus for the reception of a remarkable Flower-
pecker which is found only in Ceylon and which has hitherto been
placed either in *Prionochilus* or in *Pachyglossa*. It differs from
both these genera in possessing only nine primaries. From
Dicæum it may be recognized by its very large, coarse bill, and
from *Piprisoma* by its rounded tail and the numerous hairs which
cover the nostrils.

In *Acmonorhynchus* the sexes differ and the young bird resembles
the female. Its habits are those of the family, but nothing is
known about its nidification.

920. Acmonorhynchus vincens.   *Legge's Flower-pecker.*

Prionochilus vincens, *Sclater, P. Z. S.* 1872, p. 729; *id. Ibis,* 1874,
p. 2, pl. i, figs. i & ii; *Hume, S. F.* ii, p. 455, iv, p. 493; *id. Cat.*
no. 240 ter; *Sharpe, Cat. B. M.* x, p. 72.
Pachyglossa vincens (*Scl.*), *Legge, Birds Ceyl.* p. 577, pl. 26.

Fig. 105.—Head of *A. vincens.*

Coloration. Male. Whole upper plumage and wing-coverts and
sides of the head and neck very dark bluish ashy approaching to
black; wings and tail glossy black, all the tail-feathers except the
two median pairs broadly tipped with white and the quills with a
broad band of white on the inner webs; chin, throat, breast, and
under tail-coverts white; remainder of lower plumage bright
yellow; under wing-coverts and axillaries white.
Female. Resembles the male, but the black of the upper plumage
is replaced by greenish brown; wings and tail not quite so black.

Iris brownish red; bill black, leaden at base; legs and feet
blackish (*Legge*).

Length about 4; tail 1·3; wing 2·35; tarsus ·45; bill from
gape ·45.

Distribution. Ceylon only, in the forests of the low hills of the
Southern Provinces.

Genus PIPRISOMA, Blyth, 1844.

The genus *Piprisoma* contains two species which Sharpe places
with *Prionochilus,* but which I prefer to keep separate on the
ground of their having but nine primaries.

In *Piprisoma* the bill is of much the same shape as in *Acmonorhynchus* but proportionally shorter. Viewed from above the bill
is nearly an equilateral triangle with the two sides sinuated. In
this genus the plumage is dull and the sexes are alike. Both species
are resident over the whole area they inhabit.

Key to the Species.

a. Upper plumage and sides of the head ashy
green; lower mandible coarse .... *P. squalidum,* p. 382.
b. Upper plumage and sides of the head green;
mandible slender...................... *P. modestum,* p. 383.

921. Piprisoma squalidum.   *The Thick-billed Flower-pecker.*

Pipra squalida, *Burton, P. Z. S.* 1836, p. 113.
Fringilla agilis, *Tickell, J. A. S. B.* ii, p. 578 (1833).
Parisoma vireoides, *Jerd. Madr. Journ. L. S.* xi, p. 8 (1840).

Piprisoma agile (*Tick.*), *Blyth, Cat.* p. 228; *Jerd. B. I.* i, p. 376;
Beavan, Ibis, 1867, p. 430, pl. x; *Hume, N. & E.* p. 158; *id.
S. F.* i, p. 434; *id. Cat.* no. 240; *Legge, Birds Ceyl.* p. 579; *G. F.
L. Marsh. Birds'-nesting Ind.* p. 60, pl.; *Scully, S. F.* viii, p. 290;
Oates, B. B. i, p. 338; *Barnes, Birds Bom.* p. 139.
Prionochilus squalidus (*Burt.*), *Sharpe, Cat. B. M.* x, p. 73.
Piprisoma squalidum, *Oates in Hume's N. & E.* 2nd ed. ii, p. 277.

Chitta jitta, Tel.

Coloration. Upper plumage ashy green, purer green on the
rump and upper tail-coverts; wings and tail brown edged with
olive-green; the latter tipped with white, broadly on the outermost feathers, more narrowly on the others, the middle feathers
being almost without any white; lores, cheeks, chin, and throat
white; sides of the head and neck ashy brown; a narrow brown
streak down each side of the throat; lower plumage pale ashy yellow
streaked with greenish brown.

Iris light brick-red; bill pale plumbeous horny; legs dusky
plumbeous (*Cleveland*).

Length about 4; tail 1·3; wing 2·4; tarsus ·5; bill from
gape ·45.

Distribution. Throughout the Himalayas at low elevations from
the Sutlej valley to Sikhim, and throughout the peninsula down to
Ceylon. The western limits of this species are difficult to define
owing to want of specimens and records of occurrence. It is said
to be very common at Baroda and then there is a great gap up to
Etáwah and another up to Debra. I have seen specimens from all
three places but from no other locality west of them.

To the east it can be traced to Midnapore and Dinapore, but it is
probable that it does not pass the longitude of Calcutta. Hume,
commenting on a collection of birds made by Inglis in Cachar,
states that it occurs in that district, but the Cachar specimens in the Hume Collection that I have examined, as noted below
are referable to *P. modestum.* I formerly erroneously recorded
P. squalidum (*P. agile*) from Pegu and Tenasserim.

Habits, &c. Constructs a small purse-like bag suspended from a
horizontal twig on a tree, from February to May. The materials
are fibres and the down of flower-buds felted together into a
pliable fabric which will bear crushing in the hand and then re-
cover its shape. The eggs, two or three in number, are white or
pinkish, marked in various ways with brownish pink or claret-
colour. They measure ·63 by ·45 *.

922. Piprisoma modestum. *Hume's Flower-pecker.*

Prionochilus modestus, *Hume, S. F.* iii, p. 298; *Hume & Dav. S. F.*
vi, p. 200; *Hume, Cat.* no. 240 *sex*; *Bingham, S. F.* ix, p. 171;
Hume, S. F. x, p. 198, note; *Oates, B. B.* i, p. 340; *Sharpe, Cat
B. M.* x, p. 74.
Piprisoma agile (*Tick.*), *apud Oates, S. F.* x, p. 198.

* The *Prionochilus pipra*, Less., of Hume's Catalogue, entered with doubt and
stated by Lesson to have been received from Ceylon, is *Iodopleura pipra*, now
known to occur only in the New World.

Coloration. Resembles *Piprisoma squalidum*, but has the upper plumage a brighter green, the sides of the head greener and the streaks below darker and more distinct. The chief difference, however, lies in the form of the lower mandible of the bill, which in *P. squalidum* is very large, swollen, and much bent upwards in front of the angle of the genys, whereas in the present species it is very slender and nearly straight.

Length about 4; tail 1·2; wing 2·4; tarsus ·45; bill from gape ·4.

Distribution. Probably throughout Burma, for Davison procured it at Amherst, Mergui, and Malawún, and Bingham in the Thoungyeen valley. I obtained it several times near the town of Pegu, and I. at one time wrongly identified my specimens with *P. agile* (*P. squalidum*). I have now reexamined my birds and find them to be the present species as suggested by Hume. In the Hume Collection there is a specimen procured in Cachar. I have examined this bird with the greatest care and find it to be the present species and not *P. squalidum*, to which Hume, probably by an oversight, refers it (S. F. xi, p. 85).

Genus **PRIONOCHILUS**, Strickl., 1841.

The genus *Prionochilus* contains two Indian species, the males of which are brightly coloured. The females are not very dissimilar to the males.

In this genus the bill is of the same shape as in *Acmonorhynchus*, but the sides are slightly concave when viewed from above. The rictal bristles are entirely absent. The wing has ten primaries, the first of which is of considerable size. The tail is square.

Key to the Species.

a. Upper plumage blue	*P. ignicapillus* ♂, p. 384.
b. Upper plumage green.	
a'. Lower plumage unstreaked..........	*P. ignicapillus* ♀, p. 384.
b'. Lower plumage streaked.	
a". Crown-patch crimson	*P. maculatus* ♂, p. 385.
b". Crown-patch orange-yellow	*P. maculatus* ♀, p. 385.

923. Prionochilus ignicapillus. *The Crimson-breasted Flowerpecker.*

Dicæum ignicapilla, *Eyton, P. Z. S.* 1839, p. 105.
Prionochilus percussus (*Temm.*), *apud Blyth, Cat.* p. 227; *Horsf. & M. Cat.* ii, p. 751; *Hume & Dav. S. F.* vi, p. 196; *Hume, Cat.* no. 240 quat.; *Oates, B. B.* i, p. 339.
Prionochilus ignicapillus (*Eyt.*), *Sharpe, Cat. B. M.* x, p. 65.

Coloration. Male. The whole upper plumage, sides of the head and neck, and lesser wing-coverts dull blue; a patch of crimson on the centre of the crown; tail brown, washed with blue on the outer webs; greater wing-coverts brown, edged with dull blue;

quills brown, edged with lighter blue ; a narrow white moustachial streak down the cheeks ; point of the chin white ; under wing-coverts and axillaries pure white ; the whole lower plumage deep yellow, paler on the vent and under tail-coverts and washed with green on the sides of the body ; a large patch of crimson on the breast.

Female. Upper plumage green ; wings and tail dark brown edged with green ; a pale red patch on the crown ; sides of the head green tinged with grey ; an ashy-grey moustachial streak ; lower plumage dull ashy green, suffused with yellow on the breast and abdomen.

The young resemble the female but have no coronal patch.

Legs, feet, claws, and lower mandible dark plumbeous ; upper mandible black ; iris dark brown (*Davison*) ; iris red-brown (*Wardlaw Ramsay*).

Length 3·8 ; tail 1·2 ; wing 2·2 ; tarsus ·55 ; bill from gape ·45.

Distribution. Procured hitherto within our limits only at Bankasun at the extreme southern point of Tenasserim. The range of this bird extends down the Malay peninsula to Sumatra and Borneo.

924. Prionochilus maculatus. *The White-throated Flower-pecker.*

Pardalotus maculatus, *Temm. Pl. Col.* iii, pl. 600, f. 3 (1836).
Prionochilus maculatus (*Temm.*), *Horsf. & M. Cat.* ii, p. 752 ; *Wald. Ibis,* 1872, p. 379 ; *Hume & Dav. S. F.* vi, p. 199 ; *Hume, Cat. no.* 240 quint. ; *Oates, B. B.* i, p. 340 ; *Sharpe, Cat. B. M.* x, p. 69.

Coloration. Male. The whole upper plumage and lesser wing-coverts green ; a patch of fiery red on the crown : greater coverts, wings, and tail brown, edged with green ; sides of the head ashy green ; lores and moustachial streak greenish white ; a dull green streak below this moustachial streak ; the space between these green streaks pale yellow ; breast and sides of the neck bright yellow, streaked with brown ; abdomen, vent, and under tail-coverts bright yellow ; sides of the body dusky yellow ; under wing-coverts and axillaries pale yellow.

Female. Differs from the male merely in having the coronal patch orange-yellow.

In the males, legs and feet very dark plumbeous, in the females dirty smalt-blue ; upper mandible and the lower to the angle of gonys black ; the rest of the bill plumbeous in males, smalt-blue in females ; iris dull red (*Hume & Davison*).

Length 3·7 ; tail 1·1 ; wing 2·1 ; tarsus ·5 ; bill from gape ·5.

Distribution. Tenasserim from Mergui down to Malawún : extending down to the Malay peninsula.

Genus PACHYGLOSSA, Hodgs., 1843.

The genus *Pachyglossa* resembles *Prionochilus* in many respects, but the wing is extremely long with a much shorter first primary, and the secondaries fall short of the tip of the wing by a distance greater than the length of the tarsus. The lower edge of the inferior mandible is nearly straight.

The sexes are dissimilar, but not very much so. I have not been able to examine a young bird, but it will, without doubt, be found to resemble the adult female.

The only species of this genus inhabits the Himalayas, and nothing appears to be known about its habits. Hodgson, *fide* Jerdon, says that it feeds on small insects and viscid berries and makes a pendulous nest.

925. Pachyglossa melanoxantha. *The Yellow-bellied Flower-pecker.*

Pachyglossa melanoxantha, *Hodgs., Blyth, J. A. S. B.* xii, p. 1010 (1843) ; *Jerd. B. I.* i, p. 378 ; *Hume, N. & E.* p. 160 ; *Godw.-Aust. J. A. S. B.* xliii, pt. ii, p. 156 ; *Hume, S. F.* ii, p. 455 ; *id. S. F.* v, p. 348 ; *id. Cat.* no. 242 ; *id. S. F.* xi, p. 85 ; *Oates in Hume's N. & E.* 2nd ed. ii, p. 279.
Prionochilus melanoxanthus (*Hodgs.*), *Sclater, Ibis,* 1874, p. 3, pl. i. fig. 3 ; *Sharpe, Cat. B. M.* x, p. 71.

Coloration. Male. The whole upper plumage, wings, tail, sides of head, neck, and breast black, the outermost pair of tail-feathers, or sometimes the two outer pairs with a patch of white on the inner web near the tip ; middle of chin, throat, and breast white ; remainder of the lower plumage bright yellow ; under wing-coverts white.

Female. Resembles the male, but the black is replaced by greenish brown, becoming paler on the sides of the head, neck, and breast ; abdomen and under tail-coverts duller yellow ; middle of chin, throat, and breast greyish white ; under wing-coverts and axillaries white.

Iris red ; legs dark plumbeous (*Godwin-Austen*).

Length about 4·5 ; tail 1·7 ; wing 2·8 ; tarsus ·55 ; bill from gape ·45.

Distribution. Sikhim and probably Nepal; extending to Dibrugarh in Assam and Sopoomah in the Nága hills.

Family PITTIDÆ.

The intrinsic muscles of the syrinx fixed at or near the middle of the bronchial semi-rings ; wing of ten primaries, the first of considerable size and reaching nearly to the tip of the wing ; tarsus elongated, the anterior covering entire and smooth ; tail very short and of twelve feathers ; feathers of crown elongate and forming a conspicuous crest when erected.

The *Pittidæ* are a compact group of birds which are found over the whole of South-eastern Asia, extending to Australia; and a single species is found in Africa. They differ from all other Indian Passeres in the structure of the syrinx and also in the formation of the wing, the first primary being of large size, whereas in all the other ten-primaried Passeres the first is markedly small. Their long legs and short tails also suffice to separate them from nearly all other Passeres.

The Pittas live habitually on the ground and feed on insects: they hop and run with great facility and their flight is strong for short distances. The males have a very sweet call consisting of a double whistle, uttered from a tree. The majority of the species prefer dense jungle, but some few may be found in gardens, sparse bamboo-jungle, and even in comparatively open country. Many of the species are locally migratory, others appear to be quite stationary throughout the year.

The Pittas make large oven-shaped nests on the ground or on thick branches near the ground and lay four or five eggs which are very richly marked.

It seems quite impossible to divide the Indian Pittas into more than two genera, as they are extremely similar to each other in structure.

Key to the Genera.

a. Feathers at sides of nape long and pointed,
　　forming conspicuous aigrettes .... ANTHOCINCLA, p. 387.
b. Feathers at sides of nape not conspicuously
　　lengthened .... PITTA, p. 388.

Genus ANTHOCINCLA, Blyth, 1862.

The only species of this genus is characterized by its conspicuous aigrettes and by its elongated and compressed bill. The sexes are not very different.

926. Anthocincla phayrii. *Phayre's Pitta.*

Anthocincla phayrii, *Blyth, J. A. S. B.* xxxi, p. 343 (1862); *Hume, S. F.* iii, p. 100, pl. ii; *Blyth, Birds Burm.* p. 100; *Hume & Dav. S. F.* vi, p. 245; *Hume, Cat.* no. 346 ter; *Bingham, S. F.* ix, pp. 177, 474; *Oates, B. B.* i, p. 420; *Sclater, Cat. B. M.* xiv, p. 413; *Oates in Hume's N. & E.* 2nd ed. ii, p. 279.

Coloration. *Male.* A black band from the forehead passing over the middle of the crown and expanding to cover the nape and whole hind neck; remainder of crown and forehead rich fulvous, each feather narrowly edged with black; lores, cheeks, and ear-coverts mixed rufous and black; a broad stripe from the eye over the ear-coverts, reaching well down the neck, white, each feather margined with black; some of the longer feathers, forming aigrettes, also barred with black; whole upper plumage rufous

2 c 2

brown ; wing-coverts tipped broadly with fulvous and with a subterminal black bar on both webs ; tertiaries and tail rather duller than the back ; primaries brown, broadly tipped paler ; a large fulvous patch at the base of each feather ; secondaries brown, edged with the colour of the back ; chin and middle of the throat white ; sides of the throat fulvous, the feathers margined with black ; remainder of lower plumage fulvous ; the feathers of the breast very narrowly and indistinctly margined with black, and some of them with black spots ; the feathers of the sides of the body and flanks distinctly spotted near the tip of both webs ; under tail-coverts pink.

Female. Differs in wanting the black coronal streak and the black on the nape and hind neck, this colour being replaced by the colour of the back, but rather darker ; the feathers of the forehead and crown margined with black ; also differs in having the breast more marked with black and the spots on the sides of the body larger.

Male : bill dark horny ; iris nut-brown ; legs and feet dirty flesh-colour blotched with brown. Female : bill horny ; iris dark brown ; legs, feet, and claws fleshy white (*Bingham*).

Length about 9 inches ; tail 2·3 ; wing 3·9 ; tarsus 1·25 ; bill from gape 1·5.

Fig. 106.—Head of *A. phayrii.*

Distribution. Burma east of the Sittoung river from the Karen hills east of Toungngoo to the valley of the Thoungyeen river.

Habits, &c. Bingham found a nest of this Pitta in Tenasserim in April. It was an oven-shaped structure on the ground at the root of a tree and was composed of leaves, roots, and grass, with a small platform of twigs leading up to the entrance, which was at the side. The nest contained four eggs, which were white marked with purple and black and measured about 1·09 by ·86.

Genus **PITTA**, Vieill., 1816.

The genus *Pitta* contains those Pittas which have no aigrettes of pointed feathers, and which have a shorter and broader bill than *Anthocincla.* The tail-feathers of the birds of this genus vary considerably in shape, in some being broad and rounded, and in others narrow and pointed.

a. Lower plumage plain fulvous.
  a'. Tail brown tinged with green.
    a''. Nape and hind neck blue.......... P. nepalensis, p. 389.
    b''. Nape and hind neck fulvous ...... P. oatesi, p. 390.
  b'. Tail blue....................... P. cœrulea, p. 390.
b. Lower plumage cross-barred.
  c'. Nape red; crown with a black coronal
    band .... P. cyanea, p. 391.
  d'. Nape ferruginous like crown; no coronal
    band .... P. gurneyi ♀, p. 395.
c. Lower plumage with some brilliant crim-
  son.
  e'. Breast and abdomen fulvous or buff.
    c''. Under wing-coverts black.
      a'''. Bill from gape to tip about 1·2 .. P. cyanoptera, p. 392.
      b'''. Bill from gape to tip about 1·6 .. P. megarhyncha, p. 393.
    d''. Under wing-coverts black with a
      large patch of white .... P. brachyura, p. 393.
  f'. Breast and abdomen all crimson ...... P. coccinea, p. 394.
  g'. Breast and abdomen green .... P. cucullata, p. 395.
d. Lower plumage deep black .... P. gurneyi ♂, p. 395.

Fig. 107.—Head of P. nepalensis.

927. Pitta nepalensis. *The Blue-naped Pitta.*

Paludicola nipalensis, *Hodgs. J. A. S. B.* vi. p. 103 (1837).
Pitta nipalensis (*Hodgs.*), *Blyth, Cat.* p. 156; *Horsf. & M. Cat.* i,
  p. 182; *Sclater, Cat. B. M.* xiv, p. 411; *Oates in Hume's N. & E.*
  2nd ed. ii, p. 281.
Brachyurus nipalensis (*Hodgs.*), *Elliot, Mon. Pitt.* pl. iii; *id. Ibis,* 1870,
  p. 413.
Hydrornis nipalensis (*Hodgs.*), *Jerd. B. I.* i, p. 502; *Hume, N. & E.*
  p. 224; *Oates, S. F.* iii, p. 337; *id. B. B.* i, p. 412; *Hume, Cat.*
  no. 344.

The Large Nepal Ground-Thrush, Jerd.

Coloration. Male. Forehead and anterior half of crown rich
fulvous, shading off into blue on the nape and hind neck; upper
plumage greenish brown; tail brown tinged with green; wings

dark brown, the outer webs of all the feathers broadly margined with fulvous; sides of the head, chin, and throat rich rusty or rufous; a concealed black patch on the side of the neck; remaining lower plumage deep fulvous; the feathers of the fore neck with concealed black bases, sometimes showing through when the tips of the feathers get worn.

Female. Differs from the male in having the throat whitish and the general colour of the head duller rufous.

The young bird is blackish above, with large fulvous spots; the front of the head is tinged with pink; the lower plumage is blackish, with broad pale pink tips to all the feathers.

Bill dusky, fleshy at the base; legs ruddy flesh-colour; claws whitish; iris lightish brown (*Jerdon*).

Length about 10; tail 2·6; wing 4·8; tarsus 2·1; bill from gape 1·4.

Distribution. The Himalayas from Nepal to Assam and the countries south of Assam to Manipur on the east and to Arrakan on the west.

Habits, &c. Breeds during May and June, constructing a covered nest of grass and leaves on the ground or on a tangled mass of branches of trees a short distance above the ground. The eggs, three or four in number, are white, sparingly marked with red and purple, and measure about 1·2 by ·95.

This species and the next are found in dense forests on the hills in the neighbourhood of water.

928. Pitta oatesi. *The Fulvous Pitta.*

Hydrornis oatesi, *Hume, S. F.* i, p. 477 (1873); *Walden in Blyth's Birds Burm.* p. 98; *Wardlaw Ramsay, Ibis,* 1877, p. 463; *Hume & Dav. S. F.* vi, p. 237; *Hume, Cat.* no. 341 bis; *Oates, B. B.* i, p. 411: *Salvadori, Ann. Mus. Civ. Gen.* (2) v, p. 574.
Pitta oatesi (*Hume*), *Sclater, Cat. B. M.* xiv, p. 416.

Coloration. Resembles *P. nepalensis,* but entirely wants all traces of blue on the nape and hind neck.

Upper mandible brown, the tip and edges salmon-colour; lower mandible brown; gape salmon-colour; inside of mouth flesh-colour; iris rich brown; eyelids plumbeous; legs and claws pinkish flesh-colour. Of the same size as *P. nepalensis.*

Distribution. Karennee; Tenasserim as far south as Muleyit mountain; the evergreen forests of the hills of Pegu.

929. Pitta cærulea. *The Giant Pitta.*

Myiothera cærulea, *Raffl. Trans. Linn. Soc.* xiii, p. 301 (1822).
Pitta cærulea (*Raffl.*), *Blyth, Cat.* p. 156; *Horsf. & M. Cat.* i, p. 181; *Hume & Dav. S. F.* vi, p. 238; *Hume, Cat.* no. 344 quart.; *Sclater, Cat. B. M.* xiv, p. 416.
Brachyurus cæruleus (*Raffl.*), *Elliot, Mon. Pitt.* pls. i & ii; *id. Ibis,* 1870, p. 412.
Brachyurus davisoni, *Hume, S. F.* iii, p. 321 (1875).
Gigantipitta cærulea (*Raffl.*), *Oates, B. B.* i, p. 413.

Coloration. Male. Forehead, front and sides of head, and the ear-coverts greyish brown, each feather narrowly margined with black; crown, nape, and back of neck black; a broad supercilium produced back nearly to the end of the black on the neck, as also a broad patch below this line and separated from it by a broad black streak starting from the eye and passing over the ear-coverts, fulvescent; chin and upper throat plain fulvescent; lower throat and sides of the neck the same, but each feather slightly margined with blackish; the whole lower plumage fulvous with a tinge of green; the throat separated from the breast by a broad black collar formed by the bases of certain of the feathers; this collar is not, however, always present; wings chiefly black, all the exposed portions when closed being blue; back, upper wing-coverts, rump, tail-coverts, and tail bright blue.

Female. The whole head and nape rufous-grey, closely barred with black; a broad streak from the eye over the ear-coverts and a broad collar round the back of the neck black; a supercilium reaching to the black collar, widening as it approaches it, and half surrounding the end of the black streak just referred to, plain fulvous; upper plumage chestnut; tail blue; wing-coverts and tertiaries chestnut; primaries and secondaries brown, more or less edged with ruddy; chin and throat pale grey; sides of the head and lower throat fulvous-grey, mottled with brownish; remainder of lower plumage fulvous, with a tinge of green; a black collar between the throat and breast, but not so conspicuous as in the male.

Legs and feet bluish fleshy or dark fleshy, tinged with pale plumbeous; bill black; inside of the mouth white; eyelids and gape very dark fleshy; irides hazel-grey (*Davison*).

Length about 11·5; tail 2·5; wing 6·2; tarsus 2·4; bill from gape 1·75.

Distribution. Tenasserim, from the foot of Nwalabo mountain southwards extending down the Malay peninsula to Sumatra and Borneo.

930. **Pitta cyanea.** *The Blue Pitta.*

Pitta cyanea, *Blyth, J. A. S. B.* xii, p. 1008 (1843) : *id. Cat.* p. 157 ; *Horsf. & M. Cat.* i, p. 182 ; *Hume & Dav. S. F.* vi, p. 238 ; *Hume, Cat. no.* 344 ter ; *Sclater, Cat. B. M.* xiv, p. 417 ; *Oates in Hume's N. & E.* 2nd ed. ii, p. 282.
Brachyurus cyaneus (*Blyth*), *Elliot, Mon. Pitt.* pl. xiii ; *id. Ibis,* 1870, p. 413 ; *Hume, S. F.* iii, p. 107.
Eucichla cyanea (*Blyth*), *Oates, B. B.* i, p. 419 ; *Salvadori, Ann. Mus. Civ.* (2) v, p. 575.

Coloration. Male. The lores and a broad streak from the eye over the ear-coverts to the nape black; forehead and crown greenish grey, changing to red, and giving place entirely to red on the nape, where the feathers are long and form a crest; a black streak from the bill, over the centre of the crown, to the nape; the whole upper plumage and tail blue; quills of the wing brown, each with a white patch at the base; cheeks and ear-coverts fulvous; below these a black moustachial stripe; chin and throat whitish, mottled with black; remainder of lower plumage light blue, barred with black, and the breast washed with yellow; the abdomen and lower tail-coverts paler blue, and barely barred at all.

Female. Differs from the male in having the upper plumage brown, tinged with blue, and the lower plumage yellowish brown, barred with black.

Bill black; inside of mouth dusky fleshy; iris dark reddish brown; eyelids plumbeous; legs pinkish flesh-colour; claws whitish.

Length about 9 inches; tail 2·3; wing 4·5; tarsus 1·8; bill from gape 1·2.

Distribution. Bhutan; Hill Tipperah; Arrakan; Pegu; Tenasserim as far south as Tavoy; extending to Siam.

Habits, &c. Breeds in May, constructing a massive globular nest of earth, leaves, and twigs on the ground. The eggs, four or five in number, are white, marked with various shades of purple, and measure about 1·07 by ·84.

931. Pitta cyanoptera. *The Lesser Blue-winged Pitta.*

Turdus moluccensis *, *P. L. S. Müll. Natursyst. Suppl.* p. 144 (1776).
Pitta cyanoptera, *Temm. Pl. Col.* pl. 218 (1823); *Blyth, Cat.* p. 157; *Horsf. & M. Cat.* i, p. 183; *Sclater, Cat. B. M.* xiv, p. 420; *Oates in Hume's N. & E.* 2nd ed. ii, p. 283.
Brachyurus cyanopterus (*Temm.*), *Elliot, Mon. Pitt.* pl. iv.
Brachyurus moluccensis (*Müll.*), *Elliot, Ibis,* 1870, p. 413; *Hume, S. F.* iii, p. 106.
Pitta moluccensis (*Müll.*), *Hume & Dav. S. F.* vi, p. 240; *Hume, Cat.* no. 345 bis; *Oates, B. B.* i, p. 415.

Coloration. Crown from the nostrils to the nape fulvous-brown; lores, cheeks, ear-coverts, a stripe over the eye, and a broad band round the back of the head black; a dark brown stripe over the head from the forehead to the nape; back, scapulars, and tertiaries dull green; rump, upper tail-coverts, and the smaller upper wing-coverts bright ultramarine-blue; chin immediately near the bill blackish, remainder of chin and throat white; breast, abdomen, and flanks ruddy buff; a broad stripe down the abdomen, the vent, and under tail-coverts bright crimson; tail black, tipped with dull blue; primaries black, each feather with a large patch of white; secondaries black, edged with dull blue on the terminal half; tertiaries black, tipped and margined with bluish green; larger wing-coverts dull green, edged with bright blue; under wing-coverts black.

Young birds have the coronal streak broader, and the feathers of the crown are narrowly margined with black; the wing-coverts are dull blue, and the colours of the other parts of the body less bright than in the adult.

Iris dark brown; eyelid and ocular region plumbeous; bill black; inside of mouth flesh-colour; legs fleshy pink; claws horn-colour.

---

* This name conveys an erroneous impression of this bird's habitat, and has been very properly rejected by most authors.

Length 8; tail 1·6; wing 4·9; tarsus 1·7; bill from gape 1·2.

Distribution. The southern portion of Arrakan; Pegu; Tenasserim. This Pitta is a seasonal visitor to the northern portion of its range, visiting Arrakan, Pegu, and Northern Tenasserim in April and May, and leaving in July. It appears to be a permanent resident in Southern Tenasserim. It ranges to Siam and down the Malay peninsula to some of the islands.

Habits, &c. Breeds in May, June, and July, constructing a large oven-shaped nest of leaves, roots, and earth, matted together on the ground or on a large branch or fallen tree near the ground. The eggs, usually five in number, are white, richly marked with red and purple, and measure about 1·05 by 0·87.

932. Pitta megarhyncha. *The Larger Blue-winged Pitta.*

Pitta megarhyncha, *Schleg. Voy. Ned. Ind.,* Pitta, p. 32, pl. 4, fig. 2
    (1863); *Hume & Dav. S. F.* vi, p. 242 ; *Hume, Cat.* no. 345 ter;
    Oates, B. B. i, p. 416; *Sclater, Cat. B. M.* xiv, p. 421 ; *Oates in
    Hume's N. & E.* 2nd ed. ii, p. 285, note.
Brachyurus megarhynchus (*Schleg.*), *Elliot, Ibis,* 1870, p. 414, pl. xii.

Coloration. Very similar to *P. cyanoptera,* but differing slightly in coloration, larger, and with a much longer bill. The coronal streak is obsolete or altogether absent, the brown of the head is darker, the breast is paler, and the black collar narrower.

Bill black ; iris deep brown ; legs and feet dark fleshy.

Length 9 ; tail 1·7 ; wing 4·7 ; tarsus 1·6 ; bill from gape 1·6.

Distribution. Southern Pegu and Tenasserim from Moulmein southwards. This species appears to visit Pegu and the more northern parts of its range in Tenasserim in May, and to depart in July, but in Southern Tenasserim it is probably a resident. Its range extends down the Malay peninsula, and it is also found in Banka.

Habits, &c. The nest of this species has not yet been found within Indian limits. Further south in the Malay peninsula a nest was found in April, and appears to have been of the usual type.

933. Pitta brachyura. *The Indian Pitta.*

Corvus brachyurus, *Linn. Syst. Nat.* i, p. 158 (1766).
Corvus brachyurus, var. bengalensis, *Gmel. Syst. Nat.* i, p. 376 (1788).
Turdus coronatus, *P. L. S. Müll. Natursyst., Suppl.* p. 144 (1776).
Pitta triostegus (*Sparrm.*), *Blyth, Cat.* p. 157.
Pitta brachyura (*Linn.*), *Gould, Cent.* pl. 23; *Hume, Cat.* no. 345 ;
    Barnes, Birds Bom. p. 169; *Sclater, Cat. B. M.* xiv, p. 423;
    Oates in Hume's N. & E. 2nd ed. ii, p. 285.
Pitta bengalensis (*Gmel.*), *Horsf. & M. Cat.* i, p. 184; *Jerd. B. I.*
    i, p. 503.
Brachyurus bengalensis (*Gmel.*), *Elliot, Mon. Pitt.* pl. vi.
Pitta coronata (*P. L. S. Müll.*), *Hume, N. & E.* p. 224; *Legge, Birds
    Ceylon,* p. 687.

The Yellow-breasted Ground-Thrush, Jerd. ; *Nourang,* Hind. ; *Shumcha,*
Beng. ; *Pona-inki,* Tel. ; *Avitchia, Ayitta,* Cing.

Coloration. Forehead and crown pale fulvous, with a broad median black band from the forehead to the nape; a narrow supercilium and the feathers under the eye white; a very broad black band passing under the eye, over the ear-coverts, and meeting the median coronal band on the nape; back, scapulars, and upper rump green; lower rump, upper tail-coverts, and lesser wing-coverts shining pale blue; tail black, tipped with dull blue; median coverts and tertiaries green; greater-coverts green, with the bases of the outer feathers black; winglet and primary-coverts black; primaries black, each with a basal white patch and a grey tip; secondaries black, tipped with white, and with the terminal portion of the outer web margined with dull blue; chin and throat white; remainder of lower plumage fulvous, the middle of the lower abdomen and the under tail-coverts crimson; under wing-coverts black, with a patch of white near the edge of the wing.

Iris dark brown; bill blackish, paling to reddish brown on culmen; legs and feet pale purplish fleshy (*Butler*).

Length about 7·5; tail 1·8; wing 4·1; tarsus 1·4; bill from gape 1·1.

Distribution. The whole of India from Eastern Rajputana and Garhwál to Sikhim and Calcutta, extending south to Cape Comorin and Ceylon. This Pitta is a local migrant, being found in the southern part of its range in the winter, and in the central and northern portions in the hot weather and rains, but a certain number of birds appear to be constant residents in all parts of its range suited to its habits.

Habits, &c. Breeds in the Central Provinces of India in July and August, building a huge globular nest of twigs and leaves on the ground or on low branches. The eggs are of the usual type, glossy white, marked with maroon and purple, and measure about 1·01 by ·86.

934. Pitta coccinea. *The Malayan Scarlet Pitta.*

Pitta coccinea, *Eyton, P. Z. S.* 1839, p. 104; *Hume & Dav. S. F.* vi, p. 511; *Hume, Cat.* no. 345 quat.; *Sclater, Cat. B. M.* xiv, p. 431.
Brachyurus granatinus (*Temm.*), *Elliot, Mon. Pitt.* pl. xv (pt.); *id. Ibis,* 1870, p. 417 (pt.).
Eucichla coccinea (*Eyton*), *Oates, B. B.* i, p. 417.

Coloration. The forehead, for about a quarter of an inch or to a point well in front of the eye, the lores, a streak over the eye, and the sides of the head black; crown and nape deep crimson, bordered on either side of the nape by a streak of lavender-blue; the whole upper plumage purplish blue, most brilliant on the back and dull on the other parts and tail; wing-feathers black, the outer edges and tips more or less tinged with blue; lesser wing-coverts plain black; the greater coverts black, all broadly tipped with glistening blue; chin and throat rufous, the feathers all tipped with dark brown; the breast purple, each feather edged with crimson; sides of the body, abdomen, vent, and under tail-coverts crimson.

A young bird shot by Mr. Davison had the legs, feet, and claws pale lavender; the bill black; the gape and a spot at the base and tip of both mandibles orange-vermilion.

Length 7; tail 1·6; wing 3·5; tarsus 1·5; bill from gape 1·05.

Distribution. Tenasserim, at the foot of Nwalabo mountain, extending down the Malay peninsula.

935. Pitta cucullata. *The Green-breasted Pitta.*

Pitta cucullata, *Hartl. Rev. Zool.* 1843, p. 65; *Blyth, Cat.* p. 157; *Horsf. & M. Cat.* i, p. 390; *Jerd. B. I.* i, p. 504; *Davison, S. F.* v, p. 457; *Hume & Dav. S. F.* vi, p. 243; *Hume, Cat.* no. 346; *Oates, B. B.* i, p. 414; *Sclater, Cat. B. M.* xiv, p. 442; *Oates in Hume's N. & E.* 2nd ed. ii, p. 286.

Melanopitta cucullata (*Hartl.*), *Hume, N. & E.* p. 225.

Brachyurus cucullatus (*Hartl.*), *Elliot, Mon. Pitt.* pl. xxviii; *Hume, S. F.* iii, p. 109.

The Green-breasted Ground-Thrush, Jerd.; Phattim-pho, Lepch.

Coloration. The crown from the nostrils to the nape rich rufous-brown; lores, cheeks, ear-coverts, chin, throat, and a collar surrounding the head black; breast and sides of the body pale greenish blue; the abdomen black; lower abdomen, vent, and under tail-coverts crimson; back, scapulars, and rump dark glossy green; upper tail-coverts and all the smaller wing-coverts bright ultramarine-blue; primaries black, with a large white patch on each feather; secondaries black, with the terminal half of the outer webs edged broadly with greenish blue; tertiaries wholly dark green; the larger wing-coverts dull green; tail black, tipped with blue; under wing-coverts black.

Bill black; inside of mouth dusky fleshy; iris dark coffee-brown; eyelids plumbeous; legs and claws fleshy pink.

Length about 7·5; tail 1·6; wing 4·5; tarsus 1·7; bill from gape 1·05.

Distribution. The Himalayas from Nepal to Assam; Tipperah; the Khási hills; Manipur; Pegu and Tenasserim. Blyth records this species from Arrakan. It extends down the Malay peninsula.

Habits, &c. Breeds from April to July, making a globular nest of leaves, twigs, and fibres on the ground, and laying four eggs, which are white marked with purple and measure about 1·07 by ·84.

936. Pitta gurneyi. *Gurney's Pitta.*

Pitta gurneyi, *Hume, S. F.* iii, p. 296, pl. iii; *Hume & Dav. S. F.* vi, p. 244; *Hume, Cat.* no. 346 bis.

Eueichla gurneyi (*Hume*), *Oates, B. B.* i, p. 418; *Sclater, Cat. B. M.* xiv, p. 448.

Coloration. Male. Forehead, front of the crown, lores, cheeks, ear-coverts, a stripe over the eye continued to the back of the head as a collar, breast, abdomen, vent, and under tail-coverts deep black; remainder of crown and nape bright glistening blue, the feathers long and forming a crest; chin and throat white; sides of the neck and a broad collar on the upper breast bright yellow; sides of the body and of the breast yellow, barred with black; under wing-coverts black, a few of the feathers in the middle being white; the upper plumage with the tertiaries and upper wing-coverts light chestnut-brown; primaries and their coverts black; secondaries black, the first two or three margined with whitish near the tip, the others margined more broadly with the same colour as the back; tail bright blue, the inner webs being almost entirely black.

Female. Forehead pale, the crown and nape bright ferruginous; cheeks, ear-coverts, and a line along the neck black, a few of the feathers of the cheeks pale orange-brown; chin and throat dirty white; remainder of the lower plumage yellow closely barred with black; the yellow most bright on the breast, and the bars almost absent on the abdomen; with the exception of the head, which has been described, the whole upper plumage is like that of the male; lower tail-coverts black, tipped with dull blue; primaries and secondaries dull brown; tail as in the male.

Bill black; iris dark brown; eyelids black; gape whitish; legs and claws fleshy white (*Davison*).

Length about 8·5; tail 2·2; wing 4·1; tarsus 1·6; bill from gape 1·1.

Distribution. The extreme southern portion of Tenasserim and the Island of Tonkah.

ALPHABETICAL INDEX.

PRINTED BY TAYLOR AND FRANCIS, RED LION COURT, FLEET STREET.

UNIFORM WITH THE PRESENT VOLUME

THE FAUNA OF BRITISH INDIA,
INCLUDING CEYLON AND BURMA.

Published under the authority of the Secretary of State
for India in Council.

EDITED BY

W. T. BLANFORD.

*Stitched in Paper Covers, med. 8vo, pp. i–xii, 1–250, and
71 woodcuts. Price 10s.*

MAMMALIA.—Part I.

BY

W. T. BLANFORD, F.R.S.

---

*Bound in Cloth, Lettered, med. 8vo, with numerous woodcuts.
Price £1 each.*

FISHES.—2 Vols.

BY

FRANCIS DAY, C.I.E., LL.D., &c.

---

*Bound in Cloth, Lettered, med. 8vo, with numerous woodcuts.
Price £1.*

BIRDS.—Vol. I.

BY

E. W. OATES, F.Z.S.

---

*Bound in Cloth, Lettered, med. 8vo, with numerous woodcuts.
Price £1.*

REPTILIA AND BATRACHIA.
(1 Vol. Complete.)

BY

G. A. BOULENGER.

LONDON:
TAYLOR AND FRANCIS, RED LION COURT, FLEET STREET, E.C.

CALCUTTA: | BOMBAY:
THACKER, SPINK, & CO. | THACKER & CO., LIMITED.

BERLIN:
R. FRIEDLANDER UND SOHN, 11 CARLSTRASSE.

www.ingramcontent.com/pod-product-compliance
Lightning Source LLC
Chambersburg PA
CBHW021348210326
41599CB00011B/798